Advances in Sustainability Science and Technology

Series Editors

The book series aims at bringing together valuable and novel scientific contributions that address the critical issues of renewable energy, sustainable building, sustainable manufacturing, and other sustainability science and technology topics that have an impact in this diverse and fast-changing research community in academia and industry.

The areas to be covered are

- Climate change and mitigation, atmospheric carbon reduction, global warming
- Sustainability science, sustainability technologies
- Sustainable building technologies
- Intelligent buildings
- Sustainable energy generation
- Combined heat and power and district heating systems
- Control and optimization of renewable energy systems
- Smart grids and micro grids, local energy markets
- Smart cities, smart buildings, smart districts, smart countryside
- Energy and environmental assessment in buildings and cities
- Sustainable design, innovation and services
- Sustainable manufacturing processes and technology
- Sustainable manufacturing systems and enterprises
- Decision support for sustainability
- Micro/nanomachining, microelectromechanical machines (MEMS)
- Sustainable transport, smart vehicles and smart roads
- Information technology and artificial intelligence applied to sustainability
- Big data and data analytics applied to sustainability
- Sustainable food production, sustainable horticulture and agriculture
- Sustainability of air, water and other natural resources
- Sustainability policy, shaping the future, the triple bottom line, the circular economy

High quality content is an essential feature for all book proposals accepted for the series. It is expected that editors of all accepted volumes will ensure that contributions are subjected to an appropriate level of reviewing process and adhere to KES quality principles.

The series will include monographs, edited volumes, and selected proceedings.

Thanuja Ramachandra · Roshani Palliyaguru ·
Gayani Karunasena
Editors

Concepts, Policies and Technology for Sustainable Buildings

Springer

Editors
Thanuja Ramachandra
Department of Building Economics
University of Moratuwa
Moratuwa, Sri Lanka

Roshani Palliyaguru
School of the Built Environment
Oxford Brookes University
Oxford, UK

Gayani Karunasena
School of Architecture and Built Environment
Deakin University
Geelong, Australia

ISSN 2662-6829 ISSN 2662-6837 (electronic)
Advances in Sustainability Science and Technology
ISBN 978-981-96-3085-1 ISBN 978-981-96-3086-8 (eBook)
https://doi.org/10.1007/978-981-96-3086-8

This Springer imprint is published by the registered company Springer Nature Singapore Pte Ltd.
The registered company address is: 152 Beach Road, #21-01/04 Gateway East, Singapore 189721, Singapore

Preface

The book series *Advances in Sustainability Science and Technology (ASST)* was initiated by Springer Nature and KES International to respond to the need for a publication channel for the latest high-quality research on a broad range of sustainability topics. This book, titled *Concepts, Policies, and Technologies for Sustainable Buildings,* is published in this book series.

With the steady increase in population, natural resources are being depleted daily. The concept of 'sustainability' emerged to overcome the problems generated by growing demands to use these limited resources efficiently and effectively. To enhance the sustainability of buildings, this book seeks to disseminate knowledge through fourteen chapters, organised into two broad parts:

1. Concepts and Policies on Energy and Water Efficiency and Waste Management
2. Technologies on Energy Efficiency and Waste Management.

Accompanied by an introduction, the fourteen chapters of this book are grounded in recent research studies that explore concepts, policies, techniques, and technologies related to energy and water efficiency and waste management. The chapters are authored by a diverse group of thirty-four contributors from Australia, New Zealand, Sri Lanka, and the United Kingdom.

The fulfilment of human needs and desires, essential for their overall welfare, is intricately connected to the natural environment. The escalating demand for these needs is giving rise to significant challenges such as climate change, economic disparities, and social injustices, including malnutrition and lack of access to clean drinking water. These issues have widespread global implications and impact various facets of human life. This has led to a huge growth in research around the concepts, policies, techniques, and technologies of sustainable buildings.

This book is designed to serve not only scientists, researchers, and academics but also professionals in the building industry who are engaged in sustainability practices.

The editors extend their gratitude to the authors and reviewers for their invaluable contributions to the successful completion of this book. The guidance and support provided by Springer during the book's preparation is also gratefully acknowledged.

Moratuwa, Sri Lanka Prof. Thanuja Ramachandra
Oxford, UK Dr. Roshani Palliyaguru
Geelong, Australia Dr. Gayani Karunasena

Introduction

In the modern world, environmental issues have received grave concerns due to growing population growth and rapid urbanisation, which continuously necessitate more and more construction projects. It is in this context that the construction industry is frequently criticised for its massive direct and indirect effects on the natural environment. The construction industry is found to have increasing implications for global resource and energy consumption, accounting for one-fifth of the world's total energy consumption and one-third of global Greenhouse Gas (GHG) emissions. Against this backdrop, sustainable development in the built environment has emerged as a prime solution to address these concerns. Achieving a sustainable built environment is predominantly articulated under the three pillars of sustainability, namely social, environmental, and economic sustainability. Thus, buildings are expected to be made "sustainable" by adopting various concepts, policies, and technologies that would result in the efficient use of resources such as energy, water, building materials, and by-products.

Despite the efforts, the progress of the construction industry's transition towards creating a sustainable built environment has been slower than anticipated. The adoption of the appropriate concepts, policies, and technologies seems to vary to a greater extent in developed and developing countries due to a lack of knowledge and awareness. In addition, adoption has been hindered by a lack of commitment, misperceptions about the implications of concepts and technologies, a lack of legislative and policy provisions for adoption, financial constraints, etc.

This warrants the diffusion of knowledge across a wide range of countries on the adoption of appropriate concepts, policies, and technologies that lead to sustainable buildings. To this end, this book presents knowledge on diverse sustainability concepts, policies, and technologies adopted in both developed and developing countries under the following two parts:

Part I: Concepts and Policies on Energy and Water Efficiency and Waste Management

Increasing energy and water efficiency and waste minimisation is a fast-track roadmap for achieving sustainable buildings. Despite several emerging concepts and policies aimed at achieving this, the progress of many nations falls behind. Furthermore, the built environment sector urgently needs a sustainability transition to exploit its potential for emission reductions in existing and new buildings. To this end, this part of the book presents concepts and policies applicable for building practitioners' compliance behaviours towards energy performance, energy retrofitting in residential buildings, sustainability transition in the built environment, construction of passive solar houses, energy performance ratings of commercial buildings, the impact of certification for water efficiency in buildings, and mitigation of the negative impacts of construction and demolition waste.

The latest data shows that most buildings are still being designed at the minimal compliance level, risking the industry blocking the pathway towards a sustainable future. Amongst several reasons, recent studies elucidate that this difficulty in going beyond minimal compliance is caused by building practitioners' various compliance behaviours. Therefore, Chap. 1 in this part of the book investigates the compliance behaviour of building practitioners in the Australian context.

Australia seeks to achieve a 40% boost in energy productivity by 2030 from its 2015 baseline. Despite several emerging policy instruments aimed at enhancing the energy efficiency of existing residential buildings, Australia's progress falls behind other developed economies due to deficiencies in these policy instruments. Accordingly, Chap. 2 reviews eight current federal-level policy instruments in Australia aimed at energy retrofitting residential buildings to identify prevailing deficiencies and recommend potential policy improvements.

It is clear that a transition, at scale, across the whole of the built environment sector is yet to materialise, because it is still one of the second highest-emitting sectors. It should accelerate its rate of emissions reduction to two or three times the current rate. Therefore, Chap. 3 makes an original contribution to existing knowledge by drawing on the multi-level perspective in the sustainability transitions literature to explain the whole UK's built environment sector's lack of progress.

The passive solar housing design is a practical concept that is more suitable for equatorial countries, such as Sri Lanka. However, a lack of policies, specifically focusing on the construction of passive solar houses, has hindered its successful implementation. Thus, Chap. 4 aims to present the existing sustainable and energy efficient building construction-related policies and guidelines in support of the construction of passive solar houses in Sri Lanka to determine the probable policy enhancements.

Limited research has explored energy performance differences between rating levels and how other factors, like climate conditions, could influence them. Hence, Chap. 5 aims to investigate the ordinal energy performance of commercial buildings rated under Australia's Commercial Building Disclosure program while considering the impact of climate differences.

Green Building Rating Systems provide structures or collections of metrics to determine the water efficiency level that a building can achieve. LEED is the most adaptable system in Sri Lanka among other green rating systems due to its global recognition. In terms of water efficiency, LEED-certified buildings have been recognised for poorer performance than non-certified buildings due to the misuse of available conservation systems accredited by the LEED certification. Hence, Chap. 6 aims to compare the water efficiency of LEED-certified and non-certified hotels to determine the benefits of LEED certification for water efficiency in Sri Lanka.

The construction industry is often criticised for the generation of high amounts of construction and demolition waste and the consumption of natural resources. To this end, Chap. 7 identifies drivers for implementing circular economy-based waste management practices in Australian construction projects to mitigate the negative impacts of construction and demolition waste generation, while Chap. 8 presents a quantification of the waste generated by building defects to gain insights into its contribution to construction and demolition waste generation in residential building projects in Australia.

Part II: Technologies on Energy Efficiency and Waste Management

The escalating demand for energy and resources in the construction industry, together with poor energy performance of existing buildings, pose challenges for sustainability. Despite the incorporation of advanced technologies in modern construction to reduce energy consumption, a significant number of buildings are yet to adopt sustainable practices, owing to a lack of awareness about potential technologies, and the considerations for selection. To this end, Part II of the book presents innovative techniques and technologies applicable for assessment of energy behaviours of building occupants, the renewable energy sources in buildings, carbon emissions of off-site construction, indoor air quality of buildings, waste minimisation of building materials, and the use of Building Information Modelling (BIM) in enhancing the application of Modern Methods of Construction (MMC) in the construction industry.

The existing literature has emphasised the need to adopt a holistic approach for assessing the impact of occupant behaviour on energy consumption in the building sector. Hence, Chap. 9 is dedicated to creating an innovative tool known as the Occupant Energy Behaviour Tool, with a primary focus on its application in the context of New Zealand.

In Sri Lanka, a developing nation, considerable efforts focus on attaining sustainable objectives, primarily solar energy. To this end, Chap. 10 delves into the comparative economic aspects of two prevalent solar Photovoltaic systems: on-grid and off-grid installations within Sri Lankan communities. It provides a conceptual map outlining the economic strategies and challenges specific to each system, as well as those shared by both.

Shipping containers is one specific type of offsite construction, which is widely used for the construction of new buildings using recycled material. Given that shipping containers are considered environmentally friendly due to the use of recycled material, Chap. 11 investigates their carbon emissions and determines whether they are really a sustainable solution.

Indoor Air Quality (IAQ) is an essential aspect of the educational environment that significantly influences the well-being of students in higher education institutions. Chapter 12 aims to present the impact of IAQ on students' well-being in Higher Education Institutes in the UK, with a specific focus on Carbon dioxide, Particulate Matter less than 2.5 micrometres in diameter (PM2.5) concentrations, and the Air Quality Index.

Addressing the waste generated during construction projects requires effective techniques and technologies for its management. Among the various materials contributing to this waste stream in New Zealand, ceramic tiles stand out due to their fragile nature during handling and installation. As research on minimising ceramic tile waste in construction projects in New Zealand remains limited, Chap. 13 aims to fill this gap by investigating current waste management practices related to ceramic tiles.

A notable gap exists in understanding how BIM integration specifically contributes to the success of diverse MMC applications. Consequently, Chap. 14 bridges the identified gap by systematically analysing the integral role of BIM within various MMC, highlighting key principles, methodologies, benefits, and challenges, as well as emerging technologies.

Contents

Editors and Contributors

About the Editors

Prof. Thanuja Ramachandra is a Professor in Quantity Surveying in the Department of Building Economics University of Moratuwa, Sri Lanka. Thanuja has earned her Ph.D. in Construction Management from the School of Engineering Auckland University of Technology (AUT), New Zealand and Master of Philosophy (M.Phil.) and B.Sc. (Hons) in Quantity Surveying from the University of Moratuwa (UoM) Sri Lanka. She is Cooperate member of Institute of Quantity Surveyors Sri Lanka (IQSSL). Thanuja has supervised over 25 postgraduate and over 100 undergraduate research dissertations and made over 80 publications in internationally refereed journals and conferences. Her current research interests include sustainability, green retrofits, carbon assessment, and circular economy.

Dr. Roshani Palliyaguru is a Senior Lecturer at the School of the Built Environment at Oxford Brookes University, United Kingdom. She holds a Ph.D. degree in Construction Management from the University of Salford, United Kingdom, and B.Sc. (Hons) in Quantity Surveying from the University of Moratuwa, Sri Lanka. Her research focuses on areas such as sustainability concerns of the built environment, disaster resilience building in the built environment, governance of public-private partnership projects, and cost-value propositions of construction projects. She has published in these areas in top peer-reviewed journals, conference proceedings, and book chapters, totalling over 60 publications.

Dr. Gayani Karunasena is a Senior Lecturer attached to the School of Architecture and Built Environment in Deakin University, Geelong Waterfront Campus, Australia. She is a Member of Australian Institute of Quantity Surveying and holds a Ph.D. degree in Construction and Property Management from the University of Salford, United Kingdom, and B.Sc. (Hons) in Quantity Surveying from the University of Moratuwa, Sri Lanka. Her research focuses on waste management, energy performance, and circular economy in the built environment. Her research has been published in top peer-reviewed journals, conference proceedings, and book chapters, totalling over 130 publications.

Contributors

A. G. U. Damsari Department of Management, Monash Business School, Monash University, Caulfield, Australia

Niluka Domingo School of Built Environment, Massey University, Albany, New Zealand

Nirodha Gayani Fernando School of Computing, Engineering & the Built Environment, Edinburgh Napier University, Edinburgh, UK

Devindi Geekiyanage School of Science, Engineering and Environment, University of Salford, Manchester, UK

Buddhini Ginigaddara School of Architecture and Built Environment, University of Newcastle, Callaghan, NSW, Australia

Argaw Gurmu School of Architecture and Built Environment, Deakin University, Geelong, Australia

Ganegodage Lakna Harindi Department of Building Economics, University of Moratuwa, Moratuwa, Sri Lanka

Abid Hasan School of Architecture and Built Environment, Deakin University, Geelong, Australia

Nimasha Dilukshi Hulathoowage School of Architecture and Built Environment, Deakin University, Geelong, Australia

Gayani Karunasena School of Architecture and Built Environment, Deakin University, Geelong, Australia

Esra Kurul School of the Built Environment, Oxford Brookes University, Oxford, UK

Chunlu Liu School of Architecture and Built Environment, Deakin University, Geelong, Australia

Yi Lu School of Architecture and Built Environment, Deakin University, Geelong, Australia

Mahnoor School of Computing, Engineering & the Built Environment, Edinburgh Napier University, Edinburgh, UK

H. Mallawaarachchi Department of Facilities Management, Faculty of Architecture, University of Moratuwa, Moratuwa, Sri Lanka

Peter Nicholas School of Architecture and Built Environment, University of Newcastle, Callaghan, NSW, Australia

U. A. S. Nissanka Department of Quantity Surveying, Sri Lanka Institute of Information Technology, Malabe, Sri Lanka

Udesh Perera School of Built Environment, Massey University, Albany, New Zealand

Diyana Sheharee Athidewa Ranasinghe Department of Building Economics, University of Moratuwa, Moratuwa, Sri Lanka

Eziaku Onyeizu Rasheed School of Built Environment, Massey University, Auckland, New Zealand

Hemantha Sandaruwan Bandara Rathnaweera Patabandige Department of Civil Engineering, Faculty of Engineering, Sri Lanka Institute of Information Technology, Malabe, Sri Lanka

James Olabode Bamidele Rotimi School of Built Environment, Massey University, Auckland, New Zealand

Mo Maleki Sadabad School of Science, Engineering and Environment, University of Salford, Manchester, UK

Abdul-Manan Sadick School of Architecture and Built Environment, Deakin University, Geelong, Australia

C. Savindi Ranthika Perera School of Architecture and Built Environment, Deakin University, Geelong, Australia

Wajiha Shahzad School of Built Environment, Massey University, Albany, New Zealand

Yuchen She School of Architecture and Built Environment, Deakin University, Geelong, Australia

Salman Shooshtarian School of Property, Construction and Project Management, RMIT University, Melbourne, Australia

P. Sridarran Department of Facilities Management, University of Moratuwa, Moratuwa, Sri Lanka

Olubukola Tokede School of Architecture and Built Environment, Deakin University, Geelong, Australia

Nguyet Tong School of Built Environment, Massey University, Albany, New Zealand

Nilupa Udawatta School of Architecture and Built Environment, Deakin University, Geelong, Australia

Anuradha Samarajeewa Waidyasekara Department of Building Economics, University of Moratuwa, Moratuwa, Sri Lanka

Achini Shanika Weerasinghe School of Built Environment, Massey University, Auckland, New Zealand

Abbreviations

AAC	Autoclaved Aerated Concrete
AB	Adaptive Behaviours
ABCB	Australian Building Codes Board
AC	Accessibility to Control/Alternating Current
ACT	Australian Capital Territory
AEC	Architecture, Engineering, and Construction
AI	Artificial Intelligence
AK	Actual Knowledge
ANOV	Analysis of Variance
AQI	Air Quality Index
AR	Augmented Reality
ASBEC	Australian Sustainable Built Environment Council
AVE	Average Variance Extracted
BEPs	BIM Execution Plans
BEEC	Building Energy Efficiency Certificate
BI	Behavioural Interventions
BIM	Building Information Modelling
BMS	Building Management Systems
BPS	Building Performance Simulations
BRANZ	Building Research Association New Zealand
BREEAM	Building Research Establishment Environmental Assessment Methodology
C&D	Construction and Demolition
CAD	Computer Aided Design
CART	Classification and Regression Trees
CBD	Commercial Building Disclosure
CCC	Climate Change Committee
CDE	Common Data Environment
CE	Circular Economy
CEB	Ceylon Electricity Board
CHP	Combined Heat and Power

CI	Construction Industry
CIC	Construction Industry Council
CIOB	Chartered Institute of Building
CITB	Construction Industry Training Board
CLT	Cross-Laminated Timber
CMB	Common Method Bias
CO	Carbon Monoxide
CO_2	Carbon Dioxide
COAG	Council of Australian Governments
CPD	Continuing Professional Development
CPs	Construction Projects
CR	Composite Reliability
CSA	Case Study Analysis/Classification System Accumulation
CSs	Case Studies
CW	Construction Waste
D&C	Design and Construct
DC	Direct Current
DECs	Display Energy Certificates
DI	Direction-Based Instruments
ECIU	Energy and Climate Intelligence Unit
EDGAR	Emissions Database for Global Atmospheric Research
EEs	Embodied Emissions
EPA	Energy Performance Assessment/Environmental Protection Agency
EPBD	European Energy Performance of Buildings Directive
EPC	Energy Performance Certificate
EPD	Energy Performance Disclosure
EU	European Union
ET	Education and Training
FI	Finance and Incentives
GA	Government Authorities
GBCs	Green Building Councils
GBRS	Green Building Rating Systems
GDP	Gross Domestic Product
GDPR	General Data Protection Regulation
GEMS	Greenhouse and Energy Minimum Standards
GHG	Greenhouse Gas
HCC	Hybrid Concrete Construction
HEEIP	House Energy Efficiency Inspections Project
HEIs	Higher Education Institutions
HIP	Home Insulation Program
IAQ	Indoor Air Quality
ICE	Institution of Civil Engineers
ICF	Insulating Concrete Formwork
IEA	International Energy Agency
IEQ	Indoor Environmental Quality

IFC	Industry Foundation Classes
IoT	Internet of Things
JIT	Just-In-Time
LCA	Life Cycle Assessment
LEED	Leadership in Energy and Environmental Design
LVL	Laminated Veneer Lumber
LR	Literature Review
MDF	Medium Density Fibreboard
MEPR	Minimum Energy Performance Requirements
ML	Machine Learning
MLP	Multi-level Perspective
MMC	Modern Methods of Construction
NAB	Non-Adaptive Behaviours
NABERS	National Australian Built Environment Rating System
NBS	National Building Specification
NCC	National Construction Code
NEPS	National Energy Performance Strategy
NO_2	Nitrogen Dioxide
NPC	National Policy on Construction
NS	National Scorecard
NatHERS	Nationwide House Energy Rating Scheme
O_3	Ozone
OECD	Organization for Economic Co-operation and Development
OFGS	Off-Grid Solar
ONGS	On-Grid Solar
OS	Organisational Support
PAYG	Pay As You Go
PBC	Perceived Behavioural Control
PIs	Policy Instruments/Preliminary Interviews
PK	Perceived Knowledge
PLS-SEM	Partial Least Squares Structural Equation Modelling
PM	Particulate Matter
PN	Personal Norms
PV	Photovoltaic
PIA	Preliminary Interview Analysis
RBS	Residential Energy Baseline Study
RE	Renewable Energy
RHI	Renewable Heat Incentive
RI	Research and Innovation
RICS	Royal Institution of Chartered Surveyors
ROI	Return on Investment
SC	Solar Service Providers
SHWR	Solar Hot Water Rebate
SME	Subject Matter Expert
SN	Subjective Norms

SRET	Small Scale Renewable Energy Target
STs	Sustainability Transitions
UK	United Kingdom
UKGBC	UK Green Building Council
UN	United Nations
UN-Habitat	United Nations Agency for Human Settlements
UNOPS	United Nations Office for Project Services
US	United States
VOCs	Volatile Organic Compounds
VM	Variables Modelling
VR	Virtual Reality
WGBC	World Green Building Council
WHO	World Health Organisation
WM	Waste Management
WWF	Worldwide Fund for Nature
ZCH	Zero Carbon Homes

Part I
Concepts and Policies on Energy and Water Efficiency and Waste Management

Chapter 1
Compliance Behaviours of Building Practitioners with Energy Efficiency Standards: A Focus on the 7-Star House Energy Rating Scheme in Australia

Yi Lu, Gayani Karunasena, and Chunlu Liu

Abstract Increasing energy efficiency is a fast-track roadmap for carbon mitigation and improving buildings' energy performance for achieving sustainable buildings. In Australia, the minimum energy efficiency standards have increased from 6 to 7 stars for new housing through the Nationwide House Energy Rating Scheme (NatHERS), which has progressively become mandatory for most states since 2023. Notwithstanding this policy effort, the latest data shows that most new houses (79.76%) have been still designed at the minimal compliance level without going 7 stars and higher, risking the industry blocking the pathway towards a sustainable future. Among other reasons, recent studies elucidate that this difficulty in going beyond minimal compliance is caused by building practitioners' various compliance behaviours (i.e., attitudes towards going beyond minimal compliance, subjective norms, perceived behavioural control, and personal norms). This chapter investigates this compliance behaviour in Victoria—a pioneering Australian state in promoting sustainable buildings. From a questionnaire survey of 73 residential architects/draftspersons, builders, and thermal performance assessors, the chapter finds that building practitioners' attitudes generally exhibit the lowest score while personal norms display the highest. Significant differences in attitudes and perceived behavioural control between building practitioners who work on volume homes and those who work on custom homes are shown. Strong variations in attitudes and personal norms are also demonstrated between thermal performance assessors and other occupations. Furthermore, distinctions in subjective norms among building practitioners with various lengths of work experience are presented. To facilitate the sustainable transition, interventions at improving attitudes should be introduced in the first place.

Y. Lu (✉) · G. Karunasena · C. Liu
School of Architecture and Built Environment, Deakin University, Geelong, Australia
e-mail: yi.lu@deakin.edu.au

G. Karunasena
e-mail: gayani.karunasena@deakin.edu.au

C. Liu
e-mail: chunlu.liu@deakin.edu.au

T. Ramachandra et al. (eds.), *Concepts, Policies and Technology for Sustainable Buildings*, Advances in Sustainability Science and Technology,
https://doi.org/10.1007/978-981-96-3086-8_1

Keywords Building practitioners · Compliance behaviour · House energy rating · Residential building · Sustainable building

1 Introduction

The Australian residential building sector is a substantial contributor to the nation's energy consumption. According to the Department of Climate Change (2023), during 2020–2021, the Australian residential sector accounted for 11.7% of the national energy consumption, while the commercial sector's share was less (7%). Australia's energy needs are largely met by fossil fuels, which are by far the main source of greenhouse gas emissions. The emissions further lead to global warming, produce waste and potentially detrimental atmospheric emissions, and put the industry at risk of achieving a net-zero and sustainable outlook [45]. Recent studies have shown that sustainable buildings can effectively limit the negative impacts of buildings on the environment, economy, and society by reducing greenhouse gas emissions [4]. Hence, sustainable buildings have been a core issue for Australia for many years [30]. In recent years, energy efficiency in new buildings has been recognised as a fast-track solution for achieving sustainable buildings [31]. The realisation of building energy efficiency relies on the deployment of appropriate policy instruments [23]. Among the existing policy framework, the mandatory minimum energy performance standards are considered essential instruments for reducing energy consumption and carbon emissions from the residential sector [49]. The current minimum performance standard is to achieve a Nationwide House Energy Rating Scheme (NatHERS) 6 stars for new housing. However, this minimum standard has increased from 6 to 7 stars under changes to the National Construction Code (NCC) 2022, which will become mandatory for the State of Victoria (Australia) in May 2024. The upgrade from a 6 to a 7-star rating has seen improvements in enhancing building energy efficiency, but it is still not enough for a net-zero and sustainable residential building industry [54]. Omrany et al. [49] underlined that at least a 7.5-star rating is needed to deliver the optimal economic and environmental outcomes for Australia. Despite the urgent need to exceed the expected minimal compliance level of 7 stars, 79.76% of new Victorian housing during May and November 2023 (i.e., the transition period towards 7 stars) was designed only to achieve the minimum NatHERS 6-star standard, approximately 7.24% went beyond 7 stars and only around 1.36% went beyond the optimal 7.5 stars [16]. Given the fast-approaching mandatory commencement date of the 7 stars and the long-term sustainable transition, going beyond minimal compliance with the NatHERS becomes an urgent challenge to be tackled [34, 38].

Among all the reasons for not going beyond 7 stars, recent studies indicated that one of the arguments is the different compliance behaviours (such as willingness and ability) of building practitioners [22, 43]. Since the energy star rating is given based on the simulation of the house design, the building practitioners who engage in the design phase of a project (e.g., architects, designers/draftspersons, builders, project managers, and thermal performance assessors) are especially relevant. However, most academic works relating to building energy policies as well as sustainable building only focus on the top and bottom (i.e., policymakers and building consumers), while neglecting the middle actors' (i.e., building practitioners') willingness and capacity to make their own free choices in compliance with the energy policies [24, 56, 66]. This neglect wrongfully assumes that a policy is straightforwardly and instantly adopted by building practitioners [75]. Nevertheless, building practitioners' varying ideas, practices, expectations, and behaviours will lead to different compliance outcomes in the design stage [56]. Thus, this chapter attempts to elucidate the different compliance behaviours that have been insufficiently understood. It helps the building authorities to better understand the distinct compliance behaviours held by the building practitioners. It will further inform policymakers of areas that need interventions for the timely achievement of Australia's building energy policy intentions of promoting building practitioners to achieve higher standards beyond the mandatory minimum standards. This chapter focuses on the State of Victoria, as effectively all houses in this State are approved using the NatHERS approach [40] and also the Victorian government's significant support for pursuing high performing residential houses.

2 Literature Review

2.1 Building Energy Policy Development and the Nationwide House Energy Rating Scheme

The residential building industry in Australia has been continuously considered an important segment in ensuring environmental, economic, and societal benefits, which are the building blocks of a sustainable future. As mentioned earlier, the high energy consumption profile in the residential sector has led to enormous environmental emissions. Beyond this, the fast-increasing number of new houses along with the booming of the Australian population has also triggered a huge challenge for sustainable housing. According to [7], Australia is estimated to have more than 5.5 million new homes to be built between 2023 and 2050. Thus, to facilitate the sustainable transition, significant contributions should be made by the residential building industry towards the improvement of energy efficiency.

Building energy policies in Australia have long been recognised as critical instruments to systematically assist in increasing residential buildings' energy efficiency [41, 46, 49]. As per the national plan *Trajectory for Low Energy Buildings* [15], the delivery of low-energy, net-zero emission residential sector by 2050 is supported by all Commonwealth, state, and territory energy ministers. Relevant policies and their corresponding purposes and targets concerning an energy-efficient and/or sustainable residential building industry are illustrated in Table 1.

Within the policy framework, the minimum energy performance requirements are contained in the NCC, which is guided by the Council of Australian Governments (COAG) and the Australian Building Codes Board (ABCB). They are considered

Table 1 Relevant policies relating to energy-efficient and broader sustainable residential buildings at federal and state (Victoria) levels. Authors' own illustration

Policy	Mandatory or voluntary	Purpose of the policy	Target regarding energy-efficient and/or sustainable residential building
Australia wide—National Construction Code (NCC) 2022	Only the minimum energy performance requirements are mandatory	Setting minimum standards for buildings' performance such as sustainability, safety	Minimum performance standard for energy is established with connection to the NatHERS
Australia wide—Nationwide House Energy Rating Scheme (NatHERS) (Introduced in 1993, revised in 2022)	Mandatory	Providing energy ratings for Australian homes, and facilitating the creation of energy-efficient, comfortable, and resilient dwellings	An increase in the required minimum stringency from 6 to 7-star NatHERS rating as adopted in the NCC 2022 changes
Australia wide—Trajectory for Low Energy Buildings (Agreed in 2019, updated in 2023)	Voluntary	A national plan that sets the trajectory towards zero energy and carbon-ready residential and commercial buildings in Australia	An outline of the trajectory towards the achievement of zero energy (and carbon) ready buildings
Australia wide—Your Home Manual (sixth edition)	Voluntary	Providing a thorough guide to support the design and construction of environmentally sustainable homes	Increased sustainability in a broad sense. No specific targets set
Australia wide—National Australian Built Environment Rating System (NABERS)—Apartment buildings (Launched in 1998)	Mandatory (for some local councils, e.g., City of Yarra and City of Bendigo) and Voluntary	Offering an assessment of a dwelling's real environmental performance based on data from the past year	Improved energy consumption. Targets are defined according to the building type, intended use, and local climate zone
State (Victoria)—Residential Efficiency Scorecard 2017	Voluntary	Providing an energy star rating for homes and offering upgrade information	Reduced energy consumption and enhanced thermal comfort, with no specific target established

critically important to deliver high performing houses and make contributions to the decarbonisation of the residential building industry [9, 10].

The Nationwide House Energy Rating Scheme (NatHERS) in Australia was established in 1993 as a mechanism for assessing building energy performance. It was introduced in the Australian Building Codes, now referred to as the NCC in 2003. According to the simulation output of the house design, an energy star rating of the predicted heating and cooling loads is provided, ranging from 0 (worst) to 10 stars (best). The minimum standards approximated 4 NatHERS stars in 2003, 5 stars in 2006, and 6 stars in 2010. Then, in 2022, being a significant change over a decade, the NCC prescribed the minimum star rating from 6 to 7 stars for new homes, with an additional Whole-of-Home rating covering the energy performance of appliances. Each update of the minimum standards demonstrates significant regulatory efforts for improving building energy efficiency and broader sustainability in the residential built environment [50, 74]. The NCC provides two overarching methods to demonstrate compliance: One is the proposal of an alternative solution (performance-based solution, i.e., Verification Using a Reference Building). However, this pathway is seldom used as it demands a high knowledge level and is labour-intensive [18, 54]. The second pathway is the deemed-to-satisfy approach ABCB [2] which further includes two options: (1) energy rating method—achieving a minimum energy rating (or equivalent design under NatHERS, or (2) elemental provisions method—satisfying all the elemental provisions in the NCC, such as minimum thermal insulation R-value for the building fabric [2]. Currently, the energy rating solution is the primary form of compliance demonstration pathway in most states and territories [49]. The encompassed housing energy efficiency provisions in the NCC 2022 have different transition periods for states and territories in Australia, which progressively start from 2023 [3].

2.2 *Building Practitioners' Compliance Behaviour*

Recent studies [22, 43] highlighted that one of the prevailing reasons for the minimal compliance phenomenon with the NatHERS was the different compliance behaviours of the building practitioners in the design stage, which is usually defined as a wide range of behavioural predictions (such as willingness and ability) to understand how people respond to rules and come to adapt their behaviour to the rules in the compliance process [14, 39]. Previous literature indicated many factors that can impact building practitioners' compliance behaviour, including (1) unfavourable attitudes towards going beyond minimal compliance [22, 55], (2) how the regulated building practitioner feels they will be perceived by others (May, 2004; (3) a building practitioner's understanding of their capacity to achieve a compliant behaviour [55], and (4) moral commitment [61]. However, these variables were largely scattered in a patchwork of studies that were rarely brought together. Therefore, it was somewhat obscure: how each behavioural variable influences and shapes building practitioners' discrete compliance behaviours [22, 43]. To address this gap, [44] integrated different compliance theories in the legal and social science disciplines, e.g., the theory of planned behaviour [6], and designed a cross-theoretical compliance behaviour assessment model in going beyond minimal compliance with building energy regulations. It is the first model that uncovers a comprehensive range of behavioural variables that co-influence building practitioners' compliance behaviour in going beyond minimal compliance with building energy regulations. As per the author's model, the compliance outcome is mainly influenced by the building practitioner's behavioural intention to go beyond minimal compliance, the attitudes towards going beyond minimal compliance, and the personal norms. Then, the behavioural intention is further governed by four predictive variables that were summarised in detail in Table 2.

In addition, several research works [22, 23, 43, 46, 57] indicate that Australian residential building practitioners may exhibit various compliance behaviours depending on their occupation-related characteristics, such as length of work experience, occupation, involved project types (volume homes versus custom homes[1]), and involved project areas (urban areas versus regional areas). Although some insights were captured from the existing literature, scanty studies have demonstrated the building practitioners' compliance behaviours according to their occupational characteristics.

3 Case Description

The case investigation was conducted in the State of Victoria (Australia), which is one of the leading Australian states to introduce minimum residential thermal insulation for new dwellings. The state's government continuously supports the pursuit of high performing residential buildings by introducing a series of relevant projects such as the Net Zero Carbons project [69] and the 7 Stars Home project [70]. Under changes to the NCC 2022, Victoria agreed to increase the minimum NatHERS rating to 7 stars, which is set to have a mandatory commencement date of 1 May 2024 in this state. During May 2023 and November 2023, there were 20,320 new detached houses assessed via the NatHERS, and the average rating was only 6.3 stars [17]. As revealed in the first section, 79.76% of them were designed only to achieve minimum NatHERS 6-star standards. A merely 1.36% achieved the optimal 7.5 stars, which is essential for realising the optimal environmental and societal benefits, which is consistent with previous results [49].

The investigation was done via a questionnaire survey on the building practitioners who have more than two years of work experience in Victoria. The questionnaire collected firstly the building practitioner's occupation-related information, and, secondly, measured the compliance behaviour of each respondent according to

[1] VOLUME home sector, also referred to as project home sector, is the mainstream development type in Australia. It is a typical of modern suburban housing developments of reasonable scale (at least 30 plus allotments) where a number of builders compete on price and design features and where the blocks or allotments are developed on a phased basis over several or more years [11].

Table 2 Predictive variables co-shaping building practitioners' compliance behaviour, adapted from [44]

Variable	Indicator	Operational definitions or examples in the context of going beyond minimal compliance with the NatHERS
Attitudes towards going beyond minimal compliance	Perceived economic benefits	Perceived benefits through, e.g., getting more clients, or differentiating the delivered projects via competitive advantage
	Perceived economic costs	Perceived increased compliance costs due to high performing designs
Subjective norms	Requests from clients	Pressure from clients' requests
	Expectations from building industry colleagues and peers	Pressure due to comparing with other building practitioners, or seeking societal intangibles, including reputations or respect or approval from other industry stakeholders
Perceived behavioural control	Self-efficacy in terms of confidence to go 7 stars and higher	Understanding of compliance options in the NCC, as well as knowledge and skills possessed by building practitioners to design and deliver a beyond the 7-star house
	Self-efficacy in terms of perceived easiness to go 7 stars and higher	Perceived level of easiness or difficulty in going 7 stars and higher that can result from past experience
	Perceived controllability in going 7 stars and higher	Whether going 7 stars and higher is within control, i.e., constraints or facilitators that influence building practitioners' compliance decisions, e.g., shortage or provision of certain sustainable building materials
Personal norms	Moral agreement with environmental protection	Degree of moral alignment with the goals of environmental protection as a building practitioner
	Moral agreement with carbon emissions reduction	Degree of moral alignment with the goals of greenhouse gas emissions reduction as a building practitioner

(continued)

Table 2 (continued)

Variable	Indicator	Operational definitions or examples in the context of going beyond minimal compliance with the NatHERS
	Agreement regarding whether going beyond 7 stars is correct	Degree of agreement towards the overall assessment guidance offered by the NatHERS
	Agreement regarding whether going beyond 7 stars can lead to emissions reduction	Degree of agreement regarding the statement: Going 7 stars and higher benefits carbon reduction for society
	Agreement regarding whether going beyond 7 stars can lead to energy consumption reduction	Degree of agreement regarding the statement: Going 7 stars and higher benefits energy consumption reduction

Table 2. The indicators were operationalised in relation to going beyond minimal compliance with the NatHERS, i.e., delivering a project at 7 stars and higher. All the indicators were measured on a 5-point Likert scale, with 1 representing the lowest value and 5 the highest. As the NatHERS rating star is given according to the simulation result of the final design, the building practitioners were further limited to those who are actively involved in the design phase of the residential projects. As per the categories prescribed by the National Registration Framework [8], "building practitioners" in this chapter included architects, designers/draftspersons, builders, project managers, and residential energy efficiency consultants (thermal performance assessors). In total, there were 73 participating building practitioners.

4 Results

As an initial data processing step, factor analysis was used to assess the reliability and validity of the compliance behaviour variables and indicators in the questionnaire [71]. According to Hair et al. [32], the results of factor analysis have shown that the variables demonstrate consistency reliability with the index of composite reliability above the threshold (0.7). Convergent validity is also ensured with all the variables' AVE scores being above 0.5. Most indicators' reliability was confirmed as their factor loading is above 0.7. Finally, the discriminant validity of the variables was established as the heterotrait-monotrait ratio was below the threshold of 0.85 [35]. In addition, the value of the Kaiser–Meyer–Olkin index was 0.808 which suggests sample adequacy.

4.1 *Building Practitioners' Compliance Behaviours in General*

To differentiate compliance behaviours of building practitioners from various occupational characteristics (i.e., length of work experience, occupation, project type, and project area), descriptive analysis is carried out by comparing mean values of each behaviour variable among these occupational-related sub-groups. An overview of the average value of the four compliance behaviour variables is depicted in Fig. 1.

According to Fig. 1, the "attitudes" has the lowest value at 2.79, which is below the middle of the measuring scales. This means that the majority of building practitioners have a calculative thinking and they generally associate little economic benefits with going beyond 7 stars. This can be attributed to building practitioners' assumption of increased compliance costs with high performing designs. As reflected in an earlier report "*Understanding How the Building Industry Responds to Energy Efficiency Standards: Final Report*" that was commissioned by the then Department of Climate Change and Energy Efficiency [5], some building practitioners (especially small-volume home practitioners) were more likely to perceive high costs to achieve compliance. Furthermore, the value of "subjective norms" stands at 4.12, indicating that these building practitioners usually pay high attention to the opinions of their important stakeholders such as clients, peers, and society. Building practitioners work in a very competitive industry. Thus, whether their clients consider and request energy-efficient houses will play an influential role. Next, the "perceived behavioural control" has an average value of 3.63. This demonstrates a mediocre level of perceived self-efficacy and controllability to go 7 stars and higher. This can imply that the building practitioners' level of knowledge on energy-efficient design techniques or compliance options is not sufficient. Finally, the "personal norms" scores at 4.27 on average, indicating a high level of normative alignment with the NatHERS

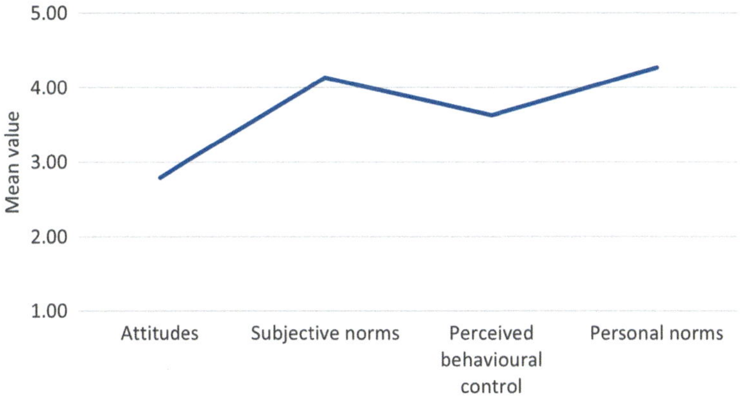

Fig. 1 Mean value of compliance behaviour predictive variables among the 73 building practitioners. Authors' own illustration

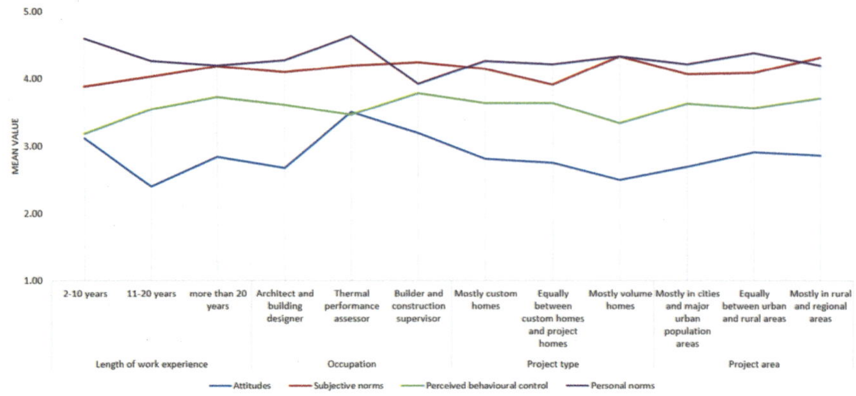

Fig. 2 Mean value of compliance behaviour variables per building practitioners' occupational characteristics. Authors' own illustration

substantive goals. With the increasing awareness of the negative impacts of greenhouse gas emissions, it is fairly common that more Australian building practitioners' personal norms are in line with their industry's low carbon commitment.

4.2 Building Practitioners' Compliance Behaviours Based on Their Occupational Characteristics

To depict different compliance behaviours of building practitioners with occupationally varying characteristics, the mean values of compliance behaviour variables of each sub-group based on occupational information are examined, as summarised in Fig. 2.

Figure 2 shows that as work experience increases, values of "subjective norms" and "perceived behavioural control" display an ascending trend. These trends indicate that experienced building practitioners generally pay more attention to the social respect and reputation of their works, and they usually possess more confidence towards beyond-minimal compliance due to extensive knowledge and experience. The value of "personal norms", on the contrary, shows a descending trend with experience increased. It can be reasonably explained as a result of boosting environmental awareness among the younger generations. The value of "attitudes" reaches a low point (2.39) among practitioners with 11–20 years of work experience, meaning this group rarely considers beyond-minimum projects profitable.

In terms of the occupations, at first glance, values of "subjective norms" and "perceived behavioural control" do not show a significant difference among the three sub-groups. However, both "attitudes" (3.50) and "personal norms" (4.64) are shown highest among the thermal performance assessors. This means that these assessors tend to think that they understand the potential economic benefits of high energy performance dwellings, and that they generally agree with the NatHERS guidelines on how to deliver such high performing projects. In contrast, the other two sub-categories of professionals who actually design the projects show considerably lower "attitudes" and "personal norms". This contrasting result is consistent with the situation highlighted by Pitt & [57] that Australian designers and builders have a low willingness to listen to their energy assessors' advice in improving design specifications for a higher compliance level.

Building practitioners who are frequently involved in volume homes also display different compliance behaviours, as compared to those who mostly engage in custom homes. Two main different compliance behavioural variables are observed between volume home practitioners and custom home practitioners. Volume home building practitioners have less favourable "attitudes" (2.50) than those who work on custom homes (2.81). This has to do with the fact that Australian volume home building practitioners are more profit-oriented so they generally care a lot about costs and benefits [22]. Lower attitudes thus infer their perceptions that going beyond minimal compliance would incur high compliance costs. The second difference lies in "perceived behavioural control", where the volume home building practitioners (3.33) had a lower value than custom home practitioners (3.64). In Australia, volume home building practitioners usually work in teams that have standardised designs that are intended at the minimal compliance level for a long period. Hence, it is without an individual's control to make further substantive changes to these set designs, which in part explains their lower perceived behavioural control. Furthermore, a lower perceived behavioural control level of volume home designers may also result from their lower confidence in delivering beyond-minimum projects. It is mostly due to a lack of up-to-date knowledge of innovative energy-efficient design techniques, or a proper understanding of the compliance options prescribed in the NCC.

Among the building practitioners from different project-area sub-groups, it seems that there exist no drastic differences in compliance behavioural variables. Victoria has three climate zones as defined by the National Bureau of Meteorology. Houses are expected to exhibit different design specifications due to the variations in their located climate zones. Contrary to this expectation, Fig. 2 shows that building practitioners working in regional or urban areas had similar compliance behaviours in all four constituent variables.

In summary, the above results unveil an overall picture of building practitioners' compliance behaviour with the NatHERS. Results also comprehensively reveal the existence of different compliance behaviours among Australian building practitioners owning various occupation-related characteristics.

5 Discussion

The overall low level of building practitioners' "attitudes" towards higher building energy performance is not exclusive to Australia. In the residential industry in the UK, small builders are found to have unfavourable attitudes towards the delivery of sustainable homes [68]. Similar passive attitudes towards green construction practices are also reported by [58] in their study on Chinese building practitioners. This consensus highlights the importance of changing attitudes. As stressed by several studies in the context of the UK, Singapore, and Indonesia [25, 37, 60, 68], stronger government incentives for going higher energy stars (e.g., subsidies to offset high premiums of designing and constructing beyond 7-star projects) are essential prerequisites for behaviour change by building practitioners. This further demonstrates the government's role in promoting building practitioners' compliance levels using different intervention tools, which is consistent with several previous studies. Doyon and Moore [20] examined the complementary effects of government interventions to facilitate the sustainable housing transition in Australia and Canada [20]. Their study showed that housing performance was typically governed by mandatory performance policies, and the use of suitable voluntary interventions can augment the implementation of the mandatory requirements if the government's willingness to make changes is unlocked. However, Australia's experience of regulating the residential building energy efficiency has shown a glacial speed of policy manifestation, even under the severe challenge of the global climate change [12]. It took the federal government more than a decade to increase the stringency of the minimum energy performance standards from 6 stars in 2010 to 7 stars in 2022, which has been well behind the other developed countries' developments [29]. Compared with the federal-level slow development, the state-level government seems to have a stronger will to expedite the progress by introducing specific policies to encourage the industry to go beyond minimal energy standards [34]. For instance, the government of the Australian Capital Territory (ACT) has established the *ACT Sustainable Energy Policy* to encourage the exceeding of minimum energy rating standards for new detached houses [36]. Nevertheless, such government support in Australia is far from sufficient. In general, the Australian government was blamed for its passive approach in encouraging net-zero transition in the residential building industry, leaving everything to the market and the industry [46].

The generally high level of "subjective norms" revealed in this study is consistent with findings in the United Arab Emirates [65]. This notes the importance of building practitioners' perceptions in terms of building up positive societal images and fulfilling customers' expectations. It further emphasises that, beyond the role of governments, consumers are also considered important in stimulating overall compliance. Thus, how to encourage more building consumers to choose an energy-efficient house requires an effective solution. An array of recent studies and reports [21, 64] has shown that Australian residential building consumers are supportive of the transition to building energy efficiency. In the US, the European Union, and other countries, building owners of energy-efficient buildings were provided with financial incentives such as tax deductions or tax payment exemption [53, 63], from which Australia can draw some lessons.

Next, this study finds a mediocre score of "perceived behavioural control" possessed by Australian building practitioners. Similar results can be seen in the Indonesian context [25]. As a consequence, bridging the skills and knowledge gap by providing relevant training and professional development is deemed essential and urgent [62]. This further underlines the significant role of the building industry's professional bodies by offering educational routes and setting the curriculum of degree and training programmes [47]. For instance, the ABCB has developed tailored NCC Continuing Professional Development (CPD) courses for architects and draftsperson, which include (1) Housing Energy Efficiency and (2) Pathways to Compliance [1]. Several research works in Australia's residential industry demonstrate that training programmes can improve building practitioners' "perceived behavioural control" [13, 67]. As with other similar training courses provided by different professional bodies in Australia (e.g., Australian Institute of Architects, Australian Building Sustainability Association, and Design Matters National), such training programmes are not compulsory. Hence, policymakers should rethink the mechanisms to drive building practitioners' participation. In addition, the level of "personal norms" also exhibits similar findings in different countries. Concurrent with this study's high degree of personal norms in Australia, many building practitioners worldwide are becoming more enthusiastic about core elements of sustainability (such as energy efficiency) in buildings [26]. In contrast to this contemporary finding, earlier studies [51, 59] revealed that building practitioners tend to have low levels of engagement with the environmental agenda. This suggests that in light of common global environmental issues such as climate change, building practitioners have gradually embraced the idea of sustainability and net zero of the built environment.

The analysis of compliance behaviour per each practitioner's characteristics highlights a notable difference in "attitudes" between thermal performance assessors (higher) and other occupations such as architects and builders (lower). Besides the "attitudes", the value of "personal norms" is also remarkably higher among thermal performance assessors than other professionals. In Australia, a thermal performance assessor determines the homes' energy rating and also provides improvement advice for the energy performance in the design phase. The divergent result between the assessors and other building practitioners in this study indicates that the professionals who actually design the homes generally possess less favourable attitudes and lower agreement with the energy efficiency requirements' substantive objectives than the thermal performance assessors. Further implications can be drawn that the role of Australian thermal performance assessors is more about data entry into energy rating tools rather than the provision of energy performance advice. The same conclusion is seen in the research work by [72]. In line of thinking, it would be helpful if architects and designers enhanced their willingness to listen to their assessors' advice. Nevertheless, a NatHERS benchmarking study performed on behalf of the Commonwealth Department of Industry reveals that 64% of thermal performance assessors have an error over 0.25 of a star [27]. Therefore, enhancing the accreditation scheme to improve the thermal performance assessor's expertise is of utmost importance.

6 Conclusions

This chapter explores building practitioners' compliance behaviours in compliance with the NatHERS in Australia. In general, the building practitioners have unfavourable "attitudes" towards going for 7 stars and higher. The high score of "subjective norms", "perceived behavioural control", and "personal norms" indicates that building practitioners are keen to enhance their technical skills and a desire to align their works with the industry's low carbon vision, but are required to consider their societal pressure from industry peers and clients. In addition, compliance behaviours further differ owing to varying occupational characteristics. There exists significant variations in "attitudes" and "perceived behavioural control" between building practitioners who engage in volume homes and those who are involved in custom homes. Strong differences in "attitudes" and "personal norms" are also visible between thermal performance assessors and other occupations like architects and designers. Then, distinctions in "subjective norms" are demonstrated within building practitioners with various lengths of work experience.

The chapter findings provide a basis for policymakers and industry advisory bodies to develop targeted response interventions, making them more aligned with the interests of building practitioners. Findings from this chapter can thus contribute to the State in terms of more meaningful interventions to augment the mandatory energy performance standard, ultimately paving the way towards sustainable buildings. Considering the four compliance behaviour variables revealed in this study, future policy design could consider (1) coupling profitability with the delivery of beyond 7-star projects (i.e., enhancing the "attitudes towards going beyond 7 stars" by lifting the economic benefits); (2) improving the accreditation scheme for thermal performance assessors (i.e., leveraging the "subjective norms" so that designers or architects are more willing to listen to assessors' advice); (3) designing incentives to encourage more clients to choose 7-star houses (i.e., taking advantage of the "subjective norms" as clients' requests are important in building practitioners' compliance decision-making); (4) developing skills and knowledge to keep pace with high energy star rating design techniques (i.e., enhancing the "perceived behavioural control" through up-to-date training and suitable mechanisms to attract training participation); and (5) refining the NatHERS framework (i.e., improving the "personal norms" by aligning the NatHERS assessment outcome with actual building energy performance).

References

1. ABCB (2022a) Build your understanding with NCC CPD [Online]. Available: https://cpd.abcb.gov.au/abcbncc/design-and-engineering. Accessed 15 Jan 2022
2. ABCB (2022b). New whole-of-home energy efficiency—what's it all about? [Online]. Available: https://ncc.abcb.gov.au/news/2022/new-whole-home-energy-efficiency-whats-it-all-about. Accessed 14 Dec 2022
3. ABCB (2023) NCC 2022 state and territory adoption dates [Online]. Available: https://abcb.gov.au/ncc-2022-state-and-territory-adoption-dates. Accessed 12 Dec 2023

4. Adebayo TS, Ullah S (2023) Towards a sustainable future: The role of energy efficiency, renewable energy, and urbanization in limiting CO_2 emissions in Sweden. Sustainable Development
5. AECOM (2012) Understanding how the building industry responds to energy efficiency standards. In EFFICIENCY, D. O. C. C. E. (ed.). Canberra, Australia
6. AJZEN I (1991) The theory of planned behavior. Organ Behav Hum Decis Process 50:179–211
7. Armstrong G, Pears A, Delafoulhouze M, Moore T (2022) 7-star housing is a step towards zero carbon-but there's much more to do, starting with existing homes. J Home Econ Inst Austr 27:54–55
8. AUSTRALIAN BUILDING CODES BOARD (ABCB) (2021) National Registration Framework for building practitioners-Model guidance on BCR recommendations [Online]. https://www.abcb.gov.au/sites/default/files/resources/2022/BCR-rec1-2-National-registration-framework.pdf. Accessed 10 Nov 2023
9. AUSTRALIAN GREENHOUSE OFFICE (2000) Impact of minimum performance requirements for class 1 buildings in Victoria. Australian Greenhouse Office, Canberra, Australia
10. AUSTRALIAN SUSTAINABLE BUILT ENVIRONMENT COUNCIL (ASBEC) & CLIMATE WORKS AUSTRALIA (2018) Built to perform—An industry led pathway to a zero carbon ready building code [Online]. Sydney: Cooperative Research Centre for Low Carbon Living. Available: https://apo.org.au/node/199346. Accessed 2 May 2022
11. Belusko M, O'Leary T (2010) Cost analyses of measures to improve residential energy ratings to 6 stars-playford North Development, South Australia. Australasian J Constr Econ Build 10:48–59
12. Berry S, Marker T (2015) Residential energy efficiency standards in Australia: where to next? Energ Effi 8:963–974
13. Bosch-Sijtsema P, Isaksson A, Lennartsson M, Linderoth HC (2017) Barriers and facilitators for BIM use among Swedish medium-sized contractors-"We wait until someone tells us to use it." Visual Eng 5:1–12
14. Bozeman B (2022) Rules compliance behavior: a heuristic model. Perspect Public Manage Governance 5:36–49
15. COMMONWEALTH OF AUSTRALIA (2023) Trajectory for low energy buildings [Online]. Available: https://www.dcceew.gov.au/energy/energy-efficiency/buildings/trajectory-low-energy-buildings. Accessed 12 Dec 2023
16. CSIRO (2023a) Australian housing data [Online]. Available: https://ahd.csiro.au/dashboards/energy-rating/states/. Accessed 30 Jan 2024
17. CSIRO (2023b) Australian housing data [Online]. https://ahd.csiro.au/. Accessed 16 September 2023
18. Daniel L, Williamson T, Soebarto V (2017) Comfort-based performance assessment methodology for low energy residential buildings in Australia. Build Environ 111:169–179
19. Department of Climate Change, Energy, the Environment and Water (2023) Australian Energy Update 2023 [Online]. Canberra: Commonwealth of Australia Canberra. Available: https://www.energy.gov.au/sites/default/files/Australian%20Energy%20Update%202023_0.pdf. Accessed 12 Jan 2024
20. Doyon A, Moore T (2020) The role of mandatory and voluntary approaches for a sustainable housing transition: evidence from Vancouver and Melbourne. Urban Policy and Research 38:213–229
21. Energy Consumer Australia (2022) Energy consumer sentiment research report [Online]. Available: https://ecss.energyconsumersaustralia.com.au/wp-content/uploads/2023/02/FINAL-Oct-2022-Qual_Essential-Report.pdf. Accessed 23 Feb 2024
22. Enker RA, Morrison GM (2019) Behavioral facilitation of a transition to energy efficient and low-carbon residential buildings. Buildings 9:226
23. Enker RA, Morrison GM (2020) The potential contribution of building codes to climate change response policies for the built environment. Energ Effi 13:789–807
24. Falana J, Osei-Kyei R, Tam VW (2023) Towards achieving a net zero carbon building: a review of key stakeholders and their roles in net zero carbon building whole life cycle. J Build Eng 108223

25. Fitriani H, Ajayi S (2023) Barriers to sustainable practices in the Indonesian construction industry. J Environ Planning Manage 66:2028–2050
26. Fitriani H, Ajayi S (2023) Investigation of requisite measures for enhancing sustainable construction practices in Indonesia. Eng Constr Archit Manag 30:2602–2620
27. Floyd W, Isaacs T, Hills R (2014) *NatHERS Benchmark Study* [Online]. Available: https://www.nathers.gov.au/sites/default/files/Research%2520Rep%2520-%2520Assessor%2520Benchmark%2520Study.pdf. Accessed 13 Dec 2023
28. Francis J, Eccles MP, Johnston M, Walker A, Grimshaw JM, Foy R, Kaner EF, Smith L, Bonetti D (2004) Constructing questionnaires based on the theory of planned behaviour: a manual for health services researchers. In: Tyne UONU (ed.). Newcastle upon Tyne, UK
29. Fuerst F, Warren-Myers G (2018) Does voluntary disclosure create a green lemon problem? Energy-efficiency ratings and house prices. Energy Economics 74:1–12
30. Gounder S, Hasan A, Shrestha A, Elmualim A (2023) Barriers to the use of sustainable materials in Australian building projects. Eng Constr Archit Manag 30:189–209
31. Hafez FS, Sa'Di B, Safa-Gamal M, Taufiq-Yap Y, Alrifaey M, Seyedmahmoudian M, Stojcevski A, Horan B, Mekhilef S (2023) Energy efficiency in sustainable buildings: a systematic review with taxonomy, challenges, motivations, methodological aspects, recommendations, and pathways for future research. Energ Strat Rev 45:101013
32. Hair JF, Ringle CM, Sarstedt M (2011) PLS-SEM: Indeed a silver bullet. J Market Theory Practice 19:139–152
33. Hair JF, Risher JJ, Sarstedt M, Ringle CM (2019) When to use and how to report the results of PLS-SEM. Eur Bus Rev 31:2–24
34. Harrington P, Toller V (2017) Best practice policy and regulation for low carbon outcomes in the built environment [Online]. CRC for Low Carbon Living. https://apo.org.au/node/232106. Accessed 1 June 2022
35. Henseler J, Ringle CM, Sarstedt M (2015) A new criterion for assessing discriminant validity in variance-based structural equation modeling. J Acad Mark Sci 43:115–135
36. Hu X, Si T, Liu C (2017) Total factor carbon emission performance measurement and development. J Clean Prod 142:2804–2815
37. Hwang BG, Tan JS (2012) Green building project management: obstacles and solutions for sustainable development. Sustain Dev 20:335–349
38. Iyer-Raniga U, Moore T, Ho O (2023) Residential building sustainability rating tools in Australia [Online]. https://www.oneplanetnetwork.org/sites/default/files/from-crm/Issue%252002_July%25202023.pdf. Accessed 20 Nov 2023
39. Langevoort DC (2018) Behavioral ethics, behavioral compliance. In: Arlen J (ed.) Research handbook on corporate crime and financial misdealing. Edward Elgar Publishing, Massachusetts, USA
40. Law T (2023) An increasing resistance to increasing resistivity. Archit Sci Rev 66:108–121
41. Li HX, Moore T, Huang J, Zhang P, Costin G (2022) Towards zero carbon housing in Victoria, Australia: a policy and incentive framework. Energ Strat Rev 40:100802
42. Lu Y, Karunasena G, Liu C (2022) Issues in compliance with low-carbon requirements in the Australian residential building industry. In: Perera S, Hardie M (eds.) 45th Australasian Universities Building Education Association (AUBEA) Conference, 2022a Penrith, Australia. Western Sydney University, Sydney, Australia, 909–918
43. Lu Y, Karunasena G, Liu C (2022) A systematic literature review of non-compliance with low-carbon building regulations. Energies 15:9266
44. Lu Y, Karunasena G, Liu C (2024) Conceptual cross-theoretical assessment model for practitioners' compliance behavior with building energy codes. J Leg Aff Disput Resolut Eng Constr 16:04523039
45. Maqbool R, Jowett E (2023) Conserving a sustainable urban environment through energy security and project management practices. Environ Sci Pollut Res 30:81858–81880
46. Martek I, Hosseini MR, Shrestha A, Edwards DJ, Durdyev S (2019) Barriers inhibiting the transition to sustainability within the Australian construction industry: An investigation of technical and social interactions. J Clean Prod 211:281–292

47. Martin S, Hall A (2002) Sustainable development and the professions. Planet 5:17–18
48. May PJ (2004) Compliance motivations: affirmative and negative bases. Law Soc Rev 38:41–68
49. Moore T, Berry S, Ambrose M (2019) Aiming for mediocrity: the case of Australian housing thermal performance. Energy Policy 132:602–610
50. Moore T, Holdsworth S (2019) The built environment and energy efficiency in Australia: current state of play and where to next. Energy Performance in the Australian Built Environment
51. Myers D (2005) A review of construction companies' attitudes to sustainability. Constr Manag Econ 23:781–785
52. Nielsen VL, Parker C (2012) Mixed motives: economic, social, and normative motivations in business compliance. Law Policy 34:428–462
53. Olubunmi OA, Xia PB, Skitmore M (2016) Green building incentives: a review. Renew Sustain Energy Rev 59:1611–1621
54. Omrany H, Soebarto V, Ghaffarianhoseini A (2022) Rethinking the concept of building energy rating system in Australia: a pathway to life-cycle net-zero energy building design. Archit Sci Rev 65:42–56
55. Pan W, Garmston H (2012) Building regulations in energy efficiency: compliance in England and Wales. Energy Policy 45:594–605
56. Parag Y, Janda KB (2014) More than filler: Middle actors and socio-technical change in the energy system from the "middle-out." Energy Res Soc Sci 3:102–112
57. PITT & Sherry (2014) National energy efficient building project report, prepared for the State of South Australia [Online]. https://energymining.sa.gov.au/__data/assets/pdf_file/0019/315415/NEEBP-final-report-November-2014.pdf. Accessed 12 March 2022
58. Qi G, Shen LY, Zeng S, Jorge OJ (2010) The drivers for contractors' green innovation: an industry perspective. J Clean Prod 18:1358–1365
59. Revell A, Blackburn R (2007) The business case for sustainability? An examination of small firms in the UK's construction and restaurant sectors. Bus Strateg Environ 16:404–420
60. Shan M, Liu W-Q, Hwang B-G, Lye J-M (2020) Critical success factors for small contractors to conduct green building construction projects in Singapore: identification and comparison with large contractors. Environ Sci Pollut Res 27:8310–8322
61. Shapiro S (2016) The realpolitik of building codes: overcoming practical limitations to climate resilience. Build Res Inform 44:490–506
62. Shari Z, Soebarto V (2012) Delivering sustainable building strategies in Malaysia: stakeholders' barriers and aspirations. ALAM CIPTA, Int J Sustain Tropical Design Res Pract 5:3–12
63. Shen L, He B, Jiao L, Song X, Zhang X (2016) Research on the development of main policy instruments for improving building energy-efficiency. J Clean Prod 112:1789–1803
64. Shooshtarian S, Hosseini MR, Martek I, Shrestha A, Arashpour M, Costin G, Seaton S (2021) Australia's push to make residential housing sustainable-Do end-users care? Habitat Int 114:102384
65. Shurrab J, Hussain M, Khan M (2019) Green and sustainable practices in the construction industry: a confirmatory factor analysis approach. Eng Constr Archit Manag 26:1063–1086
66. Simpson K, Janda KB, Owen A (2020) Preparing 'middle actors' to deliver zero-carbon building transitions. Build Cities 1:610–624
67. Steinhardt DA, Manley K, Miller W (2014) What's driving the uptake of prefabricated housing in Australia? [Online]. Available: https://eprints.qut.edu.au/216492/. Accessed 21 Feb 2023
68. Sun M, Geelhoed E, Caleb-Solly P, Morrell A (2015) Knowledge and attitudes of small builders toward sustainable homes in the UK. J Green Build 10:215–233
69. Sustainability Victoria (2022) Zero Net Carbon Homes [Online]. https://www.sustainability.vic.gov.au/energy-efficiency-and-reducing-emissions/building-or-renovating/build-for-energy-efficiency. Accessed 12 Mar 2022
70. Sustainability Victoria (2023) 7 Star homes program—participate [Online]. https://www.sustainability.vic.gov.au/energy-efficiency-and-reducing-emissions/building-or-renovating/7-star-homes-program/participate. Accessed 22 Dec 2023
71. Tabachnick BG, Fidell LS, Ullman JB (2019) Using multivariate statistics. Pearson Boston, New York, USA

72. Thoo S, Marsh A, Pitman M (2016) Improving thermal performance design outcomes through NatHERS and BIM integration. In: 50th International Conference of the Architectural Science Association, 2016 Adelaide, Australia
73. The Architectural Science Association and The University of Adelaide, 487–496
74. Wilkinson S (2013) Are sustainable building retrofits delivering sustainable outcomes? Pac Rim Prop Res J 19:211–222
75. Zapata-Lancaster G, Tweed C (2014) Designers' enactment of the policy intentions. An ethnographic study of the adoption of energy regulations in England and Wales. Energy Policy 72:129–139

Chapter 2
Navigating Progress in Policy Development for Energy Retrofitting Residential Buildings: Australian Insights

Nimasha Dilukshi Hulathoowage, Gayani Karunasena, Chunlu Liu, and Nilupa Udawatta

Abstract Australia seeks to achieve a 40% boost in energy productivity by 2030 from its 2015 baseline. Despite several emerging Policy Instruments (PIs) aimed at enhancing the energy efficiency of existing residential buildings, Australia's progress falls behind other developed economies due to deficiencies in these PIs. Therefore, this chapter reviews eight current federal-level PIs in Australia aimed at energy retrofitting residential buildings to identify prevailing deficiencies and recommend potential policy improvements. The findings reveal the need for a national plan to increase the renovation/retrofitting rate, a single national star rating system, research and innovation to integrate embodied emissions into the star rating system, and a mandatory disclosure for real estate transactions and periodic updates. The proposed financial mechanisms involve fossil fuel taxes to fund rebates for renewable energy sources. Financial incentives should be paired with educational programs and include a plan to gradually diminish incentives as the retrofit market matures. Financial incentives should target low-income groups by offering capital cost reductions rather than tax rebates. Moreover, contributing to knowledge, Table 4 offers a compendium of potential policy improvements to help Australian policymakers identify future initiatives for achieving national emission reduction targets for existing residential buildings. This study also serves as a guide for researchers to identify future research areas in policy development within this domain.

Keywords Energy retrofitting · Policy instruments · Residential buildings · Australia · National energy performance strategy

---

N. D. Hulathoowage (✉) · G. Karunasena · C. Liu · N. Udawatta
School of Architecture and Built Environment, Deakin University, Geelong, Australia
e-mail: n.hulathdoowage@research.deakin.edu.au

T. Ramachandra et al. (eds.), *Concepts, Policies and Technology for Sustainable Buildings*, Advances in Sustainability Science and Technology,
https://doi.org/10.1007/978-981-96-3086-8_2

1 Introduction

In Australia, residential buildings contribute 12% of carbon emissions—over 45 million tonnes annually [23]. While heating and cooling demands make up 40% of current household energy use [33], existing residential buildings will represent over half (55%) of the total housing stock by 2050 [14]. Therefore, the energy inefficiency of current houses significantly challenges achieving net-zero carbon residential buildings by 2050, necessitating energy retrofitting to improve their efficiency. Scientific studies have also shown the importance of energy retrofitting to meet emission reduction targets in existing houses across various climates [2, 10, 47]. Moreover, the positive impact of Policy Instruments (PIs), which promote energy retrofitting through codes, standards, and incentives, is a well-established concept [71]. However, Australian PIs often fall short of international best practices [11, 28].

Consequently, Doyon and Moore [28] examined mandatory and voluntary policies for sustainable housing by comparing Melbourne, Australia, and Vancouver, Canada. However, they did not focus on policies encouraging energy retrofitting in residential buildings. While many studies address policies for sustainable or net-zero buildings in Australia, there is a lack of comprehensive analysis of PIs, specifically aimed at energy retrofitting residential buildings [27, 28, 45]. Moreover, Zhang et al. [85] conducted a shallow examination of Australian PIs for energy retrofitting in a global study, limited to a few examples from the Australian context. Thus, this chapter provides an in-depth examination of federal-level PIs in Australia that encourage energy retrofitting in residential buildings. The primary objectives of the chapter are to (i) review the achievements made, challenges faced, and lessons learned through current PIs, and (ii) explore potential policy improvements.

2 Literature Review

Residential buildings globally account for 30% of total energy consumption, approximately three times the commercial buildings' energy consumption [49]. Moreover, projections suggest that global energy consumption in buildings will accelerate in the coming decades [48]. Thus, existing housing stock should be retrofitted to improve energy efficiency. Retrofitting generally involves modifying or updating a building or structure to enhance its performance, efficiency, or functionality. Energy retrofitting, however, focuses exclusively on improving a building's energy efficiency. Energy retrofitting encompasses three groups of retrofit measures, including passive, active, and renewable measures. While passive measures involve partially modifying the building fabric to adapt the building to passive services and to facilitate thermal comfort with minimal energy use, active measures lead to substantial repair, modification, or replacement with energy-efficient appliances [47]. In contrast, renewable measures increase the self-sufficiency of buildings in case of energy demand, and decrease emissions [47].

Various PIs have been implemented globally to increase the retrofitting rate of residential buildings. One notable example is the Building Passport introduced by the European Union (EU) through its Building Directive 2010/31/EU, which involves monitoring energy retrofitting activities throughout a building's service life [38]. Furthermore, the Clean Energy DC Omnibus Act in Washington DC focuses on iteratively retrofitting existing buildings to achieve sector-specific minimum energy star ratings [9]. In Japan, energy retrofitting is mandated through the Building Energy Efficiency Act [56]. Despite the growing number of PIs worldwide, the retrofitting rate in many developed countries remains notably low. For instance, Europe has an annual retrofitting rate of around 1%, falling significantly below the European Commission's recommended target of 3% for achieving environmental goals [37].

Numerous studies have examined PIs related to energy retrofitting in various countries. For instance, Galvin and Sunikka-Blank [40] examined energy retrofitting regulations and incentives in Germany. They proposed a new policy approach that promotes a balanced mix of partial and comprehensive energy retrofitting by leveraging household behaviour changes based on a deeper understanding of fuel-saving motivations and fuel price elasticity. Liu et al. [53] studied energy retrofitting policies implemented in China from 1996 to 2019, highlighting that existing mandatory building regulations aimed at enhancing energy efficiency are inadequate to achieve a substantial reduction in carbon emissions in the future. Bergman and Foxon [10] examined new policy framing in the UK for energy retrofitting and highlighted the government's crucial roles in a demand-driven market. These roles include ensuring stable, long-term policy frameworks, minimising financial investment risks; and fostering innovation among decentralised stakeholders.

Most of the global studies have classified PIs related to energy retrofitting to better understand and evaluate their effectiveness. Therefore, various classification methods for categorising PIs are evident in the literature. Among these classifications, seven common clusters were identified as core policy applications. Table 1 illustrates these core policy applications.

Literature indicates that DI and FI are the most common applications [45, 76]. EPA and EPD are also important, particularly in Western countries [65]. Command and control instruments are emphasised in China [53]. In addition, only one study investigates policies in Australia for energy retrofitting, and it is also limited to a few examples [85]. Moreover, despite the varied focuses of these studies, policy development processes also differ significantly across countries. Thus, investigating Australia's federal-level PIs that encourage energy retrofitting residential buildings adds new insights to the existing knowledge.

2.1 Australian Policy Instruments for Energy Retrofitting

While PIs play a crucial role in boosting the energy retrofitting rate of residential buildings by addressing market failures and barriers to the adoption of retrofit measures [10, 16, 31], Australia's PIs remain subpar compared to international best

Table 1 Core policy applications

Core policy applications	Descriptions	References
Direction-based Instruments (DI)	DI involves strategies, such as action plans, guidelines, and retrofit directives	[45, 85]
Energy Performance Assessment (EPA)	EPA evaluates the energy efficiency of buildings and appliances, often using a star rating system	[76, 83]
Minimum Energy Performance Requirements (MEPR)	MEPR regulates energy use and emissions in buildings and appliances through mandatory thresholds or efficiency requirements	[83, 65]
Energy Performance Disclosure (EPD)	EPD provides information on the energy efficiency of buildings and appliances to enhance market transparency	[83, 73]
Finance and Incentives (FI)	FI includes subsidies, low-interest loans, and tax reductions to incentivise energy retrofitting practices and emission reduction initiatives	[46, 69]
Research and Innovation (RI)	RI focuses on creating strategies and tools to improve the energy efficiency of existing residential buildings	[83, 72]
Education and Training (ET)	ET delivers the necessary knowledge to promote energy-efficient behaviour and investments in energy retrofitting and provides technical support	[69, 8]

standards [11, 28]. However, the Council of Australian Governments (COAG) Energy Council (2019) reported that enacting energy retrofitting policies for residential buildings could cut greenhouse gas (GHG) emissions by 40.3 $MtCO_2$-e by 2050, yielding a net present value of $3.4 billion. In light of this potential, Australia has a few federal-level PIs that focus on encouraging energy retrofitting in residential buildings. These PIs include National Energy Performance Strategy (NEPS), National Scorecard (NS), Greenhouse and Energy Minimum Standards (GEMS) Act 2012, Home Insulation Program (HIP), Solar Hot Water Rebate (SHWR), Small Scale Renewable Energy Target (SRET), House Energy Efficiency Inspections Project (HEEIP), and Residential Energy Baseline Study (RBS): Australia 2000–2030 (refer to Table 2).

Most PIs in Table 2 are voluntary, except for the GEMS Act. Mandatory PIs are rare due to political influences from frequent government turnovers and limited local authority [28]. Nevertheless, voluntary initiatives, such as SRET promoting rooftop solar panels, provide initial steps towards energy retrofitting by offering knowledge and resources. Moreover, PIs related to finance and incentives mainly target owner-occupied houses, resulting in a lack of federal policies for energy retrofitting rental properties, which make up 30.6% of Australian dwellings [4]. Most PIs in Table 2 follow the prescriptive path, offering explicit guidance to homeowners on retrofit measures, while the performance-based path allows more flexibility in implementation. While demand-side PIs directly encourage energy retrofitting, supply-side PIs complement these demand-side PIs by reducing the overall emission profile of existing houses.

Table 2 Federal-level policy instruments in Australia for energy retrofitting in residential buildingsPlease check the layout and modified citations of references present in all Tables, and amend if necessary.

PIs	Date/period	Mandatory/ voluntary	Housing type	Aim/purpose	Clusters of application	Demand-side/ supply-side management	Compliance path
NEPS	Proposed to develop from 2022 to 2026 [13, 23]	N/A	All housing types [22]	Accelerate nationwide energy efficiency and electrification [23]	Direction-based instruments [13]	Demand-side Management	Performance path
NS	Launched in 2021 [70]. Proposed to execute from 2021 to 2025 [58, 78]	Voluntary [70]	All housing types [70]	Encourage energy efficient improvements [28, 52, 58]	Energy performance assessment	Demand-side Management	Prescriptive, performance paths
HEEIP	Launched in 2015 [58]	Voluntary [58]	Newly built houses [3]	Evaluate construction quality for energy efficiency [3]	Research and innovation [58]	Demand-side Management	N/A
RBS	Launched in 2015 [58]	Voluntary [58]	Class1a, 1b, 2 type houses [32]	Analyse energy consumption trends in Australian houses from 2000 to 2030 [32]	Research and innovation [58]	Demand-side Management	N/A
GEMS Act 2012	Launched in 10/2012 and modified in 2013 and 2022 [GEMS Act 2012 [39]]	Mandatory [GEMS Act 2012 [39]]	New purchase of products for all housing types [GEMS Act 2012 [39]]	Assist consumers in making more informed and energy-efficient decisions [63]	Energy performance disclosure [GEMS Act 2012 [39]]	Supply-side Management	Prescriptive, performance path

(continued)

Table 2 (continued)

PIs	Date/period	Mandatory/ voluntary	Housing type	Aim/purpose	Clusters of application	Demand-side/ supply-side management	Compliance path
HIP	Proposed to execute from 2009 to 2011 [5], but cancelled in 2010 [21]	Voluntary [5]	Uninsulated owner-occupied houses [5]	Reach 2.7 million energy-efficient owner-occupied homes [5, 79]	Finance and incentives	Demand-side Management	Prescriptive path
SHWR	Proposed to execute Phase 1 from 2009 to 2012 [35]; rolled into REBS in 2010 [60] and ended in 2012 [75]	Voluntary [35]	Owner-occupied houses without solar hot water [35]	Phase out electric hot water systems to reduce national carbon emissions [74]	Finance and incentives	Demand-side Management	Prescriptive path
SRET	Initiated as RET, split as LRET and SRET, and will be continued until 2030 [18, 19]	Voluntary [19]	All housing types	Encourage the application of renewable energy sources at the domestic level [19]	Finance and incentives	Demand-side Management	Prescriptive path

Note RET, renewable energy target; LRET, large-scale renewable energy target; REGO, renewable electricity guarantee of origin; DCCEEW, the department of climate change, energy, the environment and water; N/A, Not Applicable

The next section further discusses the rationale behind choosing Australia as the case study.

3 Methodology

This chapter examines selected PIs within a specific case study (Australia). Therefore, the descriptive case study approach was used to evaluate Australia's energy retrofitting policy landscape, drawing on data from books, academic journals, government reports, and institutional websites [84]. Additionally, theme-based cross-case analysis was conducted to identify themes from the existing studies, achieving the chapter's primary objectives.

Australia was selected as the case study for several reasons. First, Australia ranks as the 14th highest emitter globally (Global Carbon Budget 2023), with a lower rate of annual decarbonisation efficiency growth in the residential building stock than other developed countries [54]. Moreover, eight million Australian homes have been constructed before implementing the National Construction Code (NCC) [41], which governs the minimum energy efficiency of new constructions and major renovations in Australia since 1996 [77].

Second, the residential building stock in Australia is projected to comprise more than half (55%) of all residential buildings by 2050 [14]. However, recent research has highlighted significant deficiencies in the energy performance of existing residential buildings [3, 7]. For instance, studies by CSIRO revealed that many houses tested for airtightness and insulation did not meet minimum standards, resembling older and less efficient dwellings [3].

Third, emerging policies in Australia are increasingly focused on promoting energy retrofitting in residential buildings. For instance, the Scorecard from the Victorian Government has now become the widely recognised energy rating program for existing homes throughout Australia [70, 81]. Moreover, mandatory minimum efficiency standards for technical equipment and appliances are a widely adopted, proven, and successful policy in Australia [15]. An extension of NatHERS star ratings for existing houses is currently being developed.

The Australian residential sector possesses a substantial inventory of energy-inefficient residential buildings, coupled with a limited number of emerging PIs to encourage energy retrofitting. This combination presents a compelling case for studying Australia's progress towards achieving the national energy retrofitting goals, especially compared to international best practices.

Australia operates under a three-tiered government system: federal, state, and municipal. The Australian Building Codes Board (ABCB), established through an inter-governmental agreement between the federal and state governments, sets minimum regulatory requirements for the built environment and maintains the NCC [28, 77]. Federal policies are crafted collaboratively by state governments, ensuring uniform adherence across states and territories [28]. Thus, energy retrofitting regulation primarily occurs at the federal level [27]. Moreover, federal PIs are crucial for addressing nationwide issues and ensuring regulatory consistency for the building products supply sector and large construction firms operating across state borders [11].

So far, this chapter has presented the background of the study and provided a brief overview of the case study and its federal-level PIs for encouraging energy retrofitting in residential buildings. The following sections will first analyse the success of these PIs and then compare them with international best practices to explore potential improvements.

4 Analysis/Results

The eight federal-level PIs outlined in Table 2 in Sect. 5.1 were further analysed in this section. The success of PIs and their implementation was assessed in Table 3 by analysing progress in achieving key objectives and how policymakers addressed challenges, providing valuable lessons for future policy development. This assessment emphasises the significance of core policy applications in the domain of energy retrofitting.

While Australia lacks a specific direction-based instrument for increasing the energy retrofitting rate, NEPS was introduced in April 2024 to boost demand-side initiatives. NEPS aims to reduce energy bills, lower GHG emissions, and enhance energy reliability and security.

As a user-friendly energy performance assessment initiative, obtaining the scorecard alone can lead to a 9% improvement in energy efficiency by enhancing homeowners' understanding of their energy usage patterns. Additionally, the scorecard facilitates the secure collection and management of building stock data, providing a valuable database for policy development and research projects [70]. However, such a program's success depends on the proficiency and expertise of assessors and their capacity to handle unusual circumstances [70]. Reliability and transparency are highly expected to boost the confidence of residents. Moreover, collaboration between certified scorecard assessors and retrofit businesses is crucial for effective and widespread energy retrofitting, as these assessors have the necessary knowledge to devise upgrade plans and must comply with standards to maintain their certification [70].

Mandatory energy performance requirements can apply to buildings, appliances, materials, or equipment [2]. In Australia, the GEMS Act 2012 sets minimum efficiency requirements for domestic and industrial appliances, which helps reduce energy demand in existing houses to some extent [63]. However, there is no direct enforcement of minimum efficiency requirements for existing houses at the federal level.

Energy performance disclosure standards can also apply to buildings, appliances, and materials to ensure market transparency. For instance, the GEMS Act 2012

Table 3 The success of policy instruments and their implementation

PIs	Success of PIs		Success of the implementation process of PIs	
	Objectives	Achievements	Challenges identified	Solutions applied
Direction-based Policy Instruments				
NEPS	Since this long-term framework was launched in April 2024, there is still no study on its success			
Energy Performance Assessment				
NS	OB01 (Deliver NS)	As of June 2022, 6800 residents have obtained scorecards [58]	Training deficiencies: evaluating chimneys, understanding star rating systems, and estimating comfort ratings [78]	Developed a new educational module to explain the estimation process of the tool to generate ratings and recommendations shown on the certificate [2]
	OB02 (Maintain high levels of benefit, quality, and trust in all jurisdictions)	Survey results in all states and territories: 85% are likely to recommend NS to others and 90% report an increased understanding of energy efficiency [78]. Proven cost savings and enhanced thermal comfort result from a better understanding of energy usage [70, 78]		
	OB03 (Collaborate with the Commonwealth to transit NS to NatHERS)	The Commonwealth allocated funds in 2023 to enhance NatHERS, integrating NS [59]. CSIRO develops RapidRate, an AI-powered tool to estimate home energy ratings with minimal input data [59]		
	OB04 (Build the national market for NS)	ACT and NSW have begun adopting the NS initiative [78]	Trust issues due to previous experiences with unsolicited callers and door-to-door salespeople [78]	N/A
	OB05 (Prepare systems and supports for mass uptake)	Two Australian banks now recognise NS for green loan eligibility, anticipating more NS usage due to green loans. Some insurance companies are also exploring its potential use [78]		

(continued)

Table 3 (continued)

PIs	Success of PIs		Success of the implementation process of PIs	
	Objectives	Achievements	Challenges identified	Solutions applied
	Others	Provide secure data storage for the building stock [70]		
Minimum Energy Performance Requirements, and Energy Performance Disclosure				
GEMS Act	OB (Encourage the creation and use of appliances and equipment that consume less energy and emit fewer GHGs)	GEMS-certified appliances save the average household energy bill between AUD140 and AUD220 per year [20, 63] (GEMS Act 2012, 2023). The product registration process received satisfactory feedback for user-friendliness and prompt inquiry resolution. In 2018, Australia achieved AUD 1.13 billion to AUD 2.15 billion in economic savings, 4.8 to 7.6 million tonnes of GHG emission savings [63]. As of 2019, GEMS regulations achieved a negative cost of AUD 200 per tonne of emissions [5]	Possibility of redundancy and confusion with existing codes and standards [62]	State and territory laws were amended to align with GEMS [62]
			Privacy concerns regarding registration data of products [63]	Enforce specific offences for sharing information [63]
			Time-consuming standard development process with technological advancements [20]	Enhanced flexibility for product registrations [20]
Finance and Incentives				
HIP	OB1 (Offer free ceiling insulation to 2.7 M homes)	Insulated about 1.2 million homes, significantly improving from the previous insulation rate of 70,000 homes per year [30, 55]	Terminated prematurely due to safety concerns: fatalities, injuries, and property damage, because of fire incidents [5, 55]	A third phase for safety remediation was introduced to inspect homes and to increase confidence and skilled installers [5]
	OB2 (Reduce annual energy expenses and GHG emissions)	Rapid implementation of large-scale decarbonisation interventions [57]. Insulated homes experienced a 35% reduction in HVAC demand and saved at least AUD 300 annually on energy bills [43]		

(continued)

Table 3 (continued)

PIs	Success of PIs		Success of the implementation process of PIs	
	Objectives	Achievements	Challenges identified	Solutions applied
Research and Innovation				
HEEIP	OB01 (Assess the validity of NatHERS standards)	Cross-validating NatHERS assessments revealed its close alignment with actual performance, though a lower outcome, ideally around 10 ACH@50 Pa, would have been preferable [3]	The challenges of recruiting households include minimising bias, ensuring a representative sample, and achieving a sufficient sample size [3]	N/A
	OB02 (Recognise improvements to existing policies to enhance energy efficiency of houses)	Identified upgrades to NatHERS: integrating certified air pressure results into calculations, offering both 'as designed' and 'as built' certifications [3]. The NCC should aim for a precise air tightness criterion of 10 ACH@50 Pa, which many houses can achieve [3]. A standard for verifying compliance through random testing [3]		
RBS	OB (Assess the validity of NatHERS standards)	Newer, higher rated NatHERS homes had increased summer cooling energy usage, contrary to expectations [32]	Complexities in estimating dynamic shell efficiency [32]	Integrating thermal modelling [32]

mandates the disclosure of star ratings for electric appliances. However, technological advancements in appliances have made the standard development process time-consuming. Consequently, policymakers introduced amendments to enhance flexibility, allowing for product-specific regulatory approaches, new regulations without legislative changes, appropriate exemptions, and ongoing assessment of regulatory requirements [20]. To avoid similar challenges in future EPD initiatives, the Commonwealth Government should engage stakeholders before setting implementation dates and collaborate closely with Standards Australia to improve determinations and related processes [20]. A centralised government portal should be established for product registration, covering multiple legislations such as the GEMS Act, electrical safety, water efficiency, and ozone protection. Moreover, education and engagement of key stakeholders are crucial for the success of disclosure programs.

Among financial initiatives, HIP was a widespread national program that completed only half of its progress due to fire hazards [5]. To prevent similar negative impacts in future initiatives, objectives should be balanced to avoid conflicts and unmet goals. Timely communication of constraints and alternatives to relevant ministers is crucial to avoiding failures. Engaging with stakeholders and experts helps understand risks and opportunities for effective planning. A robust governance framework is essential for monitoring and addressing performance deviations. Prioritising prevention through a safety inspection program, along with a testing process to tackle fraud and non-compliance, is advisable. Withholding a portion of payments to ensure quality standards or mandating a certain co-payment from program beneficiaries can enhance the effectiveness of demand-driven measures, such as incentive mechanisms [21].

Two key federal-level research and innovation programs are highlighted in Table 3. Among these two programs, RBS faced challenges in estimating unit energy consumption due to dynamic shell efficiency, which was addressed by integrating thermal modelling to predict changes in energy demand over time [32]. To prevent issues similar to those in HEEIP, conducting blower door tests before the on-site energy assessment is recommended in future projects to understand influencing factors. Allocating more on-site time allows for addressing suspicious outcomes and retesting as needed without a subsequent visit [3]. RBS further proves that AccuRate/NatHERS star ratings lack precision in indicating annual heating or cooling energy consumption in Adelaide or Brisbane, requiring further research [32].

Education and training enhance individuals' knowledge and skills in energy retrofitting residential buildings, promoting energy-efficient behaviour and investments in energy retrofitting [12, 72]. However, Australia lacks comprehensive educational programs for energy retrofitting residential buildings.

In conclusion, all PIs in Table 3 have generally accomplished their key objectives, except for HIP [5, 55]. Despite numerous challenges during the implementation of these PIs, most have been successfully overcome, as given in Table 3. Future research should address the unresolved challenges and expand knowledge on previously resolved issues. The following section establishes potential policy improvements for energy retrofitting residential buildings in Australia to address the deficiencies of the current PIs.

5 Discussion

Potential policy improvements for energy retrofitting residential buildings in Australia were explored using knowledge of current PIs and international best practices. Listed in Table 4 under seven core policy applications, these improvements aim to enhance energy retrofitting and reduce emissions holistically in existing houses. Demand-side improvements directly encourage energy retrofitting, whereas supply-side improvements target reducing emissions from energy supply and product Embodied Emissions (EEs), complementing overall emission reduction in existing houses. These improvements can also be applied to countries with similar housing stock or regulatory systems.

As a global example of direction-based instruments, the United States (US) aims to cut building emissions by 50% by 2050 [26]. The European Energy Performance of Buildings Directive (EPBD) mandates long-term energy retrofitting, with Germany targeting to increase energy retrofitting from 0.8% to 2% annually [85]. However, in Australia, energy retrofitting policies introduced by the COAG Energy Council in 2019 were slowed and abandoned due to political influences. However, the Australian federal government recently introduced the NEPS framework, focusing exclusively on demand-side management. To address such deficiencies, Table 4 proposes both demand-side and supply-side policy improvements, emphasising obligation schemes for energy suppliers. Moreover, sectorial targets are crucial to estimating budgets, defining subsidies, and identifying target groups. For instance, Europe's direction-based PIs include installing 600,000 heat pumps annually by 2028 and 13 million by 2050, aiming to reduce heat pump costs by 25–50% by 2025 [6, 80]. Moreover, the EU's Renovation Wave in 2020 aims to double renovation rates in 10 years and improve energy and resource efficiency [82]. Although EEs are often overlooked worldwide, French environmental regulations set up a national database for EEs of construction products and materials [66]. Moreover, Denmark has established a national plan to gradually introduce limits on EEs [44].

In Australia, efforts are underway to extend NatHERS to include existing homes [59]. Integrating NatHERS and the scorecard is essential for a unified national approach, addressing each system's limitations. Many countries have long-established star rating systems for existing houses, such as ENERGY STAR in the US and EnerGuide in Canada, known for their exemplary assessment tools. While the Energy Performance Certificate (EPC) in France and the UK rates buildings on an A–G scale, Germany's Efficiency Label with Heating System program labels heaters over 15 years old to encourage replacement [36]. These systems use rigorous evaluations and provide tailored recommendations to improve the energy efficiency of existing houses [25]. Moreover, an ideal rating system should provide insights into energy consumption, CO_2 emissions, recommendations for energy retrofitting with estimated costs and payback periods, and indicators of health and comfort improvements.

In Australia, efforts are underway to establish a national framework for rental property energy efficiency, with specific standards already in place in Victoria and ACT [78]. In contrast, other developed nations have stricter standards. The US's Clean Energy DC Omnibus Act 2019 in Washington, D.C. involves compliance schedules every five years for achieving sector-specific star ratings for buildings [9, 73], while New York City sets emission thresholds per square foot for various building types [73]. The UK's AECB retrofit standard targets 50 kWh/m^2/year for space heating and a maximum airtightness of 2 ACH@50 Pa [64]. Additionally, the eco-design regulation in the UK mandates that only energy-efficient products meeting defined criteria can be sold [42].

Table 4 A compendium of potential policy improvements to encourage energy retrofitting residential buildings

	Potential policy improvements for demand-side management	Potential policy improvements for supply-side management
DI	Establish economy-wide targets, i.e., national emission reduction of the existing housing stock; incorporate different policy directions for different housing categories, i.e., rental and owner-occupied houses; establish energy efficiency standards for rental properties; prioritise vulnerable groups, i.e., low-income households, people in social housing, and the least efficient and most at-risk houses; establish sectorial targets, i.e., increasing the retrofitting rate, widely implementing chosen retrofit measures; establish a comprehensive long-term retrofitting strategy and update it periodically; encourage collaborations between certified scorecard assessors and retrofit businesses; encourage one-stop retrofit shops	Establish obligation schemes requiring energy suppliers to meet emission-saving targets, i.e., integrate renewable energy sources; establish a national database for EEs of construction products and materials; establish a systematic plan to encourage reused and recycled materials and equipment, and products with low EEs on the market; establish a centralised government portal for product registration requirements
EPA	Establish a single national star rating system that represents the overall energy efficiency of a building; incorporate EEs into the star rating system; enhance assessment transparency and reliability through energy audits and household consultations	Incorporate EEs into the star rating system of appliances
MEPR	Establish minimum energy efficiency requirements for rental properties, i.e., mandatory 2-star heaters and ceiling insulation for rental dwellings, minimum emission limit for a square foot of a house; establish a recurring compliance schedule to periodically achieve mandatory performance benchmarks for houses, i.e., assessing compatibility with the current sector-specific star rating every five years; establish minimum self-sufficiency standards, such as the proportion of energy demand covered by domestic renewable energy production	Establish minimum requirements for appliances at the supply side, i.e., minimum star ratings, maximum limit for EEs; establish minimum requirements for materials at the supply side, i.e., the maximum limit for EEs
EPD	Establish mandatory disclosure during sales and lease; establish periodical disclosure for every house, i.e., every five years; incorporate building passports for the performance disclosure	Establish mandatory disclosure for EEs of appliances and materials

(continued)

Table 4 (continued)

	Potential policy improvements for demand-side management	Potential policy improvements for supply-side management
FI	Provide subsidies and tax reductions for retrofitted houses that exceed minimum energy performance standards; prioritise vulnerable groups by providing capital cost reductions rather than tax rebates; provide financial incentives with complementary education programs and customer support; establish a framework for financial incentives that decrease over time to encourage cost-effective options; establish a robust governance framework for monitoring and addressing performance deviations in financial incentives; provide green loans to encourage energy retrofitting; provide energy bill reduction schemes for vulnerable and rural communities; provide rewards for using reused and recycled building materials in energy retrofitting	Levy taxes on fossil-fuel-based appliances to provide tax rebates for renewable energy technologies; provide rewards for products with lower EEs; compelling energy suppliers to invest to reduce fuel poverty
RI	Develop a standardised method to estimate EEs of energy retrofitting; establish smart meters and online monitoring systems; investigate the accuracy of occupant behaviour datasets in the national whole-house rating system; explore methods to integrate smart-meter readings into the national whole-house rating system; incorporate both engineering algorithm modelling and thermal modelling to develop the national whole-house rating system; explore opportunities to enhance the precision of AccuRate/NatHERS star ratings for estimating cooling energy consumption	Develop a standardised method to estimate EEs of appliances; develop possible regulatory measures for reducing EEs; investigate innovative technologies with reduced EEs
ED	Develop a manual for home energy retrofitting; conduct events, workshops, and training programs to encourage energy retrofitting; drive energy retrofitting for environmental goals and thermal comfort rather than financial gains; develop educational programs about financial incentives; provide tailor-made education to different groups of people based on demographic factors, i.e., age, education, and nature of houses	Education and training on following current regulatory pathways

DI, Direction-based Instruments; EPA, Energy Performance Assessment; MEPR, Minimum Energy Performance Requirements; EPD, Energy Performance Disclosure; FI, Finance and Incentives; RI, Research and Innovation; ET, Education and Training

Australia aims to establish a national framework for Disclosure by 2025, alongside the existing scorecard [78]. Similar policies at the state level include energy assessments at the point of sale/lease in the ACT and the discontinued 'Sustainability Declaration' in Queensland [58, 68]. Globally, EPCs in Europe and similar initiatives in the US and Canada demonstrate effective mandatory disclosure, despite challenges in reliability and user acceptance [38]. To address these, the EPBD introduced the building passport in 2019, providing a digital logbook of energy retrofitting works and long-term plans based on energy audits [67]. This tool aims to offer user-friendly information on energy retrofitting benefits. In addition, commercial and multi-family residential buildings in Washington, D.C., US, must disclose energy ratings every five years to ensure compliance with sector-specific star ratings [9].

Notable federal-level subsidy programs in Australia include HIP, SHWR, and SRET. Similar to SRET, which decreases solar PV rebates as the industry matures, financial assistance should be time-bound, encouraging cost-efficient energy efficiency improvements over time. Globally, financial incentives have mostly focused on insulation, heat pumps, and solar technologies, with Australia being no exception [2, 35, 45, 73]. In Seattle, a tax on oil-based heaters funds rebates for heat pumps [50], while Boulder offers financial incentives and support for heating system upgrades [34]. The US provides tax reductions for retrofitted buildings meeting ASHRAE Standard 90.1-2001 [51]. In addition, the US government offers tax incentives and loans for those who exceed the minimum energy efficiency thresholds. However, grants and tax incentives can lead to free riding, often benefiting higher income households. Therefore, initiatives should prioritise vulnerable groups to address fuel poverty [2]. For instance, the Canadian government has implemented the 'Energy Efficiency Retrofit Program for Low-Income Households' to help low-income households and senior citizens live more comfortably [85]. In addition, utility providers and local governments usually provide financial incentives in Canada.

In Australia, consistent funding supports research and innovation to test and assess PIs. However, many energy retrofitting policies neglect EEs, highlighting the need for research on integrating EEs into star rating systems [2]. Globally, policies are emerging to address EEs, such as the Netherlands' carbon cap method, which converts whole life cycle emissions into euro-per-square-meter metrics [29]. Additionally, continuous federal funding programs in the US, like the Better Building Neighbourhood program, focus on innovative research and development for building energy efficiency improvement and fully fund low-income households [24].

The ABCB handbook in Australia offers limited guidance on energy retrofitting, demanding a comprehensive guide [77]. In contrast, European countries like Germany and Denmark have stringent energy standards and heightened awareness of energy retrofitting [17, 61]. The UK's PAS 2035:2019 provides a comprehensive guide for domestic energy retrofitting (PAS 2030: 2019, 2020). Moreover, clear and straightforward training programs are more effective in achieving energy retrofitting goals [72].

All seven policy applications must collaborate to address energy retrofitting challenges. Direction-based instruments like NEPS set national targets but face implementation hurdles and, therefore, should integrate with other applications. They can be divided into PIs for minimum energy performance requirements and energy performance disclosure, necessitating energy performance assessments. Research and innovation enhance transparency and integrity, finance and incentives reduce economic barriers, and education and training build trust. Thus, synergy among these core applications is essential to achieving the required retrofitting rate.

6 Conclusions

This chapter reviews the success of eight federal PIs in Australia, with a detailed examination in Table 3. These PIs aim to encourage energy retrofitting in residential buildings. Each PI's success was evaluated by analysing its progress towards key objectives and how policymakers overcame challenges encountered during implementation. Despite the limited number of PIs for energy retrofitting in Australia, they have been implemented successfully.

This chapter contributes to knowledge by developing a compendium of potential policy improvements under seven core policy applications, as shown in Table 4, to address the deficiencies of current PIs. These improvements aim to enhance energy retrofitting and holistically reduce emissions from existing residential buildings. The recently introduced NEPS can incorporate these policy improvements to enhance emission reductions in existing houses.

Further research is needed to assess the challenges encountered in implementing existing PIs and to develop strategies to overcome them. Since this chapter focuses on emission reductions in existing residential buildings, future studies should also extend to new constructions and other building categories.

References

1. ACT Government (2023) Rental properties subject to minimum insulation standards. https://www.act.gov.au/our-canberra/latest-news/2023/april/rental-properties-subject-to-minimum-insulation-standards
2. Alabid J, Bennadji A, Seddiki M (2022) A review on the energy-retrofit policies and improvements of the UK existing buildings, challenges and benefits. Renew Sustain Energy Rev 159, Elsevier Ltd
3. Ambrose M, James M, Law A, Osman P, White S (2013) The evaluation of 5-star energy efficiency standard for residential buildings—final report. Department of Industry, Canberra
4. Australian Bureau of Statistics (ABS) (2021) Housing: Census: Information on housing type and housing costs. https://www.abs.gov.au/statistics/people/housing/housing-census/latest-release#:~:text=Housing%20tenure%2C%202021%20Census,-This%20image%20shows&text=Of%20all%20the%20occupied%20private,and%20other%20non%2Dclassifiable%20households
5. Australian National Audit Office (ANAO) (2010) Home Insulation Program. https://www.anao.gov.au/work/performance-audit/home-insulation-program
6. BEIS (2020) Improving the energy performance of privately rented homes in England and Wales.
7. Baker M, Moore T, Daniel L, Caines R, Padilla H, Lester L (2023) Sustainable social housing retrofit? Circular economy and tenant trade-offs, AHURI Final Report No. 397, Australian Housing and Urban Research Institute Limited, Melbourne. https://www.ahuri.edu.au/research/final-reports/397, https://doi.org/10.18408/ahuri3128301
8. Beerepoot M, Beerepoot N (2007) Government regulation as an impetus for innovation: evidence from energy performance regulation in the Dutch residential building sector. Energy Policy 35(10):4812–4825. https://doi.org/10.1016/j.enpol.2007.04.015

9. Bergfeld K, Mathew P, Duer-Balkind M (2021) Making data-driven policy decisions for the nation's first building energy performance standards. Summer study on energy efficiency in buildings, 9–64. https://escholarship.org/uc/item/05m741q3
10. Bergman N, Foxon TJ (2020) Reframing policy for the energy efficiency challenge: insights from housing retrofits in the United Kingdom. Energy Res Social Sci 63 (101386). https://doi.org/10.1016/j.erss.2019.101386
11. Berry S, Marker T (2015) Residential energy efficiency standards in Australia: where to next? Energ Effi 8:963–974. https://doi.org/10.1007/s12053-015-9336-4
12. Bichard E, Thurairajah N (2013) Behaviour change strategies for energy efficiency in owner-occupied housing. Constr Innov 13(2):165–185. https://doi.org/10.1108/14714171311322147
13. Bowen E (2022) National energy performance strategy consultation opens. https://minister.dcceew.gov.au/mcallister/media-releases/national-energy-performance-strategy-consultation-opens
14. COAG Energy Council (2019) Report for achieving low energy existing homes. https://www.energy.gov.au/sites/default/files/Report%20for%20Achieving%20Low%20Energy%20Existing%20Homes.pdf
15. Cabeza LF, Bai Q, Bertoldi P, Kihila JM, Lucena AFP, Mata É, Mirasgedis S, Novikova A, Saheb Y (2022) Buildings. InL IPCC, 2022: Climate Change 2022: Mitigation of Climate Change. Contribution of Working Group III to the Sixth Assessment Report of the Intergovernmental Panel on Climate Change [Shukla PR, Skea J, Slade R, Al Khourdajie A, van Diemen R, McCollum D, Pathak M, Some S, Vyas P, Fradera R, Belkacemi M, Hasija A, Lisboa G, Luz S, Malley J (eds.). Cambridge University Press, Cambridge, UK and New York, NY, US. https://doi.org/10.1017/9781009157926.011
16. Chen S, Zhang G, Xia X, Chen Y, Setunge S, Shi L (2021) The impacts of occupant behavior on building energy consumption: a review. Sustain Energy Technol Assess 45. https://doi.org/10.1016/j.seta.2021.101212
17. Christensen TH, Gram-Hanssen K, Adjei A, de Best-Waldhober M (2011) Energy renovation practices in Danish homes: the influence of energy labels on home renovation practices. In: Proceedings of the 10th Conference of the European Sociological Association: Social Relations in Turbulent Times, Geneva, Switzerland, 7–10 September 2011
18. Clean Energy Regulator (2014) Small-scale technology certificates forecasts (2014 to 2016)
19. Clean Energy Regulator (2024) Small-scale renewable energy scheme
20. Collyer A (2019) Independent review of the greenhouse and energy minimum standards (GEMS) act 2012. Department of the Environment and Energy, Canberra
21. Commonwealth of Australia (2010) Home Insulation Program. Retrieved from https://www.anao.gov.au/work/performance-audit/home-insulation-program
22. Commonwealth of Australia (2022) National energy performance strategy: consultation paper
23. Department of Climate Change, Energy, the Environment, and Water (DCCEEW) (2023) National Energy Performance Strategy. https://www.dcceew.gov.au/energy/strategies-and-frameworks/national-energy-performance-strategy
24. Department of Energy (2024) Science & Innovation. https://www.energy.gov/science-innovation. Accessed 18 June 2024
25. Department of Energy and Environment, US (2023) Clean Energy DC Omnibus Amendment Act. https://doee.dc.gov/service/clean-energy-dc-omnibus-amendment-act
26. Department of Energy, US (2024) Decarbonizing the US economy by 2050: a national blueprint for the buildings sector. Retrived from chromeextension:// efaidnbmnnnibpcajpcglclefindmkaj/https://www.energy.gov/sites/default/files/2024-12/bto-decarbonizing-us-economy-2050-122724.pdf
27. Dowling R, McGuirk PM, Bulkeley H (2014) Retrofitting cities: local governance in Sydney, Australia. https://ro.uow.edu.au/sspapers
28. Doyon A, Moore T (2020) The role of mandatory and voluntary approaches for a sustainable housing transition: evidence from Vancouver and Melbourne. Urban Policy Res 38(3):213–229. https://doi.org/10.1080/08111146.2020.1768841

29. Dutch Green Building Council (2021). Whole life carbon: Position paper. https://www.dgbc.nl/upload/files/Circulariteit/BuildingLife/Position%20Paper%20Whole%20Life%20Carbon_EN.pdf. Accessed 18 June 2024
30. ECAR Committee (2010) Environment, communications and the arts references committee: energy efficient homes package (ceiling insulation), Canberra
31. Economidou M, Todeschi V, Bertoldi P, D'Agostino D, Zangheri P, Castellazzi L (2020) Review of 50 years of EU energy efficiency policies for buildings. Energy Build 225:110322. https://doi.org/10.1016/j.enbuild.2020.110322
32. Energy Consult PTY LTD (2015) Department of industry and science on behalf of the trans-tasman equipment energy efficiency (E3) program
33. Energy Rating (2024) Greenhouse and energy minimum standards act. https://www.energyrating.gov.au/industry-information/understand-requirements/legislative-framework
34. Energy Smart (2018) Rebates and financing. https://energysmartyes.com/rebates-financing/
35. Enviro Friendly (2024) Solar hot water rebates. https://enviro-friendly.com/solar-hot-water/rebates/
36. European Commission (2022) The energy label of existing heating appliances as a tool for energy transition of the sector. https://build-up.ec.europa.eu/en/resources-and-tools/articles/technical-article-energy-label-existing-heating-appliances-tool-energy. Accessed 18 June 2024
37. European Commission (2016) Commission Staff Working Document: Impact Assessment. Accompanying the document Proposal for a Directive of the European Parliament and of the Council amending Directive 2010/31/EU on the energy performance of buildings 2016, European Union: European Commission. p. 121
38. Fabbri M, De Groote M, Rapf O (2016) Building renovation passports: customised roadmaps towards deep renovation and better homes. Buildings Performance Institute Europe
39. GEMS Act 2012 (2022) Greenhouse and energy minimum standards act 2012. Compilation no. 12, includes amendments up to act no. 69, 2022. Retrieved from https://www.legislation.gov.au/Details/C2022C00189
40. Galvin R, Sunikka-Blank M (2013) A critical appraisal of Germany's thermal retrofit policy: turning down the heat. Springer, London
41. Get Up (2023) Submission by GetUp on the national energy performance strategy
42. Gonzalez-Torres M, Bertoldi P, Castellazzi L, Perez-Lombard L (2023) Review of EU product energy efficiency policies: what have we achieved in 40 years? J Clean Prod 421. https://doi.org/10.1016/j.jclepro.2023.138442
43. Hanger I (2014) Report of the Royal Commission into the home insulation program, Canberra. http://www.homeinsulationroyalcommission.gov.au/documentation/documents/reportoftheroyalcommissionintothehomeinsulationprogram.pdf
44. Happold B (2024) How Denmark leads the way in decarbonising the construction industry. https://www.burohappold.com/news/how-denmark-leads-the-way-in-decarbonising-the-construction-industry/. Accessed 18 June 2024
45. He Q, Wu Z, Li S, Li H, Wang Y (2023) Two decades of the evolution of China's green building policy: insights from text mining. Build Res Inform 51(2):158–178. https://doi.org/10.1080/09613218.2022.2142498
46. Hsu D (2014) How much information disclosure of building energy performance is necessary? Energy Policy 263–272. https://doi.org/10.1016/j.enpol.2013.08.094
47. Hulathdoowage ND, Karunasena G, Udawatta N, Liu C (2023) Reviewing the contribution of retrofitting for climate resilience in residential buildings. Int J Disaster Resilience Built Environ. https://doi.org/10.1108/IJDRBE-02-2023-0031
48. Ibn-Mohammed T, Greenough R, Taylor S, Ozawa-Meida L, Acquaye A (2013) Operational versus EEs in buildings—a review of current trends. Energy Build 66:232–245. https://doi.org/10.1016/j.enbuild.2013.07.026
49. International Energy Agency (IEA) (2023) Tracking Clean Energy Progress 2023, IEA, Paris, https://www.iea.org/reports/tracking-clean-energy-progress-2023

50. Khurana T (2022) Seattle's heating oil tax: a missed opportunity for environmental justice. Retrieved from https://southseattleemerald.org/feature/2022/01/08/seattles-heating-oil-tax-a-missed-opportunity-for-environmentaljustice#:~:text=In%20September%202019%2C%20Seattle%20Mayor,disproportionately%20harmed%20by%20fossil%20fuels
51. Kubba (2017) Handbook of green building design and construction: LEEF, BRREAM, and Green Globes. Elsevier
52. Li HX, Moore T, Huang J, Zhang P, Costin G (2022) Towards zero carbon housing in Victoria, Australia: a policy and incentive framework. Energ Strat Rev 40. https://doi.org/10.1016/j.esr.2022.100802
53. Liu G, Tan YT, Li XH (2020) China's policies of building green retrofit: a state-of-the-art overview. Build Environ 169. https://doi.org/10.1016/j.buildenv.2019.106554
54. Ma W, Hao JL, Guo F, Sarno LD (2022) System dynamics-life cycle assessment causal loop model for evaluating the carbon emissions of building refurbishment construction and demolition waste. Waste Biomass Valoriz 13:4099–4113. https://doi.org/10.1007/s12649-022-01796-9
55. Majumdar A (2023) Energy justice in Australia's housing decarbonisation transition: learning from community housing retrofits
56. Ministry of Land, Infrastructure and Transport (MLIT) (2015) JAPAN: act on improvement of energy consumption performance of buildings (building energy conservation act) (Act No. 53 of 2015). https://policy.asiapacificenergy.org/node/4185ACT Government. (2023). Rental properties subject to minimum insulation standards. https://www.act.gov.au/our-canberra/latest-news/2023/april/rental-properties-subject-to-minimum-insulation-standards
57. Moore T (2018) Strategic niche management and the challenge of successful outcomes. In: Moore T, de Haan F, Horne R, Gleeson BJ (eds) Urban sustainability transitions. Springer; Singapore, pp 109–126
58. Moore T, Horne R, de Kleyn L, Camboim L, Rajagopalan P, Willand N (2023) Optimising housing assessment to drive low carbon energy efficient housing upgrades
59. Nationwide House Energy Rating Scheme (2023) Existing homes. Retrieved from https://www.nathers.gov.au/existinghomes
60. Natural Solar (2024) Solar hot water rebates. Retrieved from https://naturalsolar.com.au/solar-rebates/solar-hot-water-rebates/
61. Novikova A, Vieider F, Neuhoff K, Amecke H (2011) Drivers of thermal retrofit decisions: a survey of German single- and two-family houses (CPI Report); Climate Policy Initiative Berlin, Berlin, Germany
62. Parliament of Australia (2012) Greenhouse and Energy Minimum Standards Bill 2012. Department of Parliamentary Services, Parliament of Australia
63. Parliament of Australia (2023) Greenhouse and energy minimum standards amendment (Administrative Changes) Bill 2023. Department of Parliamentary Services, Parliament of Australia
64. Passive House Plus (2019) Passive house plus: sustainable buildings. Retrieved from https://passivehouseplus.co.uk/
65. Pombo O, Rivela B, Neila J (2019) Life cycle thinking toward sustainable development policy-making: the case of energy retrofits. J Clean Prod 206:267–281. https://doi.org/10.1016/j.jclepro.2018.09.173
66. Prasad D et al. (2022) Policy pathways to a net zero carbon-built environment. In: Delivering on the climate emergency. Palgrave Macmillan, Singapore. https://doi.org/10.1007/978-981-19-6371-1_6
67. Puček MM, Khoja A, Bazzan E, Gyuris P (2023) A data structure for digital building logbooks: achieving energy efficiency, sustainability, and smartness in buildings across the EU. Buildings 13(4):1082. https://doi.org/10.3390/buildings13041082
68. Queensland Government (2009) Building and Other Legislation Amendment Bill 2009. Brisbane: Queensland Government. https://www.legislation.qld.gov.au/view/pdf/bill.first/bill-2009-1440

69. Rana A, Sadiq R, Alam MS, Karunathilake H, Hewage K (2021) Evaluation of financial incentives for green buildings in Canadian landscape. Renew Sustain Energy Rev 135. https://doi.org/10.1016/j.rser.2020.110199
70. Residential Efficiency Scorecard, (2023). Learnings from the Energy Savvy Upgrades program for vulnerable householders. https://www.homescorecard.gov.au/__data/assets/pdf_file/0031/675742/Energy-Savvy-Upgrades-evaluation-report-final-0323.pdf
71. Skillington K, Crawford RH, Warren-Myers G, Davidson K (2022) A review of existing policy for reducing embodied energy and greenhouse gas emissions of buildings. Energy Policy 168. https://doi.org/10.1016/j.enpol.2022.112920
72. Streimikiene D, Balezentis T (2019) Innovative policy schemes to promote renovation of multi-flat residential buildings and address the problems of energy poverty of aging societies in former socialist countries. Sustainability (Switzerland), 11, MDPI
73. Takahashi K, Frost J, Goldberg D, Hopkins A (2020) Survey of U.S. State and Local Building Decarbonization Policies and Programs
74. Talberg A (2010) Electric hot water systems being phased out. https://www.aph.gov.au/About_Parliament/Parliamentary_Departments/Parliamentary_Library/FlagPost/2010/October/Electric_hot_water_systems_being_phased_out#:~:text=Under%20the%20REBS%2DSHWR%2C%20eligible,program%20at%20the%20same%20address;%204%20-%20https://www.energymatters.com.au/solar-hot-water/solar-hot-water-rebates/
75. Talberg A, McCluskey I. Bills Digest 4, 2012–13—Greenhouse and Energy Minimum Standards Bill 2012. http://www.climatechange.gov.au/en/government/reduce.aspx
76. Tan Y, Liu G, Zhang Y, Shuai C, Shen GQ (2018) Green retrofit of aged residential buildings in Hong Kong: a preliminary study. Build Environ 143:89–98. https://doi.org/10.1016/j.buildenv.2018.06.058
77. The Australian Building Codes Board (ABCB) (2023) Australian Building Codes Board. https://www.abcb.gov.au/
78. The Residential Efficiency Scorecard (2022) The National Scorecard Home Energy Rating Program. https://www.homescorecard.gov.au/__data/assets/pdf_file/0032/588191/NationalScorecardReleaseEvaluation_FINAL.pdf
79. Tienhaara K (2015) Green stimulus and pink batts: The environmental politics of Australia's response to the financial crisis. J Austr Polit Economy 76:54–79
80. UK Department for Business, Energy & Industrial Strategy (2021) Heat and buildings strategy. Retrieved from https://www.gov.uk/government/publications/heat-and-buildings-strategy
81. Victoria State Government (2022) Residential Efficiency Scorecard. https://www.energy.vic.gov.au/households/save-energy-and-money/residential-efficiency-scorecard#:~:text=The%20Victorian%20Government's%20Scorecard%20is,provides%20tailored%20recommendations%20for%20improvements
82. von Malmborg F, Björklund M, Nordensvärd J (2023) Framing the benefits of European Union policy expansion on energy efficiency of buildings: a Swiss knife or a Trojan horse. Eur Policy Analysis 9(3):219–243. https://doi.org/10.1002/epa2.1184
83. Wang S, Yan C, Xiao F (2012) Quantitative energy performance assessment methods for existing buildings. Energy Build 873–888. https://doi.org/10.1016/j.enbuild.2012.08.03
84. Yin RK (2018) Case study research and applications. Sage
85. Zhang H, Hewage K, Karunathilake H, Feng H, Sadiq R (2021) Research on policy strategies for implementing energy retrofits in the residential buildings. J Build Eng 43. https://doi.org/10.1016/j.jobe.2021.103161

Chapter 3
Sustainability Transitions for Energy Efficiency in Buildings in the UK

Esra Kurul

Abstract The UK Built Environment sector urgently needs a sustainability transition in order to exploit the very high potential for emission reductions in, for example, existing residential buildings. According to the Climate Change Committee (CCC)—the UK Government's own advisory body, this potential in existing buildings is more than twice that in new buildings. However, a transition, at scale, across the whole of the UK Built Environment sector is yet to materialise. The sector remains the UK's second highest emitting sector. It should accelerate its rate of emission reduction to two or three times the current rate. This chapter makes an original contribution to existing knowledge by drawing on the multi-level perspective (MLP) in the sustainability transition (ST) literature to explain the whole sector's lack of progress, rather than one of its sub-sectors, e.g. housing. First, the ST literature is critically reviewed to identify the key characteristics of such transitions: normative directionality, which attributes public policy a central role in STs; and values, contestation, and disagreement. Second, the methodology, which is a rigorous desktop study of the UK Government policy to reduce the GHG emissions from buildings as a key enabler of sustainability transitions and of the literature on the impact of the UK policy on practice, is outlined. This study establishes that some transition has taken place with regards to reducing emissions from new buildings, but its scale and pace is limited. Moreover, there have not been radical enough changes to practices to accelerate the refurbishment of existing buildings to reduce emissions. The firms' heavy reliance on normative directionality in order to create a stable context for delivering any transformation emerges as the main barrier to transition. Novel business models that will reduce this reliance are identified as being integral to the pathway to transition, rendering this chapter's contribution significant.

Keywords Energy efficiency · Climate impact · Existing buildings · The UK · Sustainability transitions

---

Present Address:
E. Kurul (✉)
School of the Built Environment, Oxford Brookes University, Oxford, UK
e-mail: ekurul@brookes.ac.uk

T. Ramachandra et al. (eds.), *Concepts, Policies and Technology for Sustainable Buildings*, Advances in Sustainability Science and Technology,
https://doi.org/10.1007/978-981-96-3086-8_3

1 Introduction

Built environments are significant for societies across the globe. At the very basic level, they meet Maslow's physiological and safety needs by providing shelter, personal security and property. They also house most of the employment activities in cities, which generate more than 80% of the global GDP [35]. The UK's 11 Core Cities[1] deliver 26% of its economic output (Core Cities UK no date). Activities that relate to delivering and sustaining the built environments across the globe are also major contributors to the economy. For example, the construction sector, which only captures the on-site activities of contractors and sub-contractors, accounted for 14.2% of the global GDP and 7% of the UK GDP in 2022 [33]. It also provided 7% of the global employment in 2019 [27] and 9% of the UK employment in 2022 (ibid.). Moreover, two-thirds of the global net worth is locked in real estate.

Contrastingly, the built environments also have a substantial detrimental impact on the environment. The IPCC [4] highlights the continued increase between 2010 and 2019 in the total net anthropogenic GHG emissions and the cumulative net CO2 emissions since 1850. An increasing share of these emissions can be attributed to urban areas, despite the consistent expansion of policies and laws addressing mitigation. The built environments continue to emit 37–40% of the global carbon and use a third of all materials [36].

In order to reduce this impact, governments, businesses and third sector organisations across the globe have initiated a plethora of climate- and biodiversity-related actions. In the UK, these range from the 2008 Climate Change Act to industry-lead declarations of the environmental and biodiversity crises, e.g. Built Environment Declares in October 2021 just before the Conferences of the Parties (COP) 26. The UK's pioneering 2008 Act legally bound its government to reducing its GHG emissions by 80% from the 1990 levels by 2050. Its 2019 amendment raised the rate of reduction to 100%, becoming the anchor to the UK's 'net-zero' targets. Such national initiatives are coupled with global milestones such as the 2015 Paris Agreement, the UN Secretary-General's 2023 Climate Ambition Summit and the annual COPs.

By 2015, these efforts in the UK resulted in a 19% reduction, from the 1990 baseline, in direct emissions from buildings [5], mainly as a result of energy efficiency improvements in new buildings. This 19% compares with an overall reduction of 44% from UK emissions, where Energy Supply and Waste Management sectors respectively achieved 66% and 71% reductions [3]. A direct comparison of the actual reductions in different sectors does not paint a realistic picture of a sector's performance in reducing its emissions, because certain sectors inherited a bigger opportunity for reductions than others. For example, the Energy Supply sector delivered the 66% reduction through the closure of all, but one, coal powered stations in the UK.

Even so, the Built Environment sector lags behind and remains among the sectors that need to accelerate their progress, because they are not on track to meet their carbon reduction targets [6], which has been the key feature of the UK's action on

[1] Core Cities UK is an alliance of 11 cities—Belfast, Birmingham, Bristol, Cardiff, Glasgow, Leeds, Liverpool, Manchester, Newcastle, Nottingham and Sheffield.

climate change [24]. According to the UK's sixth and latest Carbon Budget [5], direct GHG emissions from buildings, which are primarily CO2 emissions, accounted for 17% of the total UK emissions, and that this share rose to 23%, if indirect emissions were included. As such, "buildings remain the UK's second highest-emitting sector" [7]. "[The sector's] rate of emissions reduction needs to accelerate two or three times the current rate" [7].

The sector has the potential to achieve this acceleration. It can become one of the locomotives for transitioning to a 'sustainable and resilient' future [26]. A 'technical potential' to reduce emissions by 40 MtCO2 in the residential sector and 11 MtCO2 in the non-domestic sector, and a 'realistic potential' to reduce emissions by 9–18 MtCO2 from existing residential buildings and an additional 4 MtCO2 from new residential buildings were established as early as 2008 [8]. A comparison of these figures with the 101.01 MtCO2 direct emissions from UK buildings in 2008 and the 86.09 MtCO2 emissions in 2018 illustrate how sizeable these opportunities are.

Given the urgency of the need to take action, this opportunity should be exploited by bringing about sustainability transitions in the sector. Pathway(s) to transitioning should be proposed, progress towards the goals of these pathways should be closely monitored, and any lack of progress should be explained. This monitoring should encompass the whole built environment sector, rather than consider its sub-sectors, e.g. housing, in isolation, because the built environments are shaped and re-shaped through activities that are undertaken by different industry sectors,[2] and by the communities that occupy them. This chapter serves this need of a whole-sector-based approach to establishing whether and to what extent sustainability transitions have been achieved in the context of improving the energy efficiency of buildings, and thus makes an original contribution to the existing literature and current debates.

2 Sustainability Transitions and Their Characteristics

Sustainability transitions are "major, long-term technological changes in the way societal functions are fulfilled. Technological transitions do not only involve changes in technology, but also changes in user practices, regulation, industrial networks, infrastructure, and symbolic meaning or culture" [19], quoted in [39]. They are "radical shifts to new kinds of socio-technical systems", which are necessary to deal with 'grand societal challenges' such as Climate Change [28]. They rely on socio-technical transformations of systems [18] in order to deal with grand challenges that "are brought about by unsustainable consumption and production patterns in systems… [e.g. buildings]" [28]. As such, buildings, heat, electricity, mobility and agro-food systems feature frequently in the sustainability transitions literature.

[2] The UK's 2007 Standard Industrial Classification (SIC) spreads these activities across industries such as construction, real estate activities and professional, scientific and technical activities which cover services such as architectural design.

Geels [19] proposed the multi-level perspective (MLP) for understanding how technological innovations drive societal transformations. MLP uses technology as "a fruitful entrance point for studying 'society in the making'" [18]. It evaluates the interactions between niche-regime, niche-landscape, and landscape-regime and posits that niche, regime and landscape evolve in tandem. Here, niches are places where innovations occur. Actors, their rules and norms, and other material and technological elements form the regimes. Landscapes are the broader context in which regimes are located.

MLP has been criticised for its lack of attention to socio-ecological systems, social sustainability and other facets of sustainability. In response, Köhler et al. [28] advocated conceptualising and explaining "how radical changes can occur in the way societal functions are fulfilled" as a central aim of transitions research. Other researchers offered conceptualisations of sustainability transitions in terms of core dimensions of transformation, kinds of systems and understanding of change processes [18]. This paper makes a contribution to the sustainability transitions literature by providing such a conceptualisation of the initiatives to reduce emissions from the UK Built Environment Sector as a whole (see Fig. 1). This conceptualisation incorporates the interactions between niche-regime, niche-landscape, and landscape-regime, and pays due attention to the actors, their rules, norms and values; and normative directionality as an important dimension of sustainability transitions.

The UK Built Environment sector is considered an appropriate context for this conceptualisation for three main reasons. First, MLP as a process theory aligns well with the processes of design, delivery and operations in the Built Environment sector. Second, MLP has global and local model components. Within the built environment context, the global component deals with the overall transition to reducing the environmental impact of built environments, recently concentrating on 'net zero'. The local component "addresses-specific activities and causal mechanisms in multi-level interactions" [21] that enable this transition. In the built environment sector, these interactions are typically between the different stakeholders of a project or through supply-chains. Third, technology as 'a fruitful entry point' reflects the policy-driven

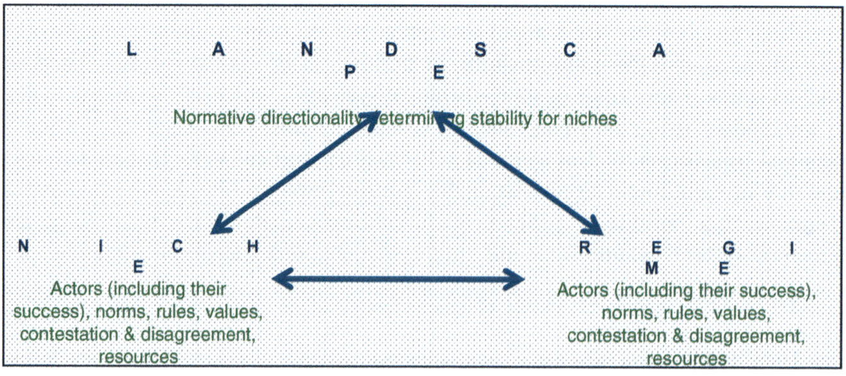

Fig. 1 Conceptual framework for sustainability transitions in the built environment

transformation initiatives within the UK built environment sector which are focussed on 'radical innovation' which centres around 'digital and manufacturing technologies' [22]. Examples of these initiatives include mandating the use of Building Information Modelling (BIM) on centrally funded Government projects from April 2016, and advocating the use of Modern Methods of Construction through government departments such as the Ministry of Defence.

Köhler et al. [28] identify several dimensions of sustainability transitions, including 'normative directionality' and 'values, contestation and disagreement'. These two characteristics can be used to monitor the progress towards exploiting the opportunity to reduce emissions from buildings and explaining the (lack of) progress. Normative directionality is concerned with using public policy to thrust transition towards a certain direction through mechanisms such as regulations, standards and taxation. It has a strong influence on the landscape which interacts with regimes and niches. It is particularly relevant in sectors such as the built environment, where private actors, e.g. firms, have limited appetite to address sustainability, which is a public good. Policies can also be used to create 'protected spaces' for niche innovations to occur and then substitute the regime as they get more widely adopted [39]. Therefore, this chapter reviews the relevant UK Government policies in order to establish whether they helped direct the private actors towards transforming their practices or encouraged them to innovate in order to reduce emissions from buildings. This review pays particular attention to 'values, contestation and disagreement'. Here, the aim is to establish *how* actors responded to the policies that were designed to direct the transition and to what extent they were concerned with protecting their vested interests. Hence, this chapter's contribution is significant because it shifts the focus of the discussion from the importance of legal instruments and government policies to firms as private actors that could bring about industry re-orientatation, which remains a nascent research area in the transitions literature.

Here, Collins et al.'s [11] dimensions for monitoring actors' progress towards sustainability transitions are relevant. They include: "a minimum level of success, [which is] interpreted as some concrete benefits for those actually delivering the projects [and those who are procuring these projects]; and the contextual factors of stability in experimental conditions over time along with adequate resources". Collins et al. (ibid) arrive at this conclusion after demonstrating that the opportunity to learn and refine innovations can be lost in 'real-world' contexts where the failure appetite of the actors is low. This is particularly the case in sectors such as the built environment, where successful project delivery is the modus operandi. When projects are deemed 'unsuccessful', there is hardly any appetite to learn as stakeholders find the experience 'painful' [11]. In addition, the tension between experimenting in a 'protected space' and the 'real world' intensifies when the niche's safety is undermined as a result of changes to the regulations, which provided safety in the first place. These niche innovations need either the stability of protection or other factors that would support the actors in their sustainability transition pathways. This paper aims to establish what these factors could be.

3 Methodology

The previous section identified the dimensions for exploring to what extent the UK Built Environment sector has transitioned to exploit the emission reduction opportunities it inherits. This section outlines and justifies the approach to this exploration in the UK, as the first country that legally bound itself to reduce GHG emissions in 2008. The UK is also home to BREEAM certification, which was the first scheme of its kind when it was launched in 1990.

A systematic review of secondary sources underpins the exploration that is presented in this chapter. These sources include annual reports on meeting the carbon budgets and the progress in reducing emissions, and academic papers that evaluate the UK Government's key policy instruments such as the Code for Sustainable Homes, the 'failed' Green Deal and the more recent Future Homes and Buildings Standards, and the actors' responses to these instruments. There is an emphasis on the Climate Change Committee's (CCC) monitoring reports, because it is the UK Government's independent, statutory body established under the Climate Change Act 2008. The CCC advises the UK Government and the devolved administrations on emission targets. It reports to the UK Parliament on progress made in reducing GHG gas emissions and preparing for and adapting to the impacts of climate change. Buildings are amongst the key areas of its reporting. The CCC makes policy recommendations to the Government. This systematic review relies on thematic analysis of these sources, which starts with the dimensions that were drawn from the sustainability transitions literature which was reported in the previous section.

4 The UK Built Environment's Journey to Sustainability Transitions

This section starts with normative directionality, in other words the policy landscape in the UK as a direct influence on the built environment sector to transition and as a mechanism to create the 'protected spaces' for niche initiatives to flourish. The policy focus on new buildings is thus demonstrated. Then, new building certification as an example of niche innovation within this landscape is reviewed. The last sub-section demonstrates that the emission reduction potential in existing buildings is big and untapped. It thus identifies a pathway to sustainability transition in the sector. As illustrated in Fig. 1, Köhler et al.'s [28] 'values, contestation and disagreement' and Collins et al.'s [11] dimensions for monitoring actors' progress towards sustainability transitions underpin these sections.

4.1 Policy Focus on New Buildings

Edmondson et al. [17] review the UK Government's policies to deliver its Zero Carbon Homes (ZCH) strategy using extensive data sources from policy documents to inquiries such as the one on 'home energy efficiency and demand reduction'. They identify the Building Regulations, National Planning Policy Framework and the Standard Assessment Procedure (SAP) as the key regulatory instruments for 2009–2016. They label the Code for Sustainable Homes, and the associated stamp duty exemption for all ZCH to be built during 2007–2011, the Eco-Towns initiative, and the Low Carbon Building Programme as economic instruments for achieving the ZCH target. All these instruments are concerned with new homes, which could be justified due to the residential sector's high impact on emissions. However, this focus has to be recognised as a key barrier to exploiting the emission reduction opportunities that are inherent in existing buildings.

Clearly, the introduction of the Code for Sustainable Homes in 2007 "as a voluntary national standard to improve the overall sustainability of new homes by setting a single framework within which the home building industry can design and construct homes to higher environmental standards" [13] was closely aligned with the normative directionality characteristic of sustainability transitions. It became mandatory for all new homes to be rated against the Code from May 2008. There was no requirement for private housing to reach a certain level of the Code, but publicly funded homes and those built on publicly owned land normally had to achieve at least Level 3. The Code also proposed that from April 2016 all new homes would be 'Zero Carbon', i.e. Code Level 6.

Hence, the Code provided a policy commitment to improving the energy efficiency of new residential buildings. Moreover, it signalled a potential market for zero or low carbon building technologies and innovations, which became an incentive for the industry, or at least the zero carbon niche, to engage in radical innovation to transform the incumbent *regime*. The introduction of the Code was thus a good example of the Government creating 'protected spaces' within the *landscape* for challenging the regime rules and for creating opportunities for niche-regime interactions.

Initially, the volume house-builders (the incumbents) that did not have the knowledge and skills to deliver ZCH were reluctant to change their practices. Edmondson et al. [17] argue that "the expectation of potential *resources* being channelled into this area seems to have provided a strong enough market signal to stimulate innovative activity amongst [some] incumbent [in other words *regime*] actors". Barratt became the notable leader amongst the volume house-builders by delivering the first Code Level 6 home at British Research Establishment's (BRE) Innovation Park, using the grant funding that was available at the time. It could be argued that Barrat's move reflects an industry that was unlikely to contest the changes which would have come about as a result of the reconfiguration of the incumbent *regime*, provided that the Government exerted *normative directionality* through policies and coupled them with *resources* in the form of grant funding.

However, these green shoots of sustainability transitions in the context of new residential buildings were fragile and short-lived. In 2010, the Conservative-Liberal Democrat Coalition Government replaced the Labour Government, which passed the 2008 Act. This new Government saw the Code as another regulation which limited the opportunities to supply new housing at scale. A series of policy moves, such as the watering down of the definition of 'zero carbon' in 2011 and the withdrawal of grant funding to build ZCH in the same year, signalled to the *regime* and the *niche* actors that the Government's commitment was weakening. The tension between experimenting in a 'protected space' and the 'real world' had intensified. As a result, the level of innovation started to decline, while the *regime* actors became more vocal in contesting the policy and influencing the Treasury to conclude that the cost of achieving the ZCH target was a 'tax on development' [17].

The Conservative Government came into power with a majority in 2015. Withdrawing the Code nine months before the April 2016 deadline when all new homes would be 'Zero Carbon' was one of its initial moves in the context of emission reductions from buildings. Hence, the Code as a pathway to sustainability transition was not given the time to unfold, contrary to the widespread acceptance in the literature that such transitions 'are long-term processes' [28]. Moreover, the policy support which, for a period, created a 'protected space' for the industry to start transitioning was removed. O'Neill and Gibbs [32] describe this move as 'policy dismantling', which is defined as "change of a direct, indirect, hidden or symbolic nature that either diminishes the number of policies in a particular area, reduces the number of policy instruments used and/or lowers their intensity." (Bauer et al., 2012, quoted in [32]).

The Conservative Government justified their dismantling of the Code by stating that elements of it were to be incorporated into the revised building regulations, which were to align with Level 4 of the Code. Given the one-year delay in implementing the 2013 increase in the energy performance standards in the Building Regulations and the smaller than expected increases in these standards, this justification was far from signalling an ambition to exercise *normative directionality*. Moreover, this policy dismantling diminished the innovation appetite amongst the *regime actors*. It is telling that the UK Green Building Council's (UKGBC) open letter to the Chancellor on the withdrawal of the Code was not signed by any of the key *actors*, i.e. the volume house-builders that were amongst its founding members.

In the event, the first set of amendments to the 2013 regulations on conservation of fuel and power in new and existing dwellings (known as Part L in England) was made in 2016. Further revisions followed in 2018 and 2022, when Future Homes Standards came into force. The 2016 and 2018 amendments to Building Regulations Part L applied to all new buildings. But, they only applied to existing buildings where new work was undertaken. Part L does not require existing buildings to be brought up to the current standards of the time. Their only stipulation is that new work that is carried out on existing buildings, e.g. extensions, comply with the relevant regulations. This is another set of evidence that the UK Government's policy instruments to reduce emissions from buildings have focused on new buildings.

The 2022 Future Homes Standard introduced phased targets for new homes to reduce their carbon emissions by 31%, in comparison to the 2018 regulations, from

June 2022 and by 75–80% from 2025. The latter target is defined as 'net-zero-ready' homes. It represents an aspiration for homes built after 2025 to operate at net zero without needing to be retrofitted. Thus, in effect, the 2016 target for net-zero homes has been delayed to at least 2025 or to 2050, representing a 9 to 34 year delay at a time when urgent action is crucially important. The 9-year delay assumes that there will not be any further delays in implementing the Future Homes Standard from 2025, and that the 'net-zero-ready' homes will be built to a standard that will not require retrofit.

Given the initial move of the target year from 2023 to 2025 during the initial consultation on the standard, further delays are expected. Moreover, the UK's past record in ensuring that homes are built to high energy performance standards during periods of policy flux does not provide confidence in homes built from 2025 to the 'net-zero-ready standard' not needing retrofit. It is estimated that about 1.5 m homes that have been built since 2015 will need to be retrofitted at a cost of £30bn to £45bn to become net zero.

This section has demonstrated that the UK Government has focused its *normative directionality* to new buildings, that the strength of this direction has fluctuated since 2008, and that these fluctuations were mainly driven by politics. The next section reviews the development of new building certifications as a niche that has the potential to reconfigure the incumbent regime. It establishes to what extent this potential has been exploited.

4.2 Building Certification as a 'Niche'

The first new 'green' building certification system in the world—Building Research Establishment Environmental Assessment Methodology (BREEAM)—was launched in 1990 [1]. This can be regarded as a pivotal point in sustainability transitions not only in the UK, but also in the built environment sectors across the globe. BREEAM has been followed by other prominent certification tools such as the US-developed Leadership in Energy and Environmental Design (LEED), and the one that is developed and administered by Germany's Deutsche Gesellschaft für Nachhaltiges Bauen (DGNB—German Sustainable Building Council).

While these tools have resulted in an interest in buildings that are designed to outperform those that only comply with the existing regulations, there are a number of fundamental problems with regards to their role in transitioning to sustainable built environments. The key practitioners who were involved in developing these tools expressed "an element of regret on how [these tools] have failed to transform the industry" [1]. This perhaps emanated from the origins of these certifications.

BREEAM was originally devised for a developer, i.e. an *actor*, to cover the design and construction phases of a commercial building. This focus remains in most certification tools. They almost exclusively deal with reducing resource use by allocating points to, for example, installing PV panels and choosing 'sustainable' materials.

So, they belong to the 'do no further harm' domain rather than that of regenerating what the sector has depleted.

These tools have many different facets such as energy efficiency, sustainable materials and waste management. Certification can be sought against different facets, resulting in buildings carrying the same certification for different reasons. For example, one building could be certified BREEAM Excellent, because its energy performance is very high, while another building could have reduced waste by design. The lack of transparency to the owners and/or end-users of certified buildings results in a situation where a certified building does not deliver the expectations of these stakeholders. The key process impact of the use of these certification methods has been the introduction of an additional role within the project team in order to 'keep on top' of the bureaucratic paperwork to achieve certification.

In addition, the market penetration of these certification tools has been limited, mainly due to their cost. Licina (2019) estimates that, by 2019, a million buildings had been certified worldwide. If all the certified buildings were in the EU, they would constitute 0.76% of the buildings.[3] The World Green Building Council [38] uses the best available data at the time of publication to state that only 856 buildings were certified by various Green Building Councils (GBCs) as operating at 'net zero', using one of the 18 GBC certification schemes. Clearly, the percentage of certified buildings worldwide forms a tiny proportion of all buildings. Moreover, certification in its own right does not result in sustainable buildings, because it typically relies on the evaluation of the design of a scheme and not necessarily on its actual performance in use [2]. There is abundant evidence that points to the 'performance gap' between the design and operational performance of green buildings, starting with de Wilde [15] and van Dronkelaar et al. [37]. Given the lack of market penetration of certification tools and the performance gap even in certified buildings, only a very small percentage of new buildings operate as low or zero carbon buildings.

So, how do these developments render the innovative behaviour of incumbents? They provide additional evidence that all incumbents in the built environment sector cannot be characterised as 'defenders' of the existing regime and that some incumbents can be innovative when the business conditions are appropriate. However, the outcomes of these innovative initiatives do not necessarily result in reductions in emissions from new buildings, as the well-established 'performance gap' in certified buildings show.

---

[3] The RICS (2020) estimates that there are approximately 131 million buildings in the member states of the European Union (EU). https://www.rics.org/news-insights/energy-efficiency-of-the-building-stock-in-the-eu

4.3 Emission Reduction Potential in Existing Buildings: An Untapped Opportunity as a Pathway

The previous two sections have shown that sustainability transitions within the sector have been focused on new buildings. The large emission reduction potential in existing buildings, which was established as early as 2008, remains untapped.

The CCC's [8] first report in 2008 identified the 'technical potential' to reduce 40 MtCO2 in the residential sector and 11 MtCO2 in the non-domestic sector. The report stated a realistic potential to reduce emissions by 9–18 MtCO2 from existing residential buildings and an additional 4 MtCO2 from new buildings in this sector. The notable focus on existing buildings is justified on two bases. First, new buildings had already become 70% more efficient by 2007 than homes built before 1990 [14]. Second, few new buildings were going to be added to the residential stock during the UK's first three carbon budgets between 2008 and 2022, because in any year new houses make up less than 1% of the stock, and demolition rates are less than 0.1% per year. The 2008 report continued to "estimate that 99% of existing homes will still exist in 2020, and that these will then form 88% of the housing stock".

The next [9] progress report acknowledged the 9% indirect emission reductions from residential buildings between 1990 and 2007 and 6% direct emissions for 2003–2007. The former was driven by the switch from coal to gas power generation, and the latter partially as a result of demand reduction due to high energy costs. The report also highlighted that the existing policy[4] at the time did not provide sufficient incentives for the reduction measures to be taken up, representing a lack of *normative directionality*. The 2010 Progress Report [10] attributes the 2009 direct and indirect emission reductions to the 2007–2008 financial crisis and the resultant recession and determines that the rate of progress in implementing energy efficiency measures in the residential sector had been slow. In the Committee's view, the policies still need strengthening, despite the Government committing to a 'Green Deal', passing early legislation to deliver a National Energy Efficiency Programme, the Renewable Heat Incentive (RHI) proposals and the wider roll-out of Energy Performance Certificates (EPCs) and Display Energy Certificates (DECs) in non-residential buildings. In contrast, 2010 saw a 7% increase in emissions from the residential sector, which are attributed to the cold weather.

The launch and termination of the Green Deal—the UK Government's flag-ship retrofit programme, in just over three years between January 2013 and March 2016 provides strong evidence that CCC's calls for both focusing on existing buildings and providing *normative directionality* by strengthening policies were overlooked. The ambition was to retrofit 14 million homes by 2020. However, by the time this pay-as-you-save financial mechanism was terminated in March 2016, only a total of approximately 14,000 homes had been retrofitted. The 'Golden Rule', that required

---

[4] Carbon Emissions Reduction Target (CERT) was the main emissions reduction policy at the time. It was introduced in 2008 as the successor to the Energy Efficiency Commitment and was to run until the end of 2012. CERT set targets for energy supply companies to implement measures in homes that will reduce emissions, with failure to meet targets resulting in fines.

projected cost savings from the measures installed to exceed the loan repayments, was one of the key reasons for the failure of the Green Deal. Also, the *resources* that the financial proposition afforded to households was not attractive to most, and there was a lack of engagement with consumers by tapping into their *values* [34].

Fast forward to 2022, buildings are still the UK's second highest emitting sector, with 17% of the total direct emissions attributed to this sector, rising to 23% when indirect emissions are included [7]. The 15% reduction between 1996, when the territorial emissions from buildings peaked, and 2008, when the Climate Change Act was passed, is recognised. However, the stalling of the progress in emission reductions from buildings since 2008 should also be recognised. For example, the sector only delivered another 4% of reductions in the six years between 2009 and 2015 [5].

Table 1 shows the average percentages of emission reductions from buildings during the different periods quoted in the CCC reports. The highest average during 1996–2008 is attributed to energy efficiency improvements in new buildings. The annual reduction average between 1990 and 2018 sits at 0.57%. It needs to increase to 3.3% per annum if the sector is to deliver 43% emission reductions from 2022 levels by 2035.

The realised emission reductions during 2009–2015 yield the lowest average of annual reductions, which are followed by a 0.97% annual increase during the three years between 2015–2018. It is notable that these annual increases come immediately after the dismantling of the Code for Sustainable Homes in 2015. The slowing down of reductions, followed by a period of increase during 2015–2018, while the Building Regulations continued to be upgraded to demand higher efficiency standards in new buildings, clearly show that the future path to net zero should shift towards existing buildings and mainly residential ones where the reduction potential is highest. In its 2023 Progress Report, the CCC continues to urge the Government to "coordinate a shift in how the UK's 28 million homes and two million non-residential buildings use energy" [7]. This report paints a government that lays broad ambitions, but does not couple them with detailed policies that are prerequisites for their achievement, once again pointing to lack of *normative directionality* hampering progress along sustainability transition pathways. Something needs to change and change soon.

Table 1 Average emission reductions

Period	Duration	Total building emission change	Average total bldg. emission change per annum during period
1996–2008	12 y	−15% (decrease)	−1.25% (decrease)
1990–2015	25 y	−19%	−0.76%
2009–2015	6 y	−4%	−0.67%
2015–2018	3 y	+2.92% (increase)	+0.97% (increase)
Average emission reduction per annum during 1990–2018			-%0.57

5 Transition Pathways for Improving Building Energy Efficiency

The previous section demonstrated that the UK Government's *normative directionality* to reduce emissions from the built environment sector targeted new buildings. It also demonstrated that directionality fluctuated as a result of changes to policy implementation instruments to deliver climate action targets. As a result, process transitions for reconfiguration within *regimes* have not been supported for long enough to deliver changes in industry *norms*. The Government's limited commitment to policy and financial instruments constrained the socio-technical transformation, which would have enabled the penetration of existing technological solutions. This situation has been well illustrated in the discussion on the Code for Sustainable Homes.

Shifts in commitment to policies regarding new buildings and the relatively few instruments targeting the emission reduction potential inherent in existing buildings result in unnecessary delays to reducing the environmental impact of buildings. For example, the uplifting of the building standards from 2022 is celebrated as progress towards net-zero homes. However, it delivers a commitment to a similar performance to Level 4 of Code for Sustainable Homes six years later than originally planned. During the nine years since the Code was scrapped in July 2015, buildings were built to poor standards with gas boilers. They will need to be retrofitted to become net zero, exacerbating the retrofit challenge. So, delays are not only delaying the achievement of targets, but they are also making the problem bigger.

It could be argued that the fluctuations in commitment to policies, and therefore the above delays, are an outcome of the financial outlook that prioritises the cost and affordability of action over the cost and risk of inaction. Despite robust scientific evidence, the fast-escalating cost of inaction continues to be overlooked. For example, the failure of the Green Deal and the continuing lack of a robust and sustained retrofit programme for existing buildings left building occupants vulnerable to energy price hikes since early 2022, and to increased energy demand for cooling to deal with overheating as the temperature rises, and the summer months in the Northern Hemisphere break peak temperature records year on year. Energy and Climate Intelligence Unit's (ECIU) [16] analysis provides robust evidence for this argument by finding "that if the UK had not delayed deploying renewables, insulation, rooftop solar panels, heat pumps and electric vehicles, some households could have saved around £1,750 on bills in 2022".

ECIU [16] also provides clear evidence that the UK lags behind in terms of the socio-technical transformation that is necessary, at least with regards to adopting heat pumps—as the key aspect of the policy for decarbonising heating systems in new buildings. Installation progress in the UK remains extremely limited. The UK installed 63 heat pumps per 100,000 people, while Estonia sold 1,583 and Poland sold 259 per 100,000 in 2021.

Given the above context, this chapter argues that relying mainly on *normative directionality* is not going to be enough to start tapping into the emission reduction potential in existing residential buildings or consistently delivering 'net-zero' new buildings as pathways to sustainability transitions. Different avenues to bring about sustainability transitions in the built environment sector should be explored. At this juncture, the key question is:

> Why have firms, as key *actors*, been so reluctant to exploit the extensive business opportunity that energy retrofit of existing residential buildings and the construction of new net-zero buildings present?

The answer may well lie in the firms' reluctance to fail, which is understandable in a sector where the norm appears to be the 'successful' delivery of projects according to pre-determined KPIs. This context preconditions the firms to only take "action to mitigate climate change if it was inherently justifiable, i.e. if climate change initiatives provided immediate and concrete benefits [to the firms delivering these projects and the clients procuring them]" [23]. As such, actors are still reliant on *normative directionality* to deliver regulations and legal obligations to change before they change their own practices.

In the main, the firms that have changed their practices, despite the lack of *normative directionality*, transitioned to a more 'resource-efficient version' of their incumbent *regimes* [20] as they responded to policy changes such as the strengthening of the Building Regulations Standards, but they slowed down or terminated their innovation initiatives as soon as the Government withdrew its support for Code for Sustainable Homes. Therefore, most firms take a 'reformist' position. They seemingly operate in a 'largely efficient' market, some aspects of which 'fail' [20]. In this context, the role of the policy is to intervene to deal with failures, mainly through radical innovation which focuses on 'digital and manufacturing technologies' [22]. An example 'failure' in the sector is excessive resource consumption that results in very high levels of waste. The UK Government advocated the adoption of digital technologies such as BIM to deal with failures such as this one. This and similar interventions have however remained largely silent on the reconfiguration of the *practices* and *norms*, which are critical for creating the necessary conditions to foster the socio-technical transformation of the built environment 'system'.

Such systems are hard to change [20]. One of the reasons is the resistance of incumbent *actors*. This was evident in the sector when the Code for Sustainable Homes was scrapped and a major house-builder admitted to a Parliamentary Select Committee that they campaigned against the Code. More recently, it has been argued that the Government 'wasted' large sums of money as it tried to provide a secure niche for modern methods of construction (MMC) to flourish. A large number of MMC manufacturers and contractors making extensive use of this funding have failed in 2023. The emerging news in early 2024 does not paint a favourable picture either. Hence, radical innovations that emerge in niches do not 'gradually develop through learning processes' and replace routines as MLP posits [20]. Therefore, the question becomes

> What needs to be in place so that radical innovations to deliver energy efficiencient buildings emerge and replace *routines*?

This chapter argues that the answer lies in business models that are fit for alternative economic models such as doughnut economics and circular economy for a post-growth, post-development future. This need is also acknowledged in the limited literature on such business models within the built environment sector. Here, the resource use and waste reduction and the significance of design for fabrication, deconstruction and recycling are also emphasised. Therefore, the emphasis on new buildings continues. The all-important refurbishment and retrofit aspects of the industry remain overlooked, waiting for the Government to publish the next policy instrument. Further research should establish ways in which this situation can be reversed through business model innovation and the changes in practice. Research in this domain is still in its exploratory stages in all sectors [31], not just in the context of buildings. The methods and approaches that firms need to innovate their business models towards a circular economy (Blomsma and Brennan, 2017, quoted in [29]) are yet to be developed.

The starting point for devising these business models has to be sustainability as an integral part of a firm's *value* proposition, *value* creation and *value* capture. There are wider *value* propositions that go beyond economic value such as social value. Experimentation and learning from the outcomes of this experimentation are integral to developing the necessary methods and approaches to bring about innovative business models that will accommodate *value* propositions other than those that are financial [29].

There is an urgent need to evaluate the current 'green', 'sustainability' practices of firms and the projects that they deliver from the perspective of business models for sustainability. Value proposition, creation and capture in the built environment sector are distributed amongst different professions that come together in, usually, short-term project coalitions. This distribution renders the, currently limited, "integration of skills and resources across built environment sectors and the professionals working within them to maximise climate change knowledge capacity and action" all the more important [25]. Moreover, different stakeholders' perceptions of value varies. For example, value for facilities management teams is driven by energy performance in use, while it is likely to focus on cost of construction for holders of capital expenditure budgets in the absence of business models that attribute accountability to different stakeholders throughout the building life cycle. This complex landscape is a fertile research context.

6 Conclusions

This chapter demonstrated that currently sustainability transition initiatives focus on reducing negative impacts of new buildings. This represents a limited understanding of the scope and scale of the necessary sector transition in three key areas [25]. First, 70% of the buildings that we are going to occupy by 2050 already exist and their energy performance is, at best, low. They are largely neglected in the current discourse on transforming the built environment. Second, the built environment sectors are known to be reluctant to embrace change, let alone radical change. Such change is absolutely critical to deliver, at scale, buildings that operate at 'net zero' or at the level of their 'green building' certification. Third, the built environment sectors need to adopt business models that align with progressive economic models, e.g. circular economy, if they are to eliminate their environmental and biodiversity impacts and start contributing to regeneration. The current discourse focuses on doing no further harm, and leaves questions associated with regeneration unanswered.

Therefore, the following recommendations are made for policy-makers, leading firms in the sector and researchers. Policy-makers should recognise the importance of normative directionality and its consistency for the sustainability transitions of the niche and regime actors. They should retain and strengthen policies for long-enough for sustainability transitions to take place. Policies should accommodate values other than financial ones such as the role a sustainable built environment plays in the health and well-being of nations. Leading firms which are navigating their sustainability transitions should explore how they can integrate these value propositions into their balance-sheets. They are also fertile grounds for research. Researchers should study these firms to establish how they deal with the interactions between niche, regime and landscape, how these interactions deliver the minimum level of success which enable these firms to progress on their transition pathways and to what extent their business models are conducive to their transition.

References

1. Ade R, Rehm M (2020) The unwritten history of green building rating tools: a personal view from some of the 'founding fathers.' Build Res Inform 48:1–17. https://doi.org/10.1080/09613218.2019.1627179
2. Afroz Z, Burak Gunay H, O'Brien W (2020) A review of data collection and analysis requirements for certified green buildings. Energy Build 226:110367. https://doi.org/10.1016/j.enbuild.2020.110367
3. BEIS and ONS (2021) 2019 UK Greenhouse Gas Emissions. London
4. Calvin K, Dasgupta D, Krinner G, et al (2023) IPCC, 2023: climate change 2023: synthesis report. Contribution of working groups I, II and III to the sixth assessment report of the intergovernmental panel on climate change [Core Writing Team, H. Lee and J. Romero (eds.)]. IPCC, Geneva, Switzerland
5. CCC (2019) The sixth carbon budget: buildings
6. CCC (2024) Future emissions targets must not be loosened. https://www.theccc.org.uk/2024/02/28/future-emissions-targets-must-not-be-loosened/. Accessed 19 Apr 2024
7. CCC (2023) Progress in reducing emissions: 2023 Report to Parliament
8. CCC (2008) Building a low-carbon economy—the UK's contribution to tackling climate change
9. CCC (2009) Meeting carbon budgets – the need for a step change
10. CCC (2010) Meeting Carbon Budgets—ensuring a low-carbon recovery
11. Collins B (2020) "It's not talked about": The risk of failure in practice in sustainability experiments. Environ Innov Soc Transit 35:77–87. https://doi.org/10.1016/j.eist.2020.02.008
12. Core Cities UK Why Core Cities? https://www.corecities.com/cities/why-core-cities. Accessed 18 Apr 2024
13. DCLG (2009) The code for sustainable homes: case studies
14. DCLG (2007) Building a greener future: towards zero carbon development

15. de Wilde P (2014) The gap between predicted and measured energy performance of buildings: a framework for investigation. Autom Constr 41:40–49. https://doi.org/10.1016/j.autcon.2014.02.009
16. ECIU (2023) Cost of Not Zero in 2022
17. Edmondson DL, Kern F, Rogge KS (2019) The co-evolution of policy mixes and socio-technical systems: towards a conceptual framework of policy mix feedback in sustainability transitions. Res Policy 48:103555. https://doi.org/10.1016/j.respol.2018.03.010
18. Geels FW (2019) Socio-technical transitions to sustainability: a review of criticisms and elaborations of the Multi-Level Perspective. Curr Opin Environ Sustain 39:187–201. https://doi.org/10.1016/j.cosust.2019.06.009
19. Geels FW (2002) Technological transitions as evolutionary reconfiguration processes: a multi-level perspective and a case-study. Res Policy 31:1257–1274. https://doi.org/10.1016/S0048-7333(02)00062-8
20. Geels FW, McMeekin A, Mylan J, Southerton D (2015) A critical appraisal of sustainable consumption and production research: the reformist, revolutionary and reconfiguration positions. Glob Environ Chang 34:1–12. https://doi.org/10.1016/j.gloenvcha.2015.04.013
21. Geels FW, Schot J (2007) Typology of sociotechnical transition pathways. Res Policy 36:399–417. https://doi.org/10.1016/j.respol.2007.01.003
22. Green S (2023) Making sense of construction improvement. Routledge, London
23. Heiskanen E, Jalas M, Rinkinen J, Tainio P (2015) The local community as a "low-carbon lab": Promises and perils. Environ Innov Soc Transit 14:149–164. https://doi.org/10.1016/j.eist.2014.08.001
24. HM Government (2023) 2030 Strategic framework for international climate and nature action
25. Hürlimann AC, Nielsen J, Moosavi S et al (2022) Climate change preparedness across sectors of the built environment—a review of literature. Environ Sci Policy 128:277–289. https://doi.org/10.1016/j.envsci.2021.11.021
26. Hurlimann A, March A, Bush J et al (2024) Climate change transformation in built environments—a policy instrument framework. Urban Clim 53:101771. https://doi.org/10.1016/j.uclim.2023.101771
27. ILO (2019) 100 Statistics on the ILO. In: 100 Statistics on the ILO
28. Köhler J, Geels FW, Kern F et al (2019) An agenda for sustainability transitions research: state of the art and future directions. Environ Innov Soc Transit 31:1–32. https://doi.org/10.1016/j.eist.2019.01.004
29. Konietzko J, Baldassarre B, Brown P et al (2020) Circular business model experimentation: demystifying assumptions. J Clean Prod 277:122596. https://doi.org/10.1016/j.jclepro.2020.122596
30. Licina D et al. (2019) Occupant health & well-being in green buildings: trends and future directions. ASHRAE Journa 61
31. Munaro MR, Tavares SF, Bragança L (2020) Towards circular and more sustainable buildings: a systematic literature review on the circular economy in the built environment. J Clean Prod 260:121134. https://doi.org/10.1016/j.jclepro.2020.121134
32. O'Neill K, Gibbs D (2020) Sustainability transitions and policy dismantling: zero carbon housing in the UK. Geoforum 108:119–129. https://doi.org/10.1016/j.geoforum.2019.11.011
33. RICS (2022) RICS calls for industry and government to futureproof UK construction. https://www.rics.org/news-insights/rics-calls-for-industry-and-government-to-futureproof-uk-construction. Accessed 18 Apr 2024
34. Rosenow J, Eyre N (2016) A post mortem of the Green Deal: austerity, energy efficiency, and failure in British energy policy. Energy Res Soc Sci 21:141–144. https://doi.org/10.1016/j.erss.2016.07.005
35. The World Bank (2023) Urban Development. https://www.worldbank.org/en/topic/urbandevelopment. Accessed 15 Mar 2024
36. UNEP (2023) Building Materials and the Climate: Constructing a New Future. Paris
37. van Dronkelaar C, Dowson M, Spataru C, Mumovic D (2016) A review of the regulatory energy performance gap and its underlying causes in non-domestic buildings. Front Mech Eng 1. https://doi.org/10.3389/fmech.2015.00017

38. WGBC (2023) Advancing net zero in numbers. In: Advancing Net Zero Status Report
39. Williams S, Robinson J (2020) Measuring sustainability: an evaluation framework for sustainability transition experiments. Environ Sci Policy 103:58–66. https://doi.org/10.1016/j.envsci.2019.10.012

Chapter 4
Enhancing Sustainability and Energy Policies for Effective Implementation of Passive Solar Housing Construction in Sri Lanka

H. Mallawaarachchi and U. A. S. Nissanka

Abstract Ecological awareness and concern for future generations promote the eco-trend and energy independence, and contribute to the improvements in living standards by reducing the consumption of conventional energy. In that regard, the concept of passive solar housing has garnered global attention for achieving sustainability in the building construction industry. Therefore, countries around the world are currently focusing on implementing passive solar housing design techniques. Specially, the passive solar housing design is a practical concept related to energy that is more suitable for equatorial countries, such as Sri Lanka. However, lack of policies specifically focusing on the construction of passive solar houses has hindered its successful implementation. Still worse, no comprehensive analysis of the existing policies and guidelines has also been conducted ensuring relevant policy enhancements. Thus, the aim of the research is to review the existing sustainable and energy-efficient building construction-related policies and guidelines in support of the construction of passive solar houses in Sri Lanka to determine the probable policy enhancements. A descriptive case study strategy was adopted, and the data were collected through a scoping review of available national policies and guidelines of sustainable and energy-efficient building construction in Sri Lanka during the last decade and through expert interviews. As this research aimed, several provisions were recognised that support passive solar housing construction. Accordingly, potential areas of enhancements to the existing sustainability-related policies and guidelines, such as providing financial incentives for passive solar residents, reduction of excessive generation and usage of non-renewable energy, adopting modern and green technology assessment, compliance, and certification, and enhancing research and

---

H. Mallawaarachchi (✉)
Department of Facilities Management, Faculty of Architecture, University of Moratuwa, Moratuwa, Sri Lanka
e-mail: harshinim@uom.lk

U. A. S. Nissanka
Department of Quantity Surveying, Sri Lanka Institute of Information Technology, Malabe, Sri Lanka

T. Ramachandra et al. (eds.), *Concepts, Policies and Technology for Sustainable Buildings*, Advances in Sustainability Science and Technology,
https://doi.org/10.1007/978-981-96-3086-8_4

development in passive solar housing construction field were identified through analysis. Further, having a separate national policy for passive solar housing design and construction, and providing financial incentives for passive solar service providers were also identified as key strategies that need to be considered. Research findings presented in this chapter lay a firm foundation for progressing in developing a good panoramic policy and regulatory framework for passive solar housing construction in Sri Lanka. Indeed, this research provides a comprehensive underpinning for encouraging researchers, policymakers, and construction professionals towards a successful implementation of passive solar housing concept in Sri Lanka.

Keywords Energy retrofit · Housing construction · Passive solar housing design · Policy enhancements · Sri Lanka

1 Introduction

The World Bank emphasises that investing on low-carbon and resilient urban development is crucial for reducing city emissions and addressing climate change (Dasgupta, Lall and Wheeler, 2022). With the fastest urbanisation expected in Asia and Africa in the coming decades, it is essential to develop and implement effective sustainable solutions to build resilient cities (Engstrom, 2023). Hence, it is essential for new buildings to be built with a high level of energy performance, since this will affect the energy consumption in the following decades [15]. The building industry has huge potential for energy reductions as worldwide awareness of energy and environmental concerns rises [1]. Furthermore, the International Energy Agency (IEA) stated that all new buildings, and 20% of the existing building stock, need to be highly energy-efficient by 2030 to achieve a net-zero emission scenario by 2050 [17]. Energy conservation in buildings refers to reducing the energy required for a building's life cycle (construction, operation, and maintenance) through pre-design measures, orientation, and use of climatic conditions, such as passive strategies, high thermal mass, renewable energy sources, and natural ventilation, which play key roles for energy conservation as well as thermal comfort [3].

Among the other sustainable building designs, passive solar housing design has gained a world concern as those buildings are designed to enhance thermal comfort of building occupants using the building's physical structure and site conditions alone [3]. This energy retrofit of residences takes advantage of a building's site, climate, and materials to minimise energy use [16, 25]. Hence, many countries have introduced national-level policies and guidelines supporting the implementation of sustainable building construction, such as National Carbon Offset Standard for Buildings in Australia, to name a few [14]. As stated by Li et al. [14], the target of carbon reduction goals begins with public buildings and gradually works incrementally up to residential buildings, even though the effectiveness of standards and policies for residential buildings is insufficient in many countries. The Energy Consumption of Buildings Standard in China, which was introduced in 2016 by the Ministry of China

Housing and Urban–Rural Development, and the National Carbon Offset Standard for Buildings, which was established in 2017 by the Australian Federal Government with the collaboration of the Green Building Council, can be identified as some examples of standards and policies that focus on the total building sector [14, 16]. Further, it is evident in previous literature that the concern given on passive solar housing construction is limited in existing policies. Similarly in Sri Lanka, no specific policies or guidelines governing passive solar housing construction even through a few of sustainability- and energy-related policies and guidelines can be found with a limited number of provisions. Further, there is a lack of comprehensive analysis of the existing sustainability- and energy-related policies and guidelines towards the adoption of passive solar housing design and construction. Hence, with the aim of proposing policy enhancements for strengthening the passive solar housing design implementation, the research presented in this chapter seeks to achieve the following research objectives, limiting the research's scope to the context of Sri Lanka.

Objective 1: Identify existing sustainability- and energy-related policies and guidelines and their provisions applicable to passive solar housing construction in Sri Lanka.

Objective 2: Propose probable enhancements of the existing policies towards passive solar housing construction in Sri Lanka.

Objective 3: Investigate the challenges that could impede the adoption of passive solar housing construction in Sri Lanka and propose probable strategies to overcome them.

The next section is to understand the concept of passive solar housing, its' design, and construction technologies prior to conducting a scoping review of existing sustainability- and energy-related policies towards adoption of passive solar housing concept.

2 Literature Review

2.1 *The Passive Solar Housing Concept*

The demand for energy is increasing exponentially due to ever-increasing innovations in technology-based industries as well as population growth. This high energy consumption has a negative impact on environmental sustainability and energy conservation [9]. For this reason, the concept of passive solar housing has been introduced as a housing concept powered by natural energy forces without using mechanical or electrical components [8]. These passive solar houses are built according to a unique architectural design based on a site analysis of the land and the benefits of climate changes. The concept of passive solar housing has been introduced worldwide and is now prevalent in many countries.

The concept of passive solar design has been introduced as a building construction concept that utilises passive energy sources without the use of additional power sources or mechanical energy [2]. Passive solar houses are so named because they do not rely on mechanical or electrical components. This concept can be applied to new as well as existing buildings by modifying them. The main concept of passive solar houses is to absorb and store energy from common building materials such as windows, walls, and floors [12]. Passive solar homes regulate indoor temperatures using sunlight, wind, and vegetation. This design maximises solar energy by strategically positioning windows, walls, and floors to collect and distribute solar radiation throughout the living area. Thermal mass, which absorbs and releases heat during the day, is also utilised in passive solar designs. Additionally, passive solar housing design incorporates insulation, ventilation, and shading devices such as overhangs and trellises. Passive solar architecture reduces the need for mechanical heating and cooling, thereby lowering the home's environmental impact and energy expenses [3]. The interior environment of a passive solar house is bright. The structure of such a building blocks 99.9% of the ultraviolet rays that are harmful to the body, which helps prolong the lifespan of interior fabrics [19]. The passive solar housing concept also helps to keep interiors cool during hot weather and warm during cold seasons. These homes are especially comfortable, as the interior temperature does not fluctuate like in a normal house. This presents an opportunity to move away from traditional energy-related devices [19].

The efficiency of passive solar houses depends on climatic conditions, the size of the building, the adopted passive solar housing system, and its design criteria [8]. When compared to the layout of another building, there is a significant reduction in electricity demand. Constructing a passive solar house costs more than a typical building. However, such buildings bring economic benefits in the long term [21]. Further, passive solar housing design generates several other benefits. Solar homes are comfortable and warm, maintaining stable temperatures throughout the year with little additional heating [6]. The greatest savings in heating and emissions can be achieved with a well-designed passive solar home, which can reduce heating costs by up to 70% (Kahoorzadeh et al. 2014) and similarly decrease emissions, benefiting both the occupants and the environment. Passive solar houses can be coupled with the use of a solar thermal system, which also utilises the sun's energy, to generate effective, minimal-cost, environmentally friendly energy [10].

The ability to store internal energy also affects the sustainability of passive solar houses. It should also be noted that the quantity of energy wastage must be minimised to maintain environmental sustainability. Since this housing concept is used to reduce non-renewable energy consumption, the level of non-renewable energy use should also be considered. Accordingly, the use of non-renewable energy is also a factor in the sustainability of these houses [25]. Environmental sustainability can be achieved through eco-friendly design and construction using the most eco-friendly materials during the housing design stage [22]. Along with changes in internal temperature, natural ventilation also affects the environmental sustainability of passive solar houses [28]. Additionally, internal and external noise can be factors in environmental sustainability (Ahmed et al., 2019). The rate of resource depletion and its short-term and long-term effects are considered [22]. In addition, high-end air filtration is considered an environmentally sustainable factor [27].

2.2 *Passive Solar Housing Design and Construction Technology*

In the case of passive solar houses, a variety of unique technologies are used. The passive solar housing design and construction technology has evolved from traditional and simple to modern and sophisticated over time. Solar housing features can be added to an existing typical house. The result is a reduction in the cost of pre-existing heating and cooling after remodelling, thus achieving energy efficiency more efficiently. A key element here is to plan the orientation of the building, especially considering the nature of the land and the trajectory of the sun. The nature of the housing construction site can also be considered a sustainable environmental factor, including the natural location of the land, and surrounding natural afforestation or vegetation, which directly affects the efficiency of these houses [22]. Separate strategies must be put in place to store solar energy during peak hours and at other times. It is also important that the shading devices are positioned and adjusted to the correct size [5]. Attention should also be paid to the directions they face. There are three ways to harness solar energy in passive solar houses: direct gain system, indirect gain system, and isolated gain system [12]. Passive solar houses should be designed based on the trajectory of the sun. The direction of sunlight from sunrise to sunset is a major environmental factor [4]. The passive sun collects heat in a house, and the sunlight coming through the south-facing windows is stored as heat mass by the heat storage material. Properly oriented windows design the ecological layout and terrain location according to mathematical methods [5]. At the same time, it eliminates the need for additional heat mass, as sufficient furniture, and drywall well-insulated houses in temperate climates. Accordingly, they use architecture and masonry to make the most of these. A solar house is designed to distribute stored heat by convection and radiation. These are executed according to special designs when applying colour to the house and its elements [10]. These houses are also affected by the gradual increase in global warming and changes in rainfall. In a tropical country like Sri Lanka, there are four seasons—spring, autumn, summer, and winter—but only two main seasons are hot and rainy [10]. Windows can be covered to reduce unwanted heat gain during the day, and vegetation can be used to achieve this objective. Landscaping methods and topography can also be utilised for this purpose [12].

Additionally, insulation materials should be selected for the construction of the roof, interior floor, walls, and other building elements. The design of a house uses materials such as bricks, roof tiles, and concrete to absorb heat from the sun during the hot season. At the same time, it eliminates the need for additional heat mass is sufficient for furniture and drywall well-insulated houses in temperate climates.

Accordingly, they use architecture and masonry to make the most of these. A solar house is designed to distribute stored heat by convection and radiation. These are executed according to special designs when applying colour to the house and its elements [10]. A key element of the design of a passive solar house is to be careful when applying the raw material to the roof, especially when the sun is shining most of the year. Also, considering the colour of the interior wall of such a house is important [10]. As shown in previous research, it is advisable to use a light colour, close to white for the exterior walls of a passive solar house. The maximum benefits of a passive solar house can be achieved through technical strategies and designs such as designing a roof arrangement to maximise structural cooling and positioning windows for natural ventilation [21].

2.3 Case Study: Passive Solar Housing Construction in Sri Lanka

The concept of passive solar housing is active in Sri Lanka over the years; however, it is still not popular. Even though many passive solar design features have been adapted to modern building designs, the construction of passive solar houses is still lacking. There is one building belonging to the passive solar housing construction in Sri Lanka which has obtained the "EnerPHit" pilot project certificate from the Passive Housing Institute of Sri Lanka. Passive Housing Institute of Sri Lanka is an independent research institute that has played an especially crucial role in the development of the passive solar housing concept. It is the first dormant housing certified building in Sri Lanka and is a garment manufacturing centre located in Colombo, the capital of the country. Accordingly, in countries such as Sri Lanka, this passive solar housing concept can be applied to overcrowded factories, conference halls, and private homes. This will greatly reduce the energy consumption of the country, which is facing an energy crisis [11].

In terms of the residential building sector, the Sri Lanka Sustainable Energy Authority [24] published the "Guideline for Sustainable Energy Residences in Sri Lanka" in 2019. This guideline primarily focuses on energy efficiency and sustainability, aimed at reducing carbon emissions from residential buildings. Furthermore, GREENSL® National Green Rating System for New Constructions published by the Green Building Council Sri Lanka also provides important provisions for the use of solar power in buildings. In addition, the Ministry of Urban Development and Housing of Sri Lanka initiated the "Sri Lanka Sustainable Housing and Construction Roadmap (2020–2050)" in collaboration with the Global ABC, UNEP, United Nations Agency for Human Settlements (UN-Habitat), United Nations Office for Project Services (UNOPS), One Planet Network DBS Programme, and Worldwide Fund for Nature (WWF). The vision of this programme is "By 2050, housing and construction in Sri Lanka are socially inclusive, environmentally sustainable, and carbon–neutral". Furthermore, the National Policy on Construction, Construction

Industry Development Act No. 33 of 2014, and the National Physical Planning Policy are other policies applicable for sustainable and energy-efficient building construction in Sri Lanka. Indeed, the National Energy Policy and Strategies of Sri Lanka and the Code of Practice for Energy Efficient Buildings in Sri Lanka are other policies supporting sustainable and energy-efficient buildings in Sri Lanka.

3 Methodology

The descriptive case study strategy was implemented by selecting in the context of Sri Lanka as the case study. The concept of passive solar housing and its design and construction technologies were initially reviewed through available literature. Within the selected context, the research was designed to conduct under two stages as stated in Table 1.

To analyse the provisions of existing sustainability- and energy-related policies and guidelines in Sri Lanka, the scoping review technique was adopted (Stage 1). As stated by Pare and Kitsiou (2017), scoping review is an ideal technique to review extant literature on an emergent topic. Further to the authors, inclusion and exclusion criteria must be established to help researchers eliminate studies that are not aligned with the research questions.

Figure 1 illustrates the scoping review process adopted in this study, detailing the inclusion and exclusion criteria.

Initially, all relevant sustainable and energy-efficient building construction-related national policies and guidelines in Sri Lanka were gathered from national ministry websites, departmental websites, and databases covering the period 2013–2024. Accordingly, 11 national policies and guidelines were found in the search. The selected policies and guidelines were undergone through a screening process based on the inclusion criteria. The screening process determined how relevant the publications were in relation to the article inclusion criteria in which the use of key terms "Renewable Energy" and "Solar" or "Passive Solar" in the title and/or the document content was considered. As a result, eight (08) sustainability- and energy-related national policies and guidelines were selected for the scoping review, while three

Table 1 Stages of research design

Stage	Research strategy	Outcome	Objectives
Stage 1	Scoping review	Identify existing provisions supporting passive solar housing construction in Sri Lanka	Objective 1
Stage 2	Expert interviews	Propose probable enhancements of the existing policies towards passive solar housing construction in Sri Lanka	Objective 2
		Investigate the challenges faced when adopting passive solar housing construction in Sri Lanka and propose probable strategies to overcome the identified challenges	Objective 3

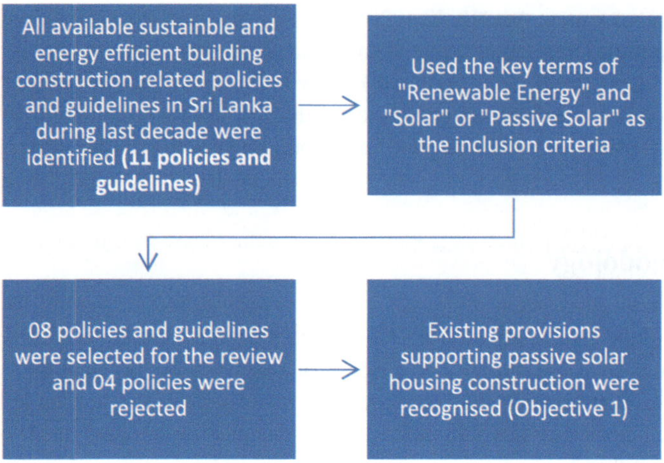

Fig. 1 Scoping review process (Author generated)

(03) guidelines that are not aligned with the research objectives were excluded. The selected policies and guidelines were then reviewed to identify existing legislative provisions supporting passive solar housing construction in Sri Lanka (Objective 1).

As the second stage, expert interviews were conducted with five (05) selected construction professionals, having more than 05 years of experience and knowledge in the fields of building construction and passive solar housing. The experts were selected through convenient sampling rather than picking experts randomly from the entire population of construction experts. The experts' profiles are presented in Table 2.

According to demographic information of experts who participated in the expert interviews (Table 1), all experts had over 05 years of experience in building construction in which more than 75% of experts had sufficient knowledge and experience in passive solar housing design and construction. It is proved that all construction professionals who participated in the expert interviews were a suitable and appropriate sample.

Table 2 Experts' profiles

Designation	Years of experience in building construction	Years of experience in passive solar housing
Architect	12	6
Quantity surveyor	12	5
Civil engineer	06	4
Project manager	10	6
Architect	10	5

The experts were asked about the possible enhancements of the existing policies and guidelines towards implementation of passive solar housing construction in Sri Lanka. Further, they were asked about the challenges faced when adopting passive solar housing construction in Sri Lanka, and probable strategies to overcome the identified challenges (RQ3). Hence, the interview guideline consists of three key sections: (i) Enhancements for existing policies, (ii) Challenges faced when adopting passive solar housing construction in Sri Lanka, and (iii) Probable strategies to overcome the identified challenges. The collected data were analysed by using template analysis technique.

Key findings are presented in the following section.

4 Results and Discussion

The scoping review outcomes, the analysis of the data derived through expert interviews, and key findings derived through analysis are presented.

4.1 Legislative Provisions Supporting Passive Solar Housing Construction in Sri Lanka

As shown in Fig. 1, out of 11 policies and guidelines in initial search, 08 policies and guidelines were selected that meet the inclusion criteria adopted. The selected policies and guidelines include Guideline for Sustainable Energy Residences in Sri Lanka of 2019, National Energy Policy and Strategies of Sri Lanka, Sri Lanka Sustainable Housing and Construction Roadmap (2020–2050), GREENSL® National Green Rating System for New Constructions (2022), National Policy on Construction (NPC), Construction Industry Development Act No. 33 of 2014, and the National Physical Planning Policy and Code of Practice for Energy Efficient Buildings in Sri Lanka. Three (03) policies and guidelines were excluded in review that are not meeting the inclusion criteria, such as Urban Development Authority Act, National Policy on Sustainable Consumption & Production for Sri Lanka and National Policy on Construction (2014).

After conducting a comprehensive review of the selected policies and guidelines, various legislative provisions were derived which support the passive solar housing construction in Sri Lanka (Objective 1). Table 3 summarises the key provisions identified by reviewing 08 policies and guidelines selected through a screening process.

It was found that there is no specific policy or guideline to govern passive solar housing construction in Sri Lanka. However, various provisions for adopting passive solar design features were recognised that can be used to promote passive solar

Table 3 Legislative provisions supporting passive solar housing construction

Policy/guideline	Legislative provisions supporting passive solar housing construction
Guideline for Sustainable Energy Residences in Sri Lanka in 2019	• Measures for integrating solar energy systems • On-site solar photovoltaic (PV) power generation provisions for residential buildings • Thermal heat storage options for homes • Solar thermal systems for residential hot water production
National Energy Policy and Strategies of Sri Lanka	• Initiating a national programme for improving energy efficiency and conservation aims to strengthen the participation of all stakeholders in the household, industrial, and commercial sectors • Launching a home productivity improvement initiative, with energy efficiency and conservation as the core focus, empowering women in the process • Implementing taxation and other incentives or disincentives to foster the adoption of energy-efficient technologies • Promoting sustainable neighbourhoods as a key element in urban planning to reduce energy demand • Making energy efficiency a primary consideration for retrofits and requiring new building designs to be evaluated for energy performance on a mandatory basis • Supporting cleaner energy sources and technologies to minimise environmental impacts locally and globally • Facilitating renewable energy investments for electricity generation through competitive schemes to lower costs and increase investor participation • Conducting research to address the challenges posed by the integration of intermittent renewable energy sources like wind and solar into the power grid • Introducing advanced forecasting technologies for wind, solar, and rainfall patterns • Promoting energy storage solutions to stabilise intermittent renewable energy sources, regulate voltage and frequency, support local grids, shave peak loads, and enhance grid resilience • Establishing innovative financing schemes and instruments, involving both public and private sectors, to accelerate renewable energy development • Offering a premium green tariff or green energy certification to interested electricity customers as an optional rate, with a portion of the proceeds directed to renewable energy development

(continued)

Table 3 (continued)

Policy/guideline	Legislative provisions supporting passive solar housing construction
Sri Lanka Sustainable Housing and Construction Roadmap (2020–2050)	• Provisions for utilising clean energy sources • Installing rooftop solar panels and benefiting from the national grid as virtual storage, with three options available: (i) Net metering (offsetting household consumption with solar energy generated, without financial transactions), (ii) Net accounting (offsetting household usage with solar generation and receiving payment for excess energy from the CEB), and (iii) Net plus (receiving payment from the CEB for all solar energy generated) • Imposing penalties for excessive energy consumption through revised energy tariffs • Provisions to increase the share of renewable energy in residential electricity supply, utilising solar, wind, and other alternative energy sources
GREENSL® National Green Rating System for New Constructions (2022)	• Reducing non-renewable energy consumption in water transmission by 50% through the use of renewable energy sources such as solar, wind, low-impact hydro, and biogas • Ensuring that at least 5% of the building's total energy consumption (measured by annual energy cost) is supplied by Solar Photovoltaic (PV) technology • Designing the building to maximise natural daylight and provide ample view opportunities • Incorporating green spaces with native or adaptive vegetation at various building levels, including green roofs, terraces, and façades • Minimising solar heat gain through the building envelope
National Policy on Construction (NPC)	• Provisions for designing and developing disaster-resilient, energy-efficient, and environmentally sustainable buildings, structures, and construction practices • Promoting the adoption of energy-efficient and eco-friendly technologies
Construction Industry Development Act No. 33 of 2014	• Encouraging sustainable growth in the construction industry with a focus on designing and developing energy-efficient buildings and structures • Ensuring compliance with fundamental public health, occupational health, environmental, and technical standards, including those related to efficient resource and energy use

(continued)

Table 3 (continued)

Policy/guideline	Legislative provisions supporting passive solar housing construction
The National Physical Planning Policy	• Promoting the development and use of alternative energy sources, including solar, wind, oceanic, mini hydro, and biogas, for domestic applications • Provisions to enhance the availability and reliability of energy supply while encouraging the adoption of renewable energy sources to reduce greenhouse gas emissions
Code of Practice for Energy-Efficient Buildings in Sri Lanka (2008)	• Provisions for incorporating solar water heating systems • Reducing solar heat gain through building design by aligning the building's long axis east–west, minimising openings on the east and west (especially the west), using light-coloured walls and roofs, applying appropriate internal and external shading for windows, maintaining moderate window-to-wall ratios, minimising air infiltration into occupied spaces, and optimising insulation of the building envelope

housing construction. By analysing the existing policies and guidelines for sustainable and energy-efficient building construction in Sri Lanka listed in Table 3, five (05) key areas were derived and that needs to be enhanced to ensure an effective implementation of passive solar housing construction in Sri Lanka. Those areas include (i) Financial incentives for passive solar residents, (ii) Reduction of excessive generation and usage of non-renewable energy, (iii) Modern and green technology adaptations, (iv) Assessment, compliance, and certification, and (v) Research and development. These areas can be considered to introduce provisions under existing policies and/or for developing a new national policy for passive solar housing construction in Sri Lanka.

4.2 Probable Enhancements of the Existing Policies and Guidelines Towards Passive Solar Housing Construction in Sri Lanka

The five (05) key areas derived through scoping review were then presented to the selected experts to identify the probable enhancements for the existing policies and guidelines on sustainable and energy efficient building construction to implement passive solar housing construction in Sri Lanka.

Many of the experts stated that, it is important to introduce strategies to reduce the excessive generation and usage of non-renewable energy in residences while promoting the use of clean energy, such as solar. Provisions for increasing the renewable energy share in the electrical energy supply of residences and for encouraging the development and use of renewable energy sources to reduce greenhouse emissions are other strategies that can be considered in reducing the excessive generation and usage of non-renewable energy in residences. Introducing taxation polices, providing incentives, and disincentives for the excess use of energy, encouraging renewable energy investments for electricity generation, introducing innovative financing schemes, introducing financial instruments to encourage renewable energy development, penalising excessive energy usage by residents through revised energy tariffs, and encouraging net metering, net accounting, and net-plus options in utilising solar energy for households can be considered under "Financial incentives for passive solar residents." Further, provisions for "Modern and green technology adaptations" can be considered by storing thermal heat in houses, effective forecasting technologies, encouraging energy storage solutions, adopting solar photovoltaic technology, maximising the use of daylighting and view opportunities, introducing provisions for green areas with native or adapted vegetation, introducing provisions for the use of solar water heating, and minimising solar gain through the building envelope.

As proposed by experts, two strategies, introducing green tariff/green energy certification packages for residences and ensuring that housing construction is complying with efficient resource and energy utilisation standards, can also be considered. It may ensure the passive solar housing construction satisfies the relevant technical standards in certification of passive solar houses. "Research and development" is another vital area expressed by most of the experts that helps inducing passive solar housing construction in Sri Lanka. Hence, providing financial assistance for conducting research on overcoming adverse impacts of renewable energy absorption, researching efficient technologies for passive solar housing construction, researching challenges for adopting passive solar housing construction, and strategies to overcome the identified challenges are important.

4.3 Challenges Faced When Adopting Passive Solar Housing Construction Concept in Sri Lanka and Strategies to Overcome

As acquired through analysis, many challenges could impede the adoption of passive solar housing construction concept in Sri Lanka. Absence of policy and regulatory framework for passive solar housing construction is a key implementation barrier as expressed by many of the experts. Further, the use of informal practices in housing construction, poor quality management of small-scale housing projects are other challenges faced in implementing passive solar housing concept in Sri Lanka. Indeed, lack of incentives for builders and house owners to utilise solar energy has hindered the successful implementation of passive solar housing construction concept in the construction industry.

Accordingly, a majority of the experts proposed that improving the implementation of passive solar housing construction in Sri Lanka will indeed require a multifaceted approach to address the identified challenges. Hence, various strategies were proposed to overcome the identified challenges as shown in Table 4.

Table 4 Strategies to overcome the identified challenges

Major area of improvement	Proposed strategies
Introducing a policy on passive solar housing construction in Sri Lanka	• Enhancing the policy by including specific targets, timelines, and incentives for builders and homeowners to encourage compliance • Establishing monitoring and enforcement mechanisms to ensure adherence to the policy
Introducing new green building certification for passive solar houses	• Collaborating with international green building organisations to develop a certification that aligns with global standards and best practices • Providing technical assistance and resources to builders and architects to meet certification requirements
Introducing a green financing process for housing	• Partnering with financial institutions to develop tailored financing options for passive solar housing projects, such as preferential interest rates or grants for energy-efficient upgrades • Simplifying the application process and providing guidance to homeowners and developers on accessing green financing options
Awareness raising programmes for service providers, house owners, and small-scale builders	• Utilising multiple channels for awareness campaigns, including workshops, seminars, social media, and community outreach programmes • Tailor messaging to address the specific needs and concerns of different stakeholder groups, emphasising the economic and environmental benefits of passive solar housing
Expert training programmes for service providers for skill development on adaptation of passive solar design techniques and technologies	• Collaborating with universities, technical institutes, and industry experts to develop comprehensive training programmes • Offering hands-on workshops and on-the-job training opportunities to ensure practical skills development • Providing ongoing support and mentorship to participants to facilitate continuous learning and improvement

Since the lack of regulations creates uncertainty for stakeholders and may hinder investment in passive solar housing projects, conducting further research to understand the specific barriers to policy implementation, engaging with policymakers to advocate for the development of robust regulations, and collaborating with international organisations with experience in sustainable housing policies can be considered to overcome such barriers. Further, informal practices can lead to substandard construction quality and undermine the effectiveness of passive solar design principles, which could be overcome by conducting training programmes and capacity-building initiatives targeting local builders and contractors to improve their understanding of passive solar techniques and encourage the adoption of formal construction practices.

5 Conclusions

Carbon reduction goals primarily focus on public buildings and then gradually widen to residential buildings, even though the effectiveness of standards and policies for residential buildings remains inadequate in many countries. Despite the lack of standard policies and procedures available on carbon reduction, the concept of passive solar houses granted the world attention in reducing the excessive generation of carbon from residential buildings. However, the absence of policies specifically targeting the construction of passive solar houses has impeded their successful implementation. This research provides a comprehensive underpinning for enhancing the policy and regulatory framework for passive solar housing construction. The identified potential areas of enhancements, such as financial incentives for passive solar residents, reduction of excessive generation and usage of non-renewable energy, modern and green technology adaptations, assessment, compliance, and certification, and research and development, lay a firm foundation for progressing in developing a policy and regulatory framework for passive solar housing construction. However, many challenges impede the successful implementation of passive solar housing construction in Sri Lanka, such as the absence of a policy and regulatory framework for passive solar housing construction, use of informal practices in housing construction, poor quality management of small-scale housing projects, and no incentives for builders and house owners for utilising solar energy. It is vital to take necessary actions to overcome the barriers to adopt passive solar housing construction and to accelerate progress towards a more sustainable built environment. Hence, this research can be applied by researchers, policymakers, and construction professionals as a way forward for the successful implementation of the passive solar housing concept in Sri Lanka.

References

1. Arora NK, Fatima T, Mishra I, Verma M, Mishra J, Mishra V (2018) Environmental sustainability: challenges and viable solutions. [online] Environmental Sustainability, Springer Singapore. https://doi.org/10.1007/s42398-018-00038-w
2. Barber S (2018) History of passive solar energy, 2–12
3. Bekele MT, Atakara C (2023) Residential building energy conservation in mediterranean climate zone by integrating passive solar and energy efficiency design strategies. Buildings 13(4):1073
4. Borzooeian M (2014) Passive solar design in ecological houses. II(Ix):1–8
5. Branz (2018) Passive solar. Design guidelines. Construction Specifier, 42(9):116–120:122

6. Energy Institute S (2000) Passive Solar Design Increase energy efficiency and comfort in homes by incorporating passive solar design features Buildings for the 21st Century. Technology Fact Sheet. http://www.nrel.gov/docs/fy01osti/29236.pdf
7. Green Building Council of Sri Lanka—GBCSL (2022) GREENSL® Rating System for New Constructions. GBCSL: Sri Lanka
8. Gupta CL (1993) Solar passive buildings for developing countries. Sadhana 18(1):77–104. https://doi.org/10.1007/BF02811388
9. He Q, Hossain MU, Ng ST, Augenbroe G (2021) Identifying practical sustainable retrofit measures for existing high-rise residential buildings in various climate zones through an integrated energy-cost model. Renew Sustain Energy Rev 151:111578
10. Jayasinghe MTR, Sujeewa LC, Fernando KKJS, WRA (1997) Passive solar techniques for Sri Lanka. Research For Industry, June. https://www.researchgate.net/publication/304247421
11. Jayasinghe MTR, Attalage R (1999) or residential buildings in low altitudes of passive techniques for residential buildings. January, January, 18–27
12. Lentz TR, Iowa SU (2010) Analysis of the passive design and solar collection techniques of the houses in the 2009 U. S. Department of Energy' s Solar Decathlon competition by Timothy Robert Lentz A thesis submitted to the graduate faculty in partial fulfillment of the requirem. Energy
13. Li K, Ma M, Xiang X, Feng W, Ma Z, Cai W, Ma X (2022) Carbon reduction in commercial building operations: a provincial retrospection in China. Appl Energy 306. https://doi.org/10.1016/j.apenergy.2021.118098
14. Li Y, Li S, Xia S, Li B, Zhang X, Wang B, Ye T, Zheng W (2023) A review on the policy, technology and evaluation method of low-carbon buildings and communities. In: Energies (Vol. 16, Issue 4). MDPI. https://doi.org/10.3390/en16041773
15. Liu C, Sharples S, Mohammadpourkarbasi H (2023) A review of building energy retrofit measures, passive design strategies and building regulation for the low carbon development of existing dwellings in the hot summer–cold winter region of China. Energies 16(10):4115
16. Liu G, Li X, Tan Y, Zhang G (2020) Building green retrofit in China: policies, barriers and recommendations. Energy Policy 139:111356
17. Maduta C, Melica G, D'Agostino D, Bertoldi P (2022) Towards a decarbonised building stock by 2050: the meaning and the role of zero emission buildings (ZEBs) in Europe. Energ Strat Rev 44:101009
18. Ministry of Urban Development and Housing. Sri Lanka Sustainable Housing and Construction Roadmap 2020–2050. Ministry of Urban Development and Housing: Sri Lanka
19. Nagarjuna Kumar B, Bommareddy SSR, Asadi SS (2018) Evaluation of factors declining cost variation for construction of endowment buildings. Int J Civil Eng Technol 96:235–245
20. Paré G, Kitsiou S (2017) Methods for literature reviews. In: Handbook of eHealth evaluation: an evidence-based approach [Internet]. University of Victoria
21. Pitroda J, Biraj L, Dhiraj N, Jay N (2016) A critical literature review on benefits due to passive solar energy system in educational building. Int J Constr Res Civil Eng 25:16–20
22. Scott J, Edge M, Laing R (2006) Passive solar design of mass housing: ensuring environmental improvements at the planning stage for suburban housing. J Build Apprais 2(3):207–222. https://doi.org/10.1057/palgrave.jba.2950041
23. Sri Lanka Sustainable Energy Authority (2008) Code of practice for energy efficient buildings in Sri Lanka. Sri Lanka Sustainable Energy Authorit, Sri Lanka
24. Sri Lanka Sustainable Energy Authority (2019) Guideline for Sustainable Energy Residences in Sri Lanka" Sri Lanka Sustainable Energy Authority, Sri Lanka
25. Tarigh AD, Tarigh FD, Nikranjbar A (2012) A survey of energy-efficient passive solar houses. In: 2012 International Conference on Clean and Green Energy, 27 February, 18–25
26. Xu W, Guo H, Ma C (2022) An active solar water wall for passive solar greenhouse heating. Appl Energy 308:118270
27. Yılmaz M, Bakış A (2015) Sustainability in construction sector. Procedia Soc Behav Sci 195:2253–2262

28. Źróbek S, Trojanek M, Źróbek-Sokolnik A, Trojanek R (2015) The influence of environmental factors on property buyers' choice of residential location in Poland. J Int Stud 83:164–174

Chapter 5
Impact of Rating Levels and Climate Zones on the Performance of Energy Star-Rated Office Buildings

Abdul-Manan Sadick and Abid Hasan

Abstract Building certification (both green and energy-efficiency ratings) can contribute to reducing the operational energy consumption of buildings. Certification schemes award ordinal rating levels, and energy performance is expected to align with the ordinal levels where performance increases from lower to higher levels. Other factors like climate conditions can distort this ordinal performance expectation. However, limited research has explored energy performance differences between rating levels and how other factors, like climate conditions, could influence them. Therefore, this research aimed to investigate the ordinal energy performance of commercial buildings rated under Australia's Commercial Building Disclosure (CBD) program while considering the impact of climate differences. A total of 2,606 buildings rated under the "whole building energy rating" category was downloaded from the CBD dataset. The results showed that energy performance followed the ordinal star ratings of the CBD program when climate differences were ignored. However, when climate differences are considered, ordinal performance differences between neighbouring rating levels (like 1star–2star, 3star–4star) are insignificant, while ordinal performance differences between distant ratings (like 1star–3star, 2star–4star) are significant. Therefore, targeted upgrade programs focused on retrofitting 1star, 2star and 3star rated building to 4star and above would contribute to reducing energy consumption and enhance the impact of the CBD program.

Keywords Office buildings · Energy consumption · Energy star rating · Climate zone

A.-M. Sadick (✉) · A. Hasan
School of Architecture and Built Environment, Deakin University, Geelong, Australia
e-mail: s.sadick@deakin.edu.au

A. Hasan
e-mail: abid.hasan@deakin.edu.au

T. Ramachandra et al. (eds.), *Concepts, Policies and Technology for Sustainable Buildings*, Advances in Sustainability Science and Technology,
https://doi.org/10.1007/978-981-96-3086-8_5

1 Introduction

Carbon dioxide (CO_2) emission is one of the significant contributors to climate change, and the built environment accounts for a significant proportion of it [10, 25]. Specifically, the energy consumed at the operational stage of buildings for several purposes, including heating in winter and cooling in summer, accounts for about 28% of global carbon emissions [14]. However, operational energy consumption is essential for ensuring that buildings meet their functional requirements, including providing safe and conducive indoor spaces for occupants' comfort and well-being. Consequently, the World Green Building Council has proposed the 2030 deadline for all buildings to be carbon-neutral while prioritising low-carbon technologies to ensure that all buildings operate near zero carbon by 2050 [29]. Reducing buildings' energy consumption at the operation stage will be vital in reducing their associated CO_2 emissions and achieving net zero for all new buildings by 2030 and near-zero carbon for all buildings by 2050 [15].

Building certification (both green and energy-efficiency ratings) is among the predominant strategies employed to reduce the 29% of global energy consumed by buildings at the operation stage [14, 18, 27, 28]. Several building certification schemes are available for rating energy performance only or other aspects, including indoor environmental quality, waste reduction and water efficiency [11, 17]. Global examples of these schemes are LEED (Leadership in Energy and Environmental Design), BREEM (Building Research Establishment Environmental Assessment Method), Passive House and WELL. Several studies have compared certified buildings' energy performance to conventional buildings, given that the former is expected to consume less energy. Scofield and Doane [26] reported that LEED-certified offices, K-12 schools and multifamily dwellings consumed less energy than conventional equivalents. Sartori and Hestnes [24] demonstrated that a passive house decreased energy demand by a factor of three compared to a conventional house over an assumed service life of 80 years [24]. These findings suggest that certified green and energy-efficient buildings consume less energy than conventional buildings. However, Vosoughkhosravi et al. [30] found that LEED-certified buildings with low energy and atmosphere scores consumed more energy than non-certified buildings. Hence, building certification does not guarantee superior energy performance, given that energy consumption would be influenced by design choices within the confines of certification schemes.

Building certification schemes generally have ordinal certification levels with defined performance expectations for each level. For example, LEED certifications include "Certified", "Silver", "Gold" and "Platinum" Hence, several studies have investigated the reliability of these performance expectations. Majcen et al. [20] revealed that homes with a low energy label consumed about 20-50% less energy than projected by the label, while buildings with high energy-efficient labels consumed about 5-10% more energy than expected. These deviations show that the predicted energy performance of certified buildings may vary from the actual performance. Alam and Devjani [3] investigated a 5-star Green Star-rated mixed-use (teaching, study, academic/administrative offices) educational building. They found that the electricity consumption was 2.4 times higher than expected, and 48% was attributable to consumption during non-working hours. This performance gap is not surprising given that students may stay longer to study while researchers may work in their offices after hours. These findings demonstrate that the expected energy performance of certified buildings may differ significantly from actual performances.

The energy performances of certified green and energy-efficient buildings are expected to align with the ordinal certification levels of certification schemes. However, limited research has focused on investigating the reliability of those expectations. In an earlier study, Newsham et al. [22] reported a minimal link between the assessed energy efficiency of LEED buildings and certification level, suggesting that certification level had minimal impact on actual energy performance. However, in a recent study, Elkhapery et al. [9] found significant differences in average electricity savings between all LEED certification levels except for LEED-Certified and LEED-Silver. These contrasting findings can be attributed to several factors, including the fact that project design teams can prioritise specific certification criteria above others based on clients' project goals. For example, a team seeking LEED certification may prioritise criteria such as "Material selection and resources used" and "Water efficiency and water use reduction" over "Energy efficiency and atmosphere". Therefore, energy performance differences between certification levels may not significantly vary between all levels, as evidenced by the findings reported above.

Australia's predominant building certification schemes, particularly for commercial office buildings, are Green Star and NABERS (National Australian Built Environment Rating System). Although most building certification schemes are voluntary, the CBD (Commercial Building Disclosure) program established by the Building Energy Efficiency Disclosure Act 2010 was mandated for Australian office buildings meeting the following criteria:

- Greater than 1000 square metres
- For administrative, clerical, professional or other information-based activities and includes any Support facilities for those activities and
- For sale, lease or sublease" (DCCEEW, 2023b)

CBD aims to increase the energy efficiency of Australia's big office buildings and educate potential purchasers and tenants (DCCEEW, 2023b). The BEEC (Building Energy Efficiency Certificate) offered under the CBD program is valid for 12 years and uses the NABERS Energy for Offices star rating to assess energy performance (DCCEEW, 2023b). The NABERS ordinal star ratings include 1star (Making a start), 2star (Opportunities for upgrades), 3star (Market standard), 4star (High performance), 5star (Superior performance) and 6star (Market leaders) [21]. Although the CBD program has been operational for more than a decade, there has been limited research focused on exploring the energy performance differences between the ordinal star ratings. However, Kim et al. [16] analysed the CBD dataset and reported that 6star NABERS-rated office spaces save less energy than 4star and 5star. It is vital to note that NABERS Energy for Offices star rating has three assessment strategies, "base building energy rating" (focused on central services like heating, lifts and lobby lights), "whole building energy rating" (focused on both central services and energy used in tenant spaces) and "tenancy energy rating" (focused energy used in tenants' space only). Kim et al. [16] focused on "tenancy energy rating" only, hence, their analysis does not account for the energy consumed by central building services. The energy performance of buildings, certified or not certified, is significantly influenced by several factors, including the properties of the building materials and climate conditions [12, 13].

Given that Australia has seven climate zones [32], it is plausible that differences in climate zones would significantly influence the energy consumption patterns of CBD-certified buildings. Additionally, the NABERS energy star rating for offices considers other unique factors, such as location, for equitable and comparable ratings [21]. This research, therefore, aimed to investigate differences in energy consumption between the different ordinal star ratings of the "whole building energy rating" category in the CBD dataset. Additionally, the differences in energy performance were explored within and between the seven Australian climate zones to examine the potential impact of climate differences on the energy performance of CDB-certified buildings.

2 Methodology

2.1 Data

Given this research's aim, data was obtained from the publicly available CBD program dataset at https://www.cbd.gov.au/registers/downloadable-cbd-program-data-set (DCCEEW, 2023a). The dataset was retrieved for this research on 17 January 2023. Additionally, the "Whole Building" subset with unique NABERS reference numbers, totalling 2,606, was filtered for this research. The original NABERS star rating in the dataset ranged from 0 to 6 with 12 subcategories that were translated to 0star (0), 1star (1, 1.5), 2star (2, 2.5), 3star (3, 3.5), 4star (4, 4.5) 5star (5, 5.5) and 6star (6). Consequently, the filtered data included 255, 180, 350, 657, 778 and 381 of 0star, 1star, 2star, 3star, 4star and 5star respectively. 6star was excluded due to the limited number of records, while 0star (i.e. conventional buildings) was also excluded because it is out of scope. These filters resulted in 2,346 Whole Building data analysed in this research.

The CDB dataset includes a wide range of information, including project location, building characteristics (number of floor levels, total floor area), assessor's name, rating scope (base building or whole building), energy performance (annual consumption), lighting power density and NABERS star rating level. Postcode climate zone information was obtained from the Australian Energy Regulator website and merged with the filtered CBD data for climate zone-related analysis (ACIL [2]. The climate zone information from the Australian Energy Regulator website did not include Western Australia postcodes,hence, missing climate zone information was

Table 1 Australian climate zones [32]

Zone	Description
zone1	Hot humid summer, warm winter
zone2	Warm humid summer, mild winter
zone3	Hot dry summer, warm winter
zone4	Hot dry summer, cool winter
zone5	Warm temperate
zone6	Mild temperate
zone7	Cool temperate

sourced from the Australian Building Code Board's interactive climate zone map [1]. Key characteristics of the Australian climate zones are summarised in Table 1.

2.2 *Analysis*

Given that this research aimed to investigate differences in energy consumption between the ordinal star rankings of whole buildings from the CBD dataset, analysis of variance (ANOVA) was the preferred method, given that the CBD rating has more than two categories [8]. The response variable in this research is annual energy consumption intensity. The NABERS energy star ratings in the CBD dataset are relative to the total floor area considered for the performance assessment. Therefore, the annual energy consumption values in the CBD dataset were standardised to annual energy consumption intensities by dividing annual energy consumption values by the NABERS-rated area. This standardisation ensured that buildings of different sizes could be fairly compared. Preliminary analysis of the standardised annual energy consumption showed that the different star ratings included outliers, as shown in Figure 1. The Leven test for homogeneity of variance was statistically significant ($F[4, 2341] = 47.2$, $p = 0.0001$), affirming non-homogeneity of variance. Additionally, the QQ plot in Figure 2 points to a violation of normality, given that not all data points are approximately close to the theoretical line.

Conventional ANOVA was deemed inappropriate, given that its core assumptions of homogeneity of variance and normality were violated. Hence, robust ANOVA was preferred in this research to eliminate or minimise the likelihood of type I and type II errors in conventional ANOVA when core assumptions are violated [19, 31]. Robust ANOVA integrates robust statistical methods, including trimmed mean and robust hypothesis testing strategies not based solely on the F-test used in conventional ANOVA [19] (Wilcox 2012). Furthermore, robust ANOVA eliminates unnecessary data transformation to yield conventional ANOVA assumptions and is suitable for unequal sample sizes and distributions [19].

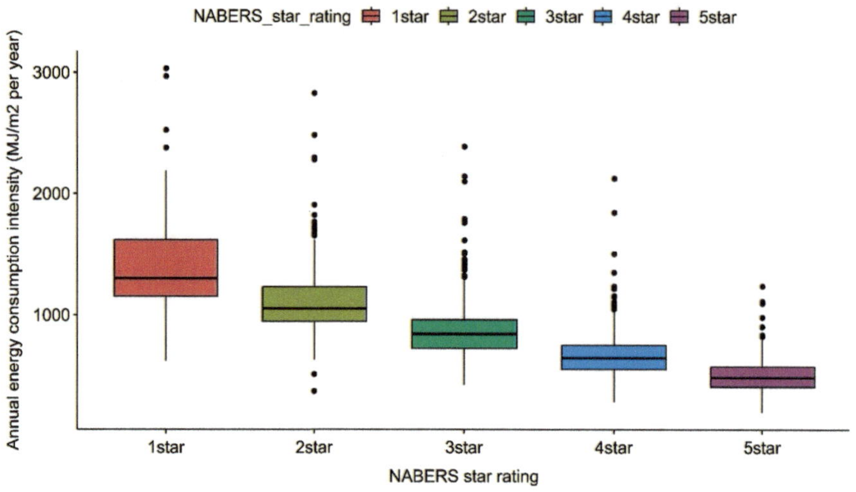

Fig. 1 Distribution of energy consumption intensities for whole building energy ratings

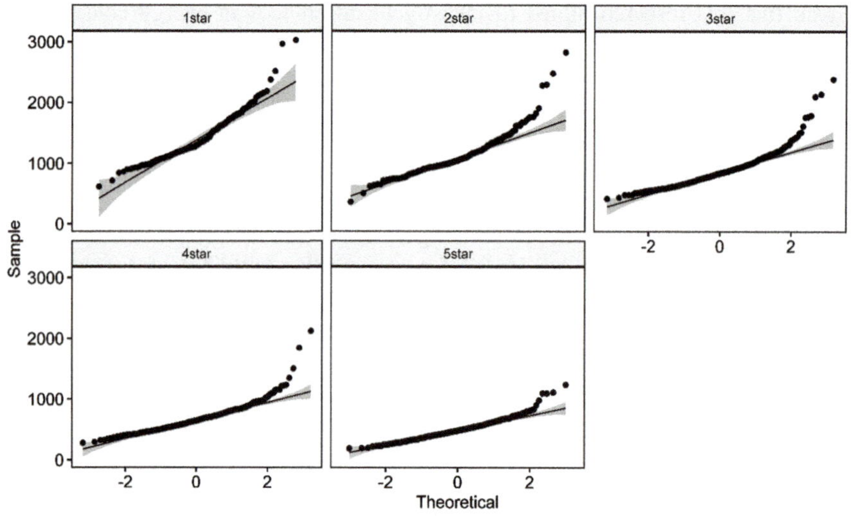

Fig. 2 QQ plot of CBD energy consumption intensity per year for NABERS whole building energy ratings

This research aimed to investigate energy consumption differences between the star ratings of the "whole building energy rating" category of CBD while considering the impact of climate differences. Four hypotheses were tested to achieve the aim of this research. Hypothesis 1 in this research states, "Annual energy consumption intensities will not vary between energy star rating levels when climate zones are ignored". This hypothesis aimed to provide a general overview of energy performance differences across Australia irrespective of climate zone. This hypothesis includes one factor, NABERS star rating, with five levels (i.e. 1star, 2star, 3star, 4star and 5star). Therefore, a robust one-way ANOVA was conducted to test hypothesis 1:

$$H_0 : \mu_{t1star} = \mu_{t2star} \cdots = \mu_{t5star}$$

where μ_t is the trimmed mean of annual energy consumption intensity for each star rating.

The Means were trimmed by the recommended proportion of 20% to minimise the likely impact of outliers [19], Wilcox, 2012).

The remaining hypotheses states that

- Hypothesis 2: Annual energy consumption intensities do not vary between climate zones when rating levels are ignored (i.e. between climate zone effects)
- Hypothesis 3: Annual energy consumption intensities do not vary across rating levels within climate zones (i.e. within climate zone effect)
- Hypothesis 4: Annual energy consumption intensities are not significantly influenced by the combined effect of climate zone differences and star rating levels (i.e. climate zone—NABERS star rating interaction effect)

The above hypotheses collectively assess differences in energy consumption intensities with NABERS star rating and climate zone as the interest factors. Hence, a two-way ANOVA was conducted to concurrently assess hypotheses 2, 3 and 4. Cross tabulation of the filtered data between climate zone and NABERS star rating excluded zone1, zone3, zone4 and zone7 due to relatively limited samples. Therefore, a 3 × 5 two-way ANOVA was conducted because the climate has three levels (zone2, zone5 and zone6) while the NABERS star rating has five levels (1star, 2star, 3star, 4star and 5star). Hypothesis 2 assesses the between-climate zone effect that

$$H_0 : \mu_{t1} = \cdots = \mu_{tJ}$$

where μ_t is the trimmed mean of annual energy consumption intensity for each climate zone, and $_J$ is the number of zones (1 to 3).

Hypothesis 3 assesses the within-climate zone effect that

$$H_0 : \mu_{t1} = \mu_{t2} \cdots = \mu_{tK}$$

where μ_t is the trimmed mean for each NABERS star rating within a climate zone, and K is the number of climate zones (1 to 5).

Hypothesis 4 assesses the climate zone—NABERS star rating interaction effect that

$$H_0 : \sum_{j=1}^{J} \sum_{k=1}^{K} (\propto \beta)_{jk}^2 = 0$$

where $(\alpha\beta)_{jk} = \mu_{tjk} - \mu_{tj} - \mu_{tk} + \mu_t$ is the interaction effect of climate zone (α) and NABERS star rating (β).

Following statistically significant hypotheses sssresults, post-hoc pairwise comparisons were conducted to identify pairs of factors/factor levels that differed significantly. At the same time, effect sizes were calculated to assess the practical significance of the results. The rule of thumb for effect size interpretation is 0.2 (small), 0.5 (medium) and 0.8 (large) [4]. The robust one-way ANOVA, 3 × 5 two-way ANOVA, and post-hoc analyses were conducted using the WRS package [19] in the R language and environment for statistical computing [23].

3 Analysis/Results

This section presents the results of the analyses conducted to achieve the aim of this research.

3.1 Robust One-Way ANOVA

The robust one-way ANOVA was conducted to test the null hypothesis that "annual energy consumption intensities do not vary between energy star rating levels". The results showed a statistically significant difference between at least a pair of star ratings, $F = 695.67$, $p = 0.0001$; hence, the null hypothesis was rejected. Additionally, the effect size for this result was 0.87 (95% CI, 0.84 to 0.91), indicating that the magnitude of the difference between at least a pair of star ratings was large. Robust pairwise comparisons were conducted to identify pairs of star ratings that differed significantly. As shown in Table 2, all pairwise comparisons were statistically significant, indicating differences in energy consumption intensities between all pairs of energy star ratings. The highest and lowest difference was respectively between 1star versus 5star (853.98 MJ/m^2 per year) and 4star versus 5star (161.46 MJ/m^2 per year), indicating that the differences increased in descending order from 5star to 1star. This result supports the thesis that the NABERS ordinal energy star ratings translate to incremental energy performance differences. Regarding the magnitude of these differences, the effect sizes in Table 2 are either medium or large and point to the practical significance of these differences.

3.2 Robust Two-Way ANOVA

Table 3 shows the frequency of whole building energy ratings for the seven climate zones across the star rating categories and shows how the number of whole building energy ratings for zone1, zone3, zone4 and zone7 are exceedingly small compared to

Table 2 Robust one-way ANOVA post-hoc pairwise comparisons of trimmed energy consumption intensity means

Comparisons	Mean difference	95% CI (lower)	95% CI (upper)	Adjusted P-value	Effect size
1star versus 2star	269.88	186.16	353.6	0.0001	0.61
1star versus 3star	500.57	421.54	579.6	0.0001	0.96
1star versus 4star	692.51	614.49	770.54	0.0001	0.89
1star versus 5star	853.98	775.18	932.77	0.0001	0.93
2star versus 3star	230.7	189.85	271.54	0.0001	0.70
2star versus 4star	422.64	383.98	461.3	0.0001	0.96
2star versus 5star	584.1	543.75	624.44	0.0001	1.00
3star versus 4star	191.94	166.07	217.82	0.0001	0.70
3star versus 5star	353.4	325.07	381.74	0.0001	0.95
4star versus 5star	161.46	136.38	186.55	0.0001	0.69

zone2, zone5 and zone6. Therefore, zone1, zone3, zone4 and zone7 were excluded from the robust two-way ANOVA analysis.

The box plot in Figure 3 shows the distribution of annual energy consumption intensities of energy star ratings across the three climate zones and shows a consistent ascending increase in annual energy consumption intensities from 1star to 5star across all regions. Irrespective of the outliers, Figure 3 suggests that annual energy consumption intensities were generally lower in zone2, while zone5 and zone6 seem to have similar consumption intensities.

Results of the robust 3 × 5 two-way ANOVA in Table 4 show that the two main effects (between climate zones, within climate zones) and the interaction effect (climate zone—star rating) were all statistically significant; therefore, null hypotheses 2, 3 and 4 were rejected. Post-hoc analysis was conducted to identify statistically significant pairs for the main and interaction effects.

Table 3 Frequency of CBD NABERS whole building energy ratings per climate zones

Star ratings	1star	2star	3star	4star	5star	Sum
zone1	2	4	7	24	16	53
zone2	39	62	104	132	125	462
zone3	0	0	0	1	1	2
zone4	0	2	6	12	7	27
zone5	55	131	294	347	150	977
zone6	70	117	184	206	66	643
zone7	14	34	62	56	16	182
Sum	180	350	657	778	381	2346

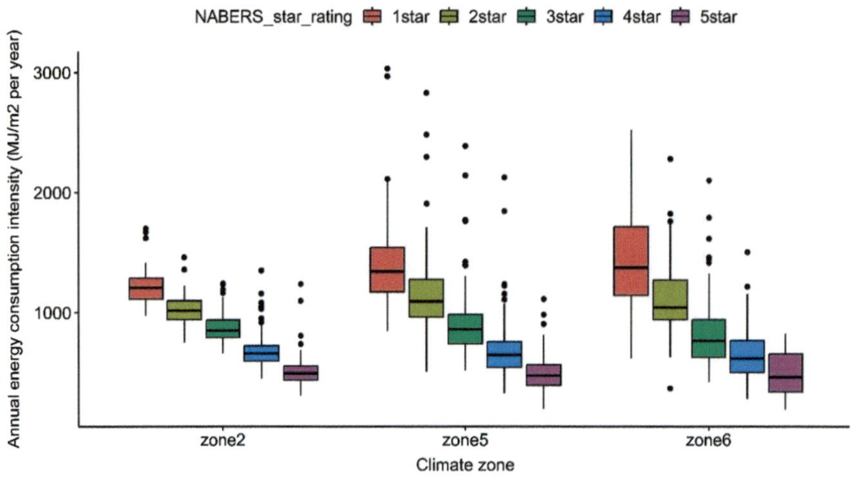

Fig. 3 Climate zone distribution of annual energy consumption intensity of CBD NABERS whole building energy ratings

Table 4 Robust 3 × 5 two-way ANOVA of annual energy consumption intensity

Factors	Test statistics	Df	P-value
Between climate zones	10.36	2	0.007
Within climate zones	2069.39	4	0.001
Climate zone–Star rating	41.71	8	0.001

3.2.1 Between-Climate Zone Effect

Table 5 shows the between-climate post-hoc pairwise comparisons test results and the contrast matrix showing the compared pairs. The between-climate zone effect assesses the collective difference in annual energy consumption intensities between pairs of climate zones, ignoring star rating levels. Only C1 (contrast 1) was statistically significant of the three pairwise comparisons. The contrast matrix shows that C1 represents the annual energy consumption intensity difference between zone2 and zone5. Although this difference was statistically significant, the effect size indicates an exceedingly low effect magnitude. The results show that when energy star rating levels are ignored, the differences in annual energy consumption intensities between the three regions are not practically significant. Additionally, this result is consistent with the pattern of energy consumption intensities depicted in Figure 3.

3.2.2 Within-Climate Zone Effect

The within-climate zone pairwise comparisons simultaneously assessed differences in annual energy consumption intensities between pairs of energy star ratings in

Table 5 Between climate zones post-hoc pairwise comparisons of CBD NABERS whole building annual energy consumption intensity

Test results					
Contrast number	Trimmed mean difference	95% CI (lower)	95% CI (upper)	Adjusted P-value	Effect size
C1	− 191.06	− 339.20	− 42.92	0.0067	0.04
C2	− 114.93	− 306.97	77.11	0.3850	0.09
C3	76.13	− 138.89	291.14	0.7766	0.06
Contrast matrix[*]					
Comparisons	C1	C2	C3		
zone2_1star	1	1	0		
zone2_2star	1	1	0		
zone2_3star	1	1	0		
zone2_4star	1	1	0		
zone2_5star	1	1	0		
zone5_1star	− 1	0	1		
zone5_2star	− 1	0	1		
zone5_3star	− 1	0	1		
zone5_4star	− 1	0	1		
zone5_5star	− 1	0	1		
zone6_1star	0	− 1	− 1		
zone6_2star	0	− 1	− 1		
zone6_3star	0	− 1	− 1		
zone6_4star	0	− 1	− 1		
zone6_5star	0	− 1	− 1		

[*]*Note* The statistically significant comparison between zone2 and zone5 has been highlighted in the table

all three climate zones. From Table 6, all the pairwise comparisons were statistically significant, indicating that annual energy consumption between all pairs of energy star ratings in each climate zone differed significantly. Based on the contrast matrix, C1 represents the annual energy consumption intensity difference between 1star and 2star in the three climate zones. The statistically significant result implies that annual energy consumption intensities for 1star versus 2star differ significantly in zone2, zone5 and zone6, and likewise all other within climate zone pairwise comparisons. Interestingly, the trimmed mean annual energy consumption intensity difference for 1star versus 2star (C1), 2star versus 3star (C5), 3star versus 4star (C8) and 4star versus 5star (C10) is 734.05, 703.98, 554.93 and 478.28 respectively. These differences imply that annual energy consumption intensity generally decreases from 1star to 5star in all three climate zones. The same ordinal pattern is observed when comparing a lower star rating to each of the ratings above it (e.g. 1star versus (2star [C1] or 3star [C2] or 4star [C3] or 5star [C4])). As shown in Table 6, the effect

Table 6 Within climate zone pairwise comparisons of CBD NABERS whole building annual energy consumption intensity

Test results					
Contrast numbers	Trimmed mean difference	95% CI (lower)	95% CI (upper)	Adjusted P-value	Effect size
C1	734.05	495.57	972.52	0.0001	0.61
C2	1438.03	1214.79	1661.27	0.0001	0.95
C3	1992.96	1773.12	2212.80	0.0001	0.90
C4	2471.24	2239.28	2703.20	0.0001	0.94
C5	703.98	576.72	831.25	0.0001	0.69
C6	1258.91	1137.96	1379.86	0.0001	0.95
C7	1737.19	1594.23	1880.15	0.0001	0.99
C8	554.93	470.10	639.76	0.0001	0.68
C9	1033.21	918.38	1148.04	0.0001	0.94
C10	478.28	370.34	586.22	0.0001	0.70

*Contrast matrix**										
Comparisons	C1	C2	C3	C4	C5	C6	C7	C8	C9	C10
zone2_1star	1	1	1	1	0	0	0	0	0	0
zone2_2star	−1	0	0	0	1	1	1	0	0	0
zone2_3star	0	−1	0	0	−1	0	0	1	1	0
zone2_4star	0	0	−1	0	0	−1	0	−1	0	1
zone2_5star	0	0	0	−1	0	0	−1	0	−1	−1
zone5_1star	1	1	1	1	0	0	0	0	0	0
zone5_2star	−1	0	0	0	1	1	1	0	0	0
zone5_3star	0	−1	0	0	−1	0	0	1	1	0
zone5_4star	0	0	−1	0	0	−1	0	−1	0	1

(continued)

Table 6 (continued)

Test results										
Contrast numbers	Trimmed mean difference		95% CI (lower)		95% CI (upper)		Adjusted P-value		Effect size	
zone5_5star	0	0	0	−1	0	0	−1	0	−1	−1
zone6_1star	1	1	1	1	0	0	0	0	0	0
zone6_2star	−1	0	0	0	1	1	1	0	0	0
zone6_3star	0	−1	0	0	−1	0	0	1	1	0
zone6_4star	0	0	−1	0	0	−1	0	−1	0	1
zone6_5star	0	0	0	−1	0	0	−1	0	−1	−1

Note For each column of the contrast matrix, the pairwise comparison is between rows of 1 and -1, as highlighted with the colour code sizes are medium or large, implying that the magnitude of the ordinal difference in annual energy consumption intensities between star ratings in each climate zone is practically significant.

3.2.3 Climate Zone—Star Rating Interaction Effect

The climate zone and star rating interaction effect aimed to determine how different climate conditions and energy star ratings influence annual energy consumption intensity differences. Out of the 30 possible pairwise comparisons for this interaction effect, only five (C6, C7, C12, C13, C15) differed significantly, as shown in the abbreviated Table 7. The effect size of each statistically significant result is medium, indicating that the magnitude of those differences has medium practical relevance. Of the five statistically significant results, C12 has the highest effect size (0.53), and the contrast matrix shows that this refers to annual energy consumption intensity for 1star versus 3star in climate zone2 and zone6. This significant result shows that climate zone differences and energy star rating level significantly influenced the energy consumption difference between 1star and 3star buildings in climate zone2 and zone6. Figure 3 shows that this interaction effect is likely due to the exceedingly high energy consumption intensities of 1star buildings in zone6, a mild temperate climate compared to warm, humid summer and mild winter in zone2. C13, with a medium effect size (0.47), follows the same pattern for 1satr versus 4star in climate zone2 and zone6. C6 and C7 also show significant interaction effects for 2star versus 4star and 2star versus 5star in climate zone2 and zone5. The last significant interaction effect, C15, is for 2star versus 3star in climate zone2 and zone6. It is imperative to note that there was no significant interaction effect for "neighbouring" climate zone5 (Mild temperate) and zone6 (Cool temperate). These findings suggest that the interaction effect of climate zone–star rating is most significant for distant zones like zone2 and zone5 and insignificant for neighbouring zones like zone5 and zone6.

4 Discussion

This research aimed to investigate differences in energy consumption in NABERS energy star-rated office buildings under the CBD program. To this effect, four hypotheses were tested in this research. Hypothesis 1 tested the idea that "annual energy consumption intensities will not vary between energy star rating levels". The test was statistically significant, and the null hypothesis was rejected, implying that annual energy consumption intensities differ between at least a pair of energy star ratings. Results of the pairwise comparisons confirmed that annual energy consumption intensity differences between all pairs of energy star ratings differed significantly with medium to large effect sizes. These findings indicate that energy consumption intensity generally decreases from lower to higher energy star rating levels, and they

Table 7 Abbreviated pairwise comparisons of "Climate zone—Star rating" interactions for annual energy consumption intensity of CBD NABERS whole building ratings

Test results					
Contrast numbers	Mean difference	95% CI (lower)	95% CI (upper)	Adjusted P-value	Effect size
C6	− 105.48	− 209.12	− 1.85	0.0418	0.33
C7	− 113.91	− 220.55	− 7.26	0.0246	0.37
C8	− 1371	− 76.83	49.41	1.0000	0.05
C9	− 2213	− 90.26	46.00	1.0000	0.10
C10	− 8.42	− 67.97	51.12	1.0000	0.04
C11	119.67	341.12	101.79	0.9041	0.27
C12	− 259.48	− 470.42	− 48.54	0.0047	0.53
C13	− 210.78	− 418.61	− 2.95	0.0437	0.47
C14	− 188.72	− 411.00	33.56	0.1897	0.44
C15	− 139.82	− 256.27	− 23.36	0.0054	0.38

Contrast matrix*										
Comparisons	C6	C7	C8	C9	C10	C11	C12	C13	C14	C15
zone2_1star	0	0	0	0	0	1	1	1	1	0
zone2_2star	1	1	0	0	0	− 1	0	0	0	1
zone2_3star	0	0	1	1	0	0	− 1	0	0	− 1
zone2_4star	− 1	0	− 1	0	1	0	0	− 1	0	0
zone2_5star	0	− 1	0	− 1	− 1	0	0	0	− 1	0
zone5_1star	0	0	0	0	0	0	0	0	0	0
zone5_2star	− 1	− 1	0	0	0	0	0	0	0	0
zone5_3star	0	0	− 1	− 1	0	0	0	0	0	0
zone5_4star	1	0	1	0	− 1	0	0	0	0	0
zone5_5star	0	1	0	1	1	0	0	0	0	0
zone6_1star	0	0	0	0	0	− 1	− 1	− 1	− 1	0
zone6_2star	0	0	0	0	0	1	0	0	0	− 1
zone6_3star	0	0	0	0	0	0	1	0	0	1
zone6_4star	0	0	0	0	0	0	0	1	0	0
zone6_5star	0	0	0	0	0	0	0	0	1	0

* Statistically significant comparison rows have been colour-coded for ease of identification

agree with Elkhapery et al. [9], who reported significant electricity consumption differences between LEED certification levels, except the bottom two levels.

Additionally, these results justify the cost and effort of achieving higher energy star ratings given that it would reduce energy consumption, building operational costs and carbon emissions. In the NABERS energy star rating nomenclature 3star is "Market standard" while 4star is "High performance". The data in this research indicate 4star as the dominant rating in the whole building category, constituting about 33% of all whole building energy ratings, while 1star to 3star is approximately 51%. Therefore, if most offices under the CBD program are upgraded to 4star, the general energy efficiency of Australian offices would increase multiple fold and contribute to achieving the 2050 net zero emissions target [5].

The two-way ANOVA test simultaneously assessed hypotheses 2 (between climate zone effect), 3 (within climate zone effect) and 4 (climate zone—star rating interaction effect). All three hypotheses' tests were statistically significant and resulted in rejecting the null hypothesis. Post-hoc pairwise comparisons showed that the between-climate zone effect was not practically significant, thus suggesting that when rating levels are ignored, differences in energy consumption intensity between the climate zones are insignificant. The climate conditions of zone2 are warm humid summer, mild winter, while zone5 and zone6 are warm temperate and mild temperate, respectively. Plausibly, the high energy consumption due to air conditioning in the warm humid summer in zone2 balances out the equally high energy consumption due to heating in the mild to cool winter of zone5 and zone6.

Given that only three zones (zone2, zone5, zone6) were considered for the two-way ANOVA due to the limited number of rated whole buildings in zone1, zone3, zone4 and zone7, future replication of this research for all seven climate zones may yield different results. All the pairwise comparisons for within climate zone effect were statistically significant, with large or medium effect sizes. Additionally, the annual energy consumption intensity differences between pairs of energy star ratings increased from lower to higher, indicating that lower ratings consumed more energy per unit area than higher ratings. This pattern is consistent in all climate zones and shows that the efficacy of the ordinal star rating is similar in all three climate zones; therefore, it supports the robustness of the NABERS energy star ratings. Furthermore, the within climate zone effect results agree with the results of hypothesis 1 in supporting the idea that the ordinal energy star ratings translate to real-world energy performance, where higher ratings consume less energy compared to lower ratings.

For the interaction effect of climate zone and star rating levels, only five of the 30 pairwise comparisons were statistically significant. The fact that most of the interaction pairwise comparisons were not statistically significant would suggest that climate zone and star rating level interaction have less influence on energy consumption intensities. However, the five significant pairwise comparisons provide insights that can inform future rendering of the CBD program. As shown in Table 8, all the significant interaction effects included zone2 on one hand and zone5 or zone6 on the other. At the same time, the rating levels consistently compared between either 1star or 2star and higher star rating levels. This pattern suggests that the interaction effect is most influential on the bottom two ratings and is logical given that they would have less stringent star rating requirements than 3star and above.

Additionally, the fact that none of the significant interaction effects involved zone5 and zone6 suggests that the interaction effect has minimal impact on rating levels of energy consumption differences between climate zones with less substantial climate condition differences. Another critical observation from Table 8 is the lack of significant interaction effect involving pairs of star ratings from 3star to 5star. A plausible

Table 8 Statistically significant pairwise comparisons for interaction effect

Contrast	Climate zones	Star rating levels
C6	zone2–zone5	2star–4star
C7	zone2–zone5	2star–5star
C12	zone2–zone6	1star–3star
C13	zone2–zone6	1star–4star
C15	zone2–zone6	2star–3star

explanation for this pattern is the relatively higher rating requirements for 3star and above that seem to neutralise the climate zone—star rating interaction effect. Given that climate zones 1, 3, 4 and 7 were not analysed in this research (due to sample limitations), it will be interesting to see if future replications of this research will find non-significant interaction effects involving climate zone pairs like zone1-zone2, zone4-zone4 and zone6-zone7.

5 Practical Implications

The findings of this research have local and global implications. First, for CBD program administrators and Australian office building owners, this research strengthens the case for prioritising financial incentives and technical support for CBD program participants to upgrade their buildings from lower to higher energy star ratings. The analysis in this research showed that 1star, 2star and 3star constituted 51% of all ratings in the whole building category, while climate conditions most impacted the energy consumption of 1star and 2star buildings. Therefore, prioritising financial incentives and technical support for 1star to 3star buildings to upgrade to 4star and above would reduce the operational energy large office buildings consume. The CBD program should consider developing guidance documents tailored to each climate zone to facilitate these upgrades, outlining optimal upgrade strategies for buildings at different star rating levels. This guidance document would provide building owners with actionable steps to maximise energy savings and cost-effectiveness within their climate.

Second, given that half of rated buildings are below 3star, Australian national, state and local governments should consider introducing mandatory minimum energy star ratings for all commercial buildings, regardless of size. However, climate zone2 showed the most potential for significant energy savings and can catalyse the reduction of commercial building energy consumption. Targeted grant programs specifically for upgrading projects in climate zone2 and for buildings with 1 and 2star ratings would maximise the impact of public funds on achieving national net zero goals. Moreover, increasing public awareness of the NABERS rating system and its role in realising energy-efficient buildings can encourage tenant demand for sustainably rated office spaces and incentivise building owners to participate in the CBD program.

Last, in the global context, the observed interaction between climate zone and star rating efficacy, while specific to the analysed climate zones in Australia, highlights the importance of considering regional climate variations when designing and implementing building energy rating systems in other countries. This climate zone and energy star rating interaction underscore the need for adaptable rating systems for local climatic contexts. Other countries can leverage the findings of this research to inform the development or refinement of their building rating systems, considering factors like granularity, regional adjustments and adaptability to diverse building types and climates. For targeted policies, incentives and support for lower-rated buildings can be a valuable model for policy development in other countries. Lastly, collaborative research efforts and knowledge sharing between countries can advance understanding of building energy performance across different climates and contexts. This global effort can lead to the development of best practices and international standards for designing and implementing effective building rating systems.

6 Conclusions

This research aimed to investigate energy consumption differences between the star ratings of the "whole building energy rating" category of the Australian CBD program while considering the impact of climate. The analysis yielded several key findings as follows:

- When climate zones are ignored, energy consumption intensity significantly decreases from lower to higher energy star ratings. Additionally, differences in energy consumption intensity between all pairs of star rating levels are statistically significant, aligning with the intended purpose of the NABERS rating system and supporting its efficacy in driving energy efficiency improvements.
- When star rating levels are ignored, differences in energy consumption between climate zones are not practically relevant, suggesting that climate differences alone may not be a significant determinant of energy performance within the analysed star rating category.
- The interaction effect of climate zone and star rating levels significantly accounts for energy consumption differences between specific climate zone and star rating combinations. Specifically, five interaction effects were observed involving climate zone2 (warm humid summer, mild winter), zone5 (warm temperate) and zone6 (mild temperate) on one hand, and star ratings 1, 2, 3, 4, 5 on the other hand. The most significant interaction effect indicated that the climate differences significantly influence energy consumption differences between 1star and 3star in zone2 and zone6. This result suggests that the efficacy of the star rating system in achieving energy savings may vary depending on the climate context and the building's current rating level.

These findings present valuable implications for stakeholders involved in improving building energy efficiency. For building owners and CBD program participants, prioritising upgrades for lower-rated buildings, particularly in climate zone 2 due to the significant climate interaction effect (Table 8), can maximise energy savings potential. Additionally, targeted technical support tailored to specific climate zones can further accelerate the CBD program's impact. Given that half of the rated buildings are from 1star to 3star, national, state and local governments should consider providing grant programs focused on upgrading existing commercial buildings to 4star and above. This scheme would reduce the energy consumption of commercial buildings and contribute to achieving net zero targets. However, this research has some limitations. Due to limited CBD-rated buildings in climate zones 1, 3, 4 and 7, the analysis only focused on three Australian climate zones. Future studies encompassing all zones could provide a more comprehensive understanding of regional variations in energy performance. Additionally, the research relied on existing CBD-rated buildings, which do not capture commercial buildings with a total floor area of less than 1000 m^2. Including additional data points for smaller commercial buildings in future studies could strengthen the analysis and refine the conclusions. By addressing these limitations and building upon the current findings, future research can further advance the understanding of energy performance in buildings and inform effective strategies for achieving sustainable and energy-efficient building operations across diverse climate zones.

Acknowledgements The authors would like to acknowledge the Department of Climate Change, Energy, the Environment and Water for the publicly accessible database. This research was not funded.

References

1. ABCB (2023) Climate zone map [Online]. Australian Builfding Code Board. Available: https://www.abcb.gov.au/resources/climate-zone-map. Accessed 17 Jan 2023
2. Allen A (2017) Electricity and gas bill benchmarks for residential customers 2017 [Online]. Australian Energy Regulator. Available: https://www.aer.gov.au/retail-markets/guidelines-reviews/electricity-and-gas-bill-benchmarks-for-residential-customers-2017. Accessed 17 Jan 2023
3. Alam M, Devjani MR (2021) Analyzing energy consumption patterns of an educational building through data mining. J Build Eng 44
4. Algina J, Keselman H, Penfield RD (2005) An alternative to Cohen's standardized mean difference effect size: a robust parameter and confidence interval in the two independent groups case. Psychol Methods 10:317
5. DCCEEW (2022) Australia submits new emissions target to UNFCCC [Online]. Available: https://www.dcceew.gov.au/about/news/australia-submits-new-emissions-target-to-unfccc. Accessed 2 Feb 2023
6. DCCEEW (2023a) Downloadable CBD program data set [Online]. Department of Climate Change, Energy, the Environment and Water. Available: https://www.cbd.gov.au/registers/downloadable-cbd-program-data-set. Accessed 17 Jan 2023

7. DCCEEW (2023b) Quick guide to the application of the CBD Program [Online]. Australian Government Department of Climate Change, Energy, the Environment and Water. Available: https://www.cbd.gov.au/quick-guide-application-cbd-program. Accessed 25 Jan 2023
8. Doncaster CP, Andrew JHD (2007) Analysis of variance and covariance: how to choose and construct models for the life sciences. Cambridge University Press, Cambridge
9. Elkhapery B, Kianmehr P, Doczy R (2021) Benefits of retrofitting school buildings in accordance to LEED v4. J Build Eng 33:101798
10. Farahzadi L, Kioumarsi M (2023) Application of machine learning initiatives and intelligent perspectives for CO_2 emissions reduction in construction. J Clean Prod 384:135504
11. Giama E, Papadopoulos A (2012) Sustainable building management: overview of certification schemes and standards. Adv Build Energy Res 6:242–258
12. Hegarty RO, Amedeo G, Kinnane O (2024) The impact of compromised insulation on building energy performance. Energy Buildings 114337
13. Hosseini M, Javanroodi K, Nik VM (2022) High-resolution impact assessment of climate change on building energy performance considering extreme weather events and microclimate–Investigating variations in indoor thermal comfort and degree-days. Sustain Cities Soc 78:103634
14. Kang X, An J, Yan D (2023) A systematic review of building electricity use profile models. Energy Buildings 281:112753
15. Kang Y, Xu W, Wu J, Li H, Liu R, Lu S, Rong X, Xu X, Pang F (2022) Study on comprehensive whole life carbon emission reduction potential and economic feasibility impact based on progressive energy-saving targets: A typical renovated ultra-low energy office. J Build Eng 58:105029
16. Kim S, Lim BTH, Oo BL (2022) Energy consumption and carbon emissions of mandatory green certified offices in Australia: evidence and lessons learnt across 2011 and 2020. Sustainability 14:13773
17. Li Y, Kubicki S, Guerriero A, Rezgui Y (2019) Review of building energy performance certification schemes towards future improvement. Renew Sustain Energy Rev 113:109244
18. Liu T, Chen L, Yang M, Sandanayake M, Miao P, Shi Y, Yap P-S (2022) Sustainability considerations of green buildings: a detailed overview on current advancements and future considerations. Sustainability 14:14393
19. Mair P, Wilcox R (2020) Robust statistical methods in R using the WRS2 package. Behav Res Methods 52:464–488
20. Majcen D, Itard LCM, Visscher H (2013) Theoretical versus actual energy consumption of labelled dwellings in the Netherlands: discrepancies and policy implications. Energy Policy 54:125–136
21. NABERS (2020) Office buildings [Online]. NABERS. Available: https://www.nabers.gov.au/sites/default/files/2022-11/Fact%20Sheet%20-%20Office%20Buildings.pdf. Accessed 24 Jan 2023
22. Newsham GR, Mancini S, Birt BJ (2009) Do LEED-certified buildings save energy? Yes, but…. Energy Buildings 41:897–905
23. R CORE TEAM 2021. R: a language and environment for statistical computing. R Foundation for Statistical Computing, Vienna, Austria
24. Sartori I, Hestnes AG (2007) Energy use in the life cycle of conventional and low-energy buildings: a review article. Energy Buildings 39:249–257
25. Scherz M, Wieser AA, Passer A, Kreiner H (2022) Implementation of life cycle assessment (LCA) in the procurement process of buildings: a systematic literature review. Sustainability 14:16967
26. Scofield JH, Doane J (2018) Energy performance of LEED-certified buildings from 2015 Chicago benchmarking data. Energy Buildings 174:402–413
27. Shen Y, Pan Y (2023) BIM-supported automatic energy performance analysis for green building design using explainable machine learning and multi-objective optimization. Appl Energy 333:120575

28. Su Y, Miao Z, Wang L, Wang L (2022) Energy consumption and indoor environment evaluation of large irregular commercial green building in Dalian, China. Energy Buildings 276:112506
29. Too J, Ejohwomu OA, Hui FKP, Duffield C, Bukoye OT, Edwards DJ (2022) Framework for standardising carbon neutrality in building projects. J Clean Prod 373:133858
30. Vosoughkhosravi S, Dixon-Grasso L, Jafari A (2022) The impact of LEED certification on energy performance and occupant satisfaction: a case study of residential college buildings. J Build Eng 59:105097
31. Wilcox RR (2012) Introduction to robust estimation and hypothesis testing. Academic Press
32. Your Home (2023) Australian climate zones [Online]. Available: https://www.yourhome.gov.au/getting-started/australian-climate-zones. Accessed 25 Jan 2023

Chapter 6
Water Efficiency in Hotel Buildings: A Focus on LEED Certification in Sri Lanka

A. G. U. Damsari and P. Sridarran

Abstract Existing water efficiency practices directly influence the water consumption in hotels. Green Building Rating Systems (GBRS) provide structures or collections of metrics to determine the water efficiency level that a building can achieve. LEED is the most adaptable system in Sri Lanka among other green rating systems due to its global recognition. LEED-certified hotels are more attractive to hoteliers because of water and energy saving mechanisms and a better environment. However, LEED-certified buildings have been recognised for poor performance than non-certified buildings due to misuse of available conservation systems accredited by the LEED certification. The study aims to compare the water efficiency of LEED certified and non-certified hotels to determine the benefits of LEED certification for water efficiency. Four case studies were conducted, comprising two LEED certified and two non-certified hotels, which are prominent five-star establishments in Sri Lanka. Semi-structured interviews were conducted with ten participants actively involved in water management within these hotels. This contextual study shows that both hotels use same efficient water fixtures and equipment, while the LEED certified hotels tend to adopt more water efficiency mechanisms like water management plans, water audits, BMS monitoring and awareness programmes. Hence, the water efficiency of the LEED certified hotels was more efficient than the non-certified hotels. Finally, a framework was developed to compare and contrast the water efficiency practices in LEED certified and non-certified hotels. It is useful to hotel industry people to identify the water efficiency practices used in LEED hotels.

Keywords Water consumption · Water efficiency · Green building rating system (GBRS) · Leadership in energy and environment (LEED) · Hotels

---

A. G. U. Damsari (✉)
Department of Management, Monash Business School, Monash University, Caulfield, Australia
e-mail: udathara.angulugahagamage@monash.edu

P. Sridarran
Department of Facilities Management, University of Moratuwa, Moratuwa, Sri Lanka
e-mail: psridarran@uom.lk

T. Ramachandra et al. (eds.), *Concepts, Policies and Technology for Sustainable Buildings*, Advances in Sustainability Science and Technology,
https://doi.org/10.1007/978-981-96-3086-8_6

1 Introduction

Water is the world's most precious resource, vital for humanity and the remainder of the world [12]. It is a fundamental requirement for living beings in the biosphere, natural procedures, communities, society, the country's economy and future generations [36]. Approximately two-thirds of the global population experiences water scarcity at least once a year [10]. Among them, 4 billion people currently face severe water scarcity and the figure that may rise to nearly 6 billion by 2050 [5]. Notably, 73% of those affected reside in Asia. Accordingly, developing countries are most affected by water shortages, flooding and poor water quality. Up to 80% of illnesses in the developing countries are linked to inadequate water and sanitation.

In Sri Lanka, water stress is already at 90.8%, which means that the country is consuming 90.8% of its total available renewable freshwater resources at present apart from environmental needs. Thus, Sri Lanka is categorised as a 'highly water stressed country' as per the global climate risk index [28]. Despite seemingly sufficient water availability at the national level, certain districts face significant constraints that could impede future social and economic development. Furthermore, this water scarcity threat could worsen in the future due to changes in seasonal rainfall, groundwater depletion, water pollution and mismanagement.

Water usage in the Sri Lankan hotel industry constitutes a significant portion of the country's overall water consumption pattern, accounting for approximately 1% of Sri Lanka's water resources. To reduce water consumption in the hotel sector, hoteliers should pay more attention [2]. Hoteliers can use several instruments, such as green rating schemes, to encourage hoteliers to maintain water efficiency to prevent the water shortage issue in the future [1]. Green Building Rating System (GBRS) evaluates the relative level of enforcement or efficiency by specific environmental priorities and requirements [19]. Several GBRS focus on water efficiency in buildings, including Leadership in Energy and Environmental Design (LEED), Energy Star, Green Globes, Living Build Challenge, NZEB, Passive House Institute US, SITES, Green SL, Building Research Establishment Environmental Assessment Method (BREEAM) [6].

Among these rating systems, LEED is the most broadly adopted GBRS in the world, with more than 79,000 programmes in over 160 countries and regions [30]. To achieve water efficiency through LEED certification guidelines, several requirements must be fulfilled by the hotel building. According to [29], the water efficiency section discusses strategies for outdoor water usage reduction, indoor water usage reduction, building level water metering, cooling tower water use reduction, water efficient landscaping, innovative wastewater technologies and plumbing fixtures and fitting efficiency.

The efficacy of LEED certification in promoting water efficiency within buildings has been a subject of debate within the construction industry [11]. Orr (2014) challenges the notion that LEED-certified buildings inherently outperform non-certified counterparts, suggesting that non-LEED buildings may, in fact, demonstrate superior performance [14]. Moreover, non-LEED buildings often employ alternative water saving practices or tools such as life cycle assessment, life cycle costing and performance evaluation to achieve water efficiency [11].

The above declaration, therefore, argues that water efficiency in LEED-certified buildings is more efficient than non-certified buildings. Several studies have been conducted regarding water saving strategies in the hotel industry and water efficiency in green and non-green buildings in the apparel sector. Yet, there is a lack of research comparing water efficiency between LEED certified and Non-certified hotel buildings in Sri Lanka. Therefore, this study aims to develop a framework to identify the benefits of LEED certification in terms of water efficiency in hotel industry by comparing water efficiency practices used in LEED certified and non-certified hotels.

2 Literature Review

2.1 Water Efficiency in Sri Lankan Hotels

Every person in the world needs on average 5 L of water for daily drinking purposes and to survive in a moderate climate [25]. In the Asia–Pacific region, some 90% of freshwater is consumed in agricultural activities alone, considerably more than the 75% used by agriculture, on average, worldwide (Food and Agricultural Organisation, 2023). Like in many other countries, agriculture is the main user of water in Sri Lanka, which consumes nearly 85% of freshwater, while total consumption for both industrial and domestic sectors, is close to 12%. Tourism is the third largest foreign exchange earning industry in the country, with a GDP share of 4.5%. The industry needs a considerable amount of water for hotel operations. Tortella and Tirado [27] indicated that the water requirement of a hotel is built upon the type of lodging and tourist activities during their stay and behaviour of the hotel. The Sri Lankan hotel industry plays a significant role in the country's overall water consumption pattern. The hotel industry consumes approximately 1% of Sri Lanka's water resources. Generally, water usage in the Sri Lankan hotel industry is nearly around 800L, making it the third largest water user in the Asia Pacific region [20]. Within these establishments, water consumption is distributed across various departments, with notable portions allocated to guest rooms and kitchen facilities. According to Macaskill et al. [18], the average water consumption per guest per day was 226 L. Additionally, kitchen operations contribute significantly, comprising 21% of total water usage within these establishments [15].

Hoteliers have adopted various water conservation strategies to enhance water efficiency. These include implementing a Water Management Plan that details current usage and sets improvement targets [29] and pursuing Green Building Rating Systems (GBRS) that incentivise water efficiency for certification [32]. The Water Hierarchy, a guideline for prioritising process changes, consists of five levels from source elimination to using fresh water [34]. Water Metering improves resource management by increasing transparency and consumer awareness [4]. Water Audits survey water use and recommend efficiency improvements [21]. Additionally, installing Faucet Aerators can reduce usage by 4%, saving over 500 gallons annually [23], and Low Flow Taps reduce water flow by mixing in air, maintaining the sensation of full flow [3]. Among the above water saving strategies, GBRS is intended to encourage and incorporate whole-building design while decreasing water efficiency and conservation effects on the environment [31].

2.2 *Green Building Rating Systems (GBRS)*

GBRS aims to address the increasingly pressing issue of water scarcity by promoting effective management of water use within buildings. Consequently, numerous GBRS have been developed with a specific focus on enhancing water efficiency. Each GBRS typically includes a dedicated section for water management, which assigns points or credits to incentivize and reward measures that contribute to improved water efficiency. According to Waidyasekara and Silva [32], Table 1 provides an overview of these points' allocation systems.

Water efficiency stands as a central category assessed in most green building rating systems [33]. Among these systems, LEED is highlighted as particularly suitable for evaluating water efficiency as it is a globally recognised green rating system. Efficient water technologies under LEED certification can be implemented at various stages of the design and construction process, with some requiring early planning and interdisciplinary coordination [13]. Moreover, the profound positive influence of LEED certification extends to the national level, highlighting its ongoing enhancements to align with advancing sustainable development needs. Within the Sri Lankan context, both LEED and Green SL certifications offer pertinent guidelines [8]. Nevertheless, LEED stands out prominently due to its dual benefits of international acclaim and

Table 1 Weightage allocation for water efficiency in GBRSs

Green rating system	Country	Points allocated for water efficiency	(%) Out of total
BREEAM	UK	9	6.8
LEED	US	10	9.1
HK-BEAM	Hong Kong	14	7.2
GREEN STAR-AUS	Australia	12	8.1
BCA Green Mark	Singapore	17	8.9
Green Star-NZ	New Zealand	12	7.9
GBI	Malaysia	10	10.0
GRIHA	India	15	14.4
Green Star-SA	South Africa	15	14.0
Green SL	Sri Lanka	14	14.0
Pearl- BRS	Abu-Dhabi	39	21.7

suitability for Sri Lanka's specific circumstances. Thus, LEED certification system is considered for comparison in this study.

The LEED certification system encompasses various categories, including LEED-NC for new buildings, LEED-CS for collaborative development by owners and tenants, LEED-CI for interior decoration of commercial buildings, LEED-ND for community development, LEED-EB for existing building operation management and specialised sub-evaluation systems such as LEED-FOR SCHOOLS [17].

LEED certification encompasses six key aspects to promote sustainability in building practices. i. Sustainable Sites strategies focus on mitigating environmental impacts and emphasise the connections between structures, ecosystems and services. ii. Water efficiency aims to reduce both indoor and outdoor water usage through specialised applications and metering methods. iii. Energy efficiency is achieved by reducing energy consumption, prioritising renewable energy sources and setting new benchmarks for conservation. iv. The Materials category considers the entire life cycle of a building, from raw material extraction to its end-of-life stage. v. Indoor Environment Quality strategies ensure good air quality and comfort in thermal, visual and acoustic aspects. Finally, vi. Innovations involve the adoption of new techniques and the latest scientific research to develop advanced design strategies [33].

A building can receive one of four LEED rating levels according to the number of points attained such as Certified (Required 40–49 points), Silver (Required 50–59 points), Gold (Required 60–79 points) and Platinum (Required 80+ points). As per Wu et al. [35], a significant positive difference was achieved at country level by getting the LEED certification and it is one of the most successful green building rating systems which is continuously improved to meet the constantly changing requirements of sustainable development.

3 Brief Description of the Methodology

The empirical study focused on four hotels, two LEED certified and two non-certified to compare the water efficiency practices. Four case studies were selected from the hotel industry including two LEED certified and two non-certified five-star hotels. Under the four case studies, semi-structured interviews were carried out with 10 experts who are engaged in the water management side at the hotel to identify the water efficient methods used in LEED certified and Non-certified hotels.

The number of respondents was determined based on data saturation and unanimity among the participants. The thematic analysis method was employed to analyse the collected data. Based on the analysis data, the results can be analysed as discussed in Section 4.

4 Analysis

4.1 Existing Water Efficiency Practices in LEED Certified and Non-Certified Hotels

Table 2 provides a comprehensive summary of the specific water efficiency practices utilised in both LEED certified and non-certified hotels. The table categorises these practices into those commonly used by both types of hotels, as well as those unique to certified or non-certified hotels. By highlighting the range of water efficiency practices, the table reveals notable similarities and differences in their approaches to water conservation. This comparison offers insights into the potential advantages of LEED certification and identifies opportunities for non-certified hotels to adopt more advanced water efficiency measures in their hotels.

The empirical findings demonstrated that both LEED certified and Non-certified hotels contributed to high quality water efficiency practices as summarised in Table 4. Tube well water is a common alternative water resource used by both hotel types, while municipal water resources are specified only for LEED certified hotels. LEED certified hotels found that, in the pursuit of LEED certification, adherence to water efficiency measures recommended by the LEED team was imperative. It was observed that, to accrue points for water efficiency, the implementation of alternative water resources such as harvested rainwater, treated wastewater and tube well water was beneficial for effectively managing the hotel's water requirements. Thus, the current water efficiency system in the LEED certified hotels is prosperous with alternative water resources including tube well water and municipal water as freshwater resources, harvested rainwater and treated wastewater for irrigation, vehicle washing, toilet flushing and spraying water on roads to reduce the dust hazard. On the other hand, respondents from non-certified hotels conveyed that their primary water source is tube well water, with rainwater and treated wastewater being employed for irrigation purposes. According to the above facts, both LEED certified and Non-certified hotels have utilised more than one alternative water resource to reduce the burden of water extraction. Moreover, water efficient fixtures and water efficiency mechanisms play a significant role in achieving the water efficiency targets in the building. When considering both methods, the usage of water efficient fixtures and equipment has been identified as cost consuming process, while water efficiency mechanisms tend to reduce water consumption cost effectively by utilising the given water fixtures. Figure 1 shows the water efficient fixtures and water efficient mechanisms that have been followed by the LEED certified hotels.

Additionally, the empirical findings demonstrated that the most efficient water fixtures are water aerators, sensor taps and dual flushing systems. The respondents in LEED certified hotels mentioned that they had installed water aerators to bathroom fittings in guest rooms and kitchen areas a few months ago. According to them, this resulted in a significant reduction in water consumption, as the aerator effectively decreases the amount of water flowing through the tap without compromising water pressure. Furthermore, a dual flushing system is also a useful fixture since it

Table 2 Water efficiency practices used in LEED certified and non-certified hotels

Category	Water efficiency practices
Water efficiency practices used both in LEED certified and non-certified hotels	Use alternative water resources
	Awareness programme
	Water aerators
	Sensor taps
	Dual flushing system
	Use sewerage water treatment plant
	Sub-metering the water distribution system
	Sprinkler system for irrigation
	Sustainability team
	Rainwater harvesting system
	Use PRV
	Use dishwashing machine
Water efficiency practices used only in LEED certified hotels	Use native plants
	Training programmes
	Water management plan
	Foreign visits
	BMS for monitoring
	Air cool chillers
	7R principal
	Water management policies
	Knowledge sharing sessions
	New management
	Set water saving targets
	Innovative technologies
	Induction programmes
Water efficiency practices used only in non-certified hotels	Reuse of condensed steam water
	Use atomizer for boilers to reduce the hardness of the water

can manage flushing volumes effectively through two types of valves fixed in the flushing system. When users need a low water amount of water for flushing, they can choose the small switch valve, while the other switch valve is used for large water requirements. Moreover, the watering time for the irrigation system is scheduled before 10.00 am and after 4.00 pm to reduce the effect of evaporation and water sprinkler system is used for watering plants. To reduce the outdoor water usage, LEED hotels have used native plants for their landscaping to minimise the irrigation water requirement. To reduce the indoor water consumption, which is the second requirement under the water efficiency category in LEED certification, LEED hotels

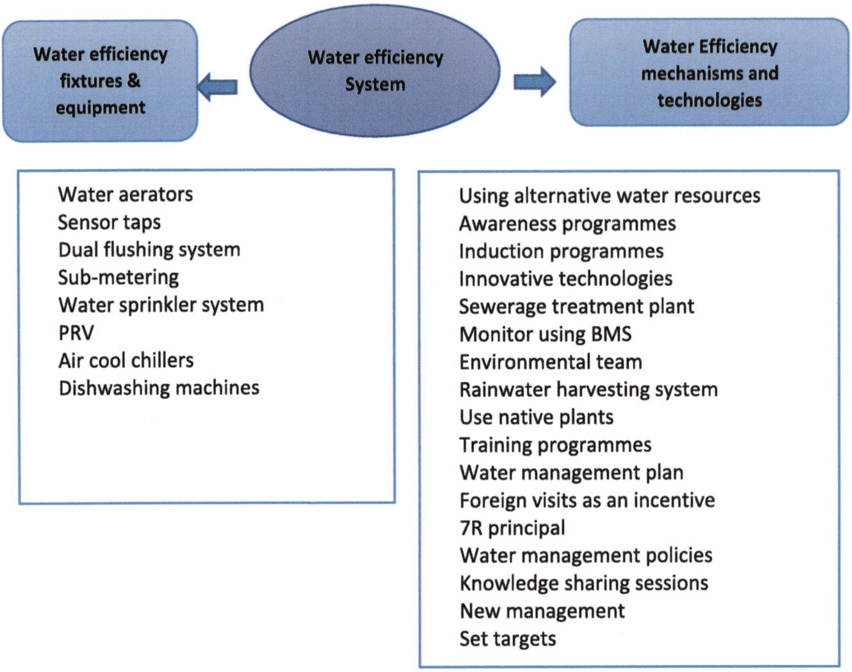

Fig. 1 Water efficient fixtures and water efficient mechanisms used in LEED certified hotels

have used sensor taps, PRVs and dishwashing machines. The third requirement of the water efficiency category under the LEED certification is cooling tower water usage reduction. According to the empirical findings, LEED hotels have used air cooled chillers instead of using water cooled chillers. Therefore, there are no cooling towers and no water evaporation loss for the hotel. However, not all the LEED certified hotels have considered installing sub-metres (which is the final primary criterion for LEED certification) for separate activities, but they have installed sub-metres in the hotel areas separately.

Within the realm of water efficiency mechanisms and technologies, LEED certified hotels have implemented awareness programmes aimed at both employees and guests. According to respondents from LEED certified hotels, employee awareness is fostered through regular monthly meetings. Furthermore, hotel management actively educates employees on water management practices as part of the employee induction process. During these induction programmes, employees are advised to utilise cutlery for their meals, serving as a practical example of water conservation efforts within the establishment. Additionally, guest awareness initiatives are carried out through the distribution of informative leaflets by hotel concierge services. These efforts aim to educate both staff and visitors about water conservation practices, fostering a culture of sustainability within the hotel environment. Moreover, LEED-certified hotels employ a range of highly efficient technologies and systems to minimise water consumption. These include advanced dishwashing machines, irrigation sprinkler systems, rainwater harvesting systems and sewage water treatment systems. Additionally, these hotels utilise Building Management Systems (BMS) to closely monitor both day-to-day water usage and fire protection systems, such as fire sprinklers, fire hydrants and fire hose reels. Through continuous monitoring via the BMS, any potential water leakages or issues within the hotel's water distribution system can be promptly identified and addressed, contributing to further water conservation efforts. To minimise outdoor water consumption, native plants have been strategically incorporated into the hotel's landscaping areas. These indigenous plant species are naturally adapted to the local environment, requiring minimal additional watering due to their inherent resilience. In LEED certified hotels, comprehensive training programmes and knowledge-sharing sessions are organised to raise awareness about innovative water management practices and cutting-edge equipment. A notable case study underscores the effectiveness of these initiatives, revealing how participants are provided with opportunities for international visits to learn firsthand about globally acclaimed water efficiency technologies. This proactive approach not only fosters a culture of continuous learning but also ensures that best practices from around the world are leveraged to enhance water conservation efforts within the hospitality sector. LEED certified hotels adhere to the esteemed 7R principle (Rethink, Refuse, Reduce, Reuse, Repair, Regift and Recycle) and these principles serve as guiding pillars, ensuring reliable management of resource consumption. In an unwavering commitment to sustainability, hoteliers have set ambitious targets to further curtail water consumption. To spearhead this vital initiative, a dedicated management team has been assembled, tasked with implementing innovative strategies to achieve heightened water efficiency throughout the hotel. Through strategic planning and dedication, these hotels not only strive for environmental friendliness but also demonstrate responsible business practices, setting a standard for sustainable hospitality that resonates far beyond their walls.

The LEED rating system incorporates various criteria aimed at enhancing environmental sustainability, including reductions in outdoor water use, indoor water use, cooling tower water use and the implementation of sub-metering. LEED-certified hotels have adopted diverse water efficiency strategies within these parameters to meet the stringent requirements set forth by the LEED certification process. Table 3 provides a summary of the water efficiency practices implemented to fulfil the specified LEED requirements.

Non-certified hotels have also demonstrated a commitment to water conservation through the implementation of water-efficient fixtures and mechanisms to a certain degree. Figure 2 shows the water efficient fixtures and water efficient mechanisms, which have been followed by the non-certified hotels.

When comparing Figs. 1 and 2, LEED certified hotels use more water efficiency mechanisms than non-certified hotels. Awareness about water efficiency and water saving practices is one of the important factors affecting water management in hotels. As a result, LEED certified hotels have focused significantly on employee and guest awareness about water efficiency practices. Employee awareness can be achieved through induction programmes, training programmes, foreign visits and

Table 3 Water efficiency practices under the LEED certification

LEED requirement	Used strategies
Outdoor water use reduction	Use native plants, reduce the artificial irrigation systems, 100% rainwater and treated wastewater used for irrigation and gardening, use water sprinklers to spray water
Indoor water use reduction	Use water efficient fixtures; sensor taps, dual flushing systems, aerators, PRV, dishwashing machines, low flow shower heads
Cooling tower water usage reduction	Use air cool chillers instead of using water cool chillers
Water metering	Sub-metering all the Sections, install BMS to monitor the system

Table 4 Comparison of water efficiency practices in LEED certified and non-certified hotels

Criteria	LEED certified hotels	Non-certified hotels
Outdoor water use reduction	Use of native plants and reduced artificial irrigation systems	Less emphasis on native plants or efficient irrigation
	100% use of rainwater and treated wastewater for irrigation	Reliance on tube well and rainwater for irrigation
	Use of water sprinklers for efficient irrigation	
Indoor water use reduction	Use of many water-efficient fixtures: sensor taps, dual flushing systems, aerators, PRV, dishwashing machines, low-flow shower heads	Use few water efficient fixtures: aerators, dual flush systems
Cooling tower water usage reduction	Use of air-cooled chillers to avoid cooling tower water loss	Likely use of water-cooled chillers, leading to water evaporation loss
Water metering	Sub-metering in different hotel areas and use of BMS	Limited sub-metering (e.g. laundry areas)
Recycling and reuse	Treat and reuse water for irrigation, vehicle washing, toilet flushing, etc	Less extensive water recycling and reuse practices
Awareness and training	Conducts regular training, induction and knowledge-sharing programmes for staff and guests	Less emphasis on awareness initiatives for staff and guests
Satisfaction with water efficiency systems	High satisfaction with water efficiency systems and effective mechanisms	General dissatisfaction with current water usage patterns

knowledge sharing sessions. Respondents in LEED certified hotels reported that they organise knowledge-sharing sessions with other hotel professionals to decrease water consumption by discovering new technologies. Additionally, respondents in LEED certified hotels consistently conveyed that their hotel facilitates international visits to

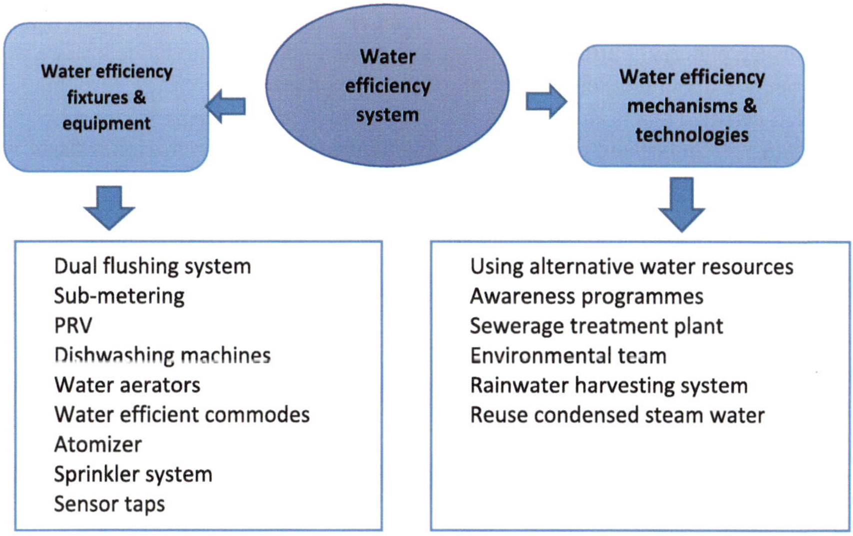

Fig. 2 Water efficient fixtures and mechanisms in non-LEED hotels

explore and learn about innovative water-efficient technologies through discussions with foreign hoteliers. Furthermore, LEED hotels have established a sustainability team to prepare water management plans and water management policies relevant to their hotel facilities. The sustainability team is also responsible for carrying out annual water audits to identify the issues of the current water efficiency system and check whether the existing water efficiency practices comply with LEED requirements. Apart from that, LEED certified hotels have implemented sewage water treatment plant to treat the wastewater coming from the hotel. In order to reduce the indoor and outdoor water consumption, LEED certified hotels recycle and reuse water for several activities. Grey water is treated using sewage treatment plants and used for gardening, vehicle washing and toilet flushing. LEED certified hoteliers reported that the balance water, which was served for the customer has been used to wash cutleries. According to the above findings, LEED hoteliers focus on every aspect of water efficiency. Even though it is a very simple mechanism, its value is highly affected by the hotel's success. Therefore, LEED certified hotels have been able to enhance their water efficiency performance as expected by LEED certification.

According to the analysis, non-certified hotels were not very vigilant in metering and sub-metering individual water use activities. Respondents in non-certified hotels declared that they have used sub-metering only for the laundry area in the hotel. Thus, individual water consumption figures are estimated based on the main meter reading. As per the research framework, unlike LEED certified hotels, there is no motive factor or documentary requirement for each water consuming data point in non-certified hotel.

According to Fig. 2, non-certified hotels have focused more on water efficient fixtures than on water efficiency mechanisms and technologies. Although they use more water efficient fixtures, non-certified hoteliers are not satisfied with their current water usage patterns. Furthermore, all respondents in non-certified hotels share the same opinion about their current water efficiency practices. In contrast, respondents in LEED certified hotels are satisfied with their current water consumption and water efficiency system because they are not solely dependent on water efficiency fixtures; they use proper water saving mechanisms to reduce water usage in the hotel. The enhancement of water efficiency performance is affected by parameters such as the total amount of water usage in the hotel, the amount of new water usage, the amount of recycled or reused water usage, the amount of water discharge from the hotel and the amount of water used during various purposes. Based on the above comparative analysis, water efficiency practices in LEED certified and non-certified hotels can be summarised as follows in Table 4.

This comparison based on LEED criteria outlines the different approaches and practices adopted by LEED-certified and non-LEED-certified hotels in Sri Lanka. LEED-certified hotels tend to employ more advanced practices and show higher satisfaction with their water efficiency systems.

4.2 *Importance of Water Efficiency System*

The empirical findings explain that water efficiency systems are essential for both LEED certified hotels and non-certified hotels. Figure 3 summarises the importance of having a water efficiency system for both LEED certified and non-certified hotels.

Based on the empirical results, it has been discovered that water conservation strategies can save money and divert less water from lakes, bays and waterways, which helps sustain a healthy climate. As per the analysis, all the respondents identified and explained the importance of having a water efficiency system for the hotel industry. All the respondents in both hotels discovered that a water efficiency system

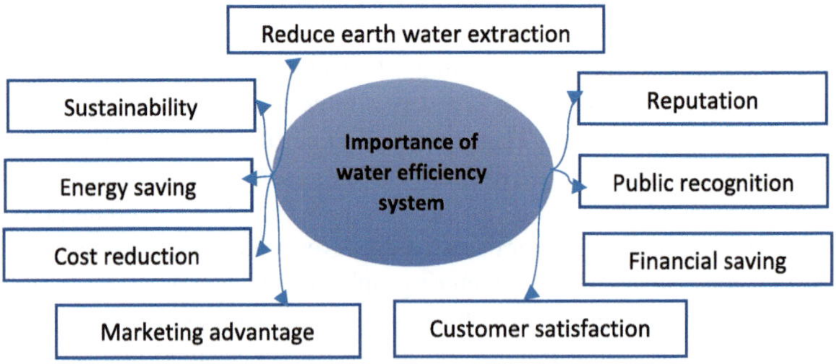

Fig. 3 Importance of water efficiency system

is important to reduce the operational cost of the hotel. Moreover, respondents in non-certified hotel stated that the expense associated with chemicals used for water treatment in the sewerage treatment plant and water treatment plant could be minimised through the implementation of appropriate water efficiency practices. Furthermore, non-LEED certified hoteliers elaborated that the cost of water heating had been decreased as a result of the efficient utilisation of hot water.

The purpose of the water efficiency system is to save water for future generations' requirements, while fulfilling the current generations' needs. Through sustainable water management, it involves solving the current world water crisis as well. According to the literature findings, water is a scarce resource; however, the hotel industry consumes more water compared to other industrial sectors. Therefore, as a hotel, there should be more concern with reducing water consumption. Moreover, respondents in LEED certified hotels emphasised that the current water efficient system helps to maintain the certifications they have earned to enhance the reputation of the hotel. Furthermore, they mentioned that water efficiency practices support to hotel marketing sector as well, since there are guests who are willing to stay at destinations that are sustainable and nature friendly. Based on the above findings, both LEED certified hotels and Non-certified hotels have identified the importance of having a water efficiency system in the hotel.

4.3 Responsibilities Toward Water Efficiency

The population has only gradually increased in recent years, but water demand continues to rise sharply. According to theoretical findings, water is a limited resource and water conservation is a good defence against the water scarcity problem. Water conservation is the responsibility of everyone in the world. Due to higher water consumption in the hotel industry, hotels should pay more attention to reduce the water wastage and save water. Therefore, the responsibilities towards water efficiency can be discussed at the organisational level, management level and individual level. Both LEED certified hotels and Non-certified hotels have identified the responsibilities under the above-mentioned levels. In the organisational level, this involves overarching strategies and policies implemented by the hotel management to promote water efficiency. At the managerial level, responsibility lies with mid-level managers and department heads who oversee day-to-day operations. Lastly, at the individual level, every staff member and guest in the hotel plays a role in conserving water.

Table 5 presents a comprehensive overview of the respective roles and responsibilities at organisational, managerial and individual levels concerning water efficiency.

At the organisational level, hotels have significant responsibilities in promoting water efficiency and addressing the global water crisis. These responsibilities include raising awareness among employees and guests about the importance of water conservation. Furthermore, participants in LEED certified hotels stated that conducting water audits within the hotel is an organisational-level responsibility. Apart from

Table 5 Responsibilities towards water efficiency

Levels	Responsibilities
Organisational level	Employee awareness
	Contribute to solve the global water crisis
	Community awareness
	Contribute to sustainable development
	Conduct water audit
	Purchase water efficient equipment
	Find innovative ideas
Management level	Prepare water management policies
	Employee awareness
	Arrange induction programmes for new employees
	Guest awareness
	Develop water management plans
	On-time maintenance
	Focus to innovations
Individual level	Reduce water wastage
	Self-discipline
	Identify and rectify the water leakages

that, both LEED certified and non-LEED hotels recognise the importance of these responsibilities, with LEED certified hotels prioritising employee and guest awareness as fundamental organisational duties. Moreover, there is a recognised responsibility among hoteliers to protect environmental water bodies, emphasising the interconnectedness of hotel operations with broader environmental conservation efforts. These organisational-level responsibilities highlight the pivotal role hotels play in driving sustainable water management practices and contributing to global water conservation efforts.

When addressing the managerial level within the hotel industry, it is widely recognised that implementing policies for water conservation is a crucial responsibility. The majority of the respondents have identified water-saving policy formulation as one of the primary duties of hotel management. This sentiment is echoed particularly strongly among LEED certified hoteliers, who emphasise the pivotal role of robust water management policies. They underscore that while establishing such systems might be relatively straightforward, the real challenge lies in consistently maintaining and monitoring them. Moreover, LEED certified hoteliers stress the importance of ongoing vigilance and adherence to water efficiency practices. They have employed various strategies to raise guest awareness, such as providing detailed building user guides that outline how to use water-efficient fixtures effectively, without compromising the integrity of the system. Additionally, both LEED certified and non-certified hotels have prioritised staff engagement through induction programmes and briefing sessions. These initiatives serve to educate employees about existing water efficiency practices, highlight current challenges and foster collaborative problem-solving. In light of these findings, it becomes evident that the management level of hotels shoulders a significant responsibility in overseeing the entire water efficiency process.

As per the factual data, both LEED-certified hotel respondents and non-LEED-certified hotel respondents declared that every individual in the hotel has a responsibility to protect water resources. Respondents emphasised that there is a collective responsibility for people worldwide to protect water resources. It was mentioned that individuals should exercise self-discipline to adhere to the rules established within the hotel and follow the provided instructions without violating them.

In addition to the organisational and managerial levels, individual responsibilities play a crucial role in the broader framework of water efficiency within hotels. The primary responsibilities for individuals include reducing water wastage, practicing self-discipline in water usage, and promptly identifying and rectifying water leakages. Reducing water wastage requires conscious efforts from every individual within the hotel. This entails being mindful of water consumption habits, such as turning off taps when not in use, using water-efficient fixtures, and avoiding unnecessary water usage wherever possible. By actively participating in water-saving practices, individuals contribute significantly to the overall conservation efforts. Self-discipline is another essential aspect of individual responsibility in water conservation. This includes being mindful of the amount of water used for activities such as bathing, washing dishes and laundry. By practicing self-discipline, individuals can significantly reduce their water footprint and contribute to overall conservation efforts. It is evident that every individual within the hotel, regardless of their role or position, bears a responsibility to adhere to water-saving policies and plans. While organisational and management levels handle the overarching water management planning and procedures, individual contributions are integral to the successful implementation of water conservation measures. Therefore, fostering a culture of water consciousness and accountability among all individuals in the hotel is paramount to achieving sustainable water management practices.

4.4 Weaknesses of Current Water Efficiency System

As described in Sect. 4.1.1, both LEED certified hotels and non-certified hotels use water efficiency systems to manage the water consumption in the hotel. However, several issues have arisen due to inefficiencies in the current water efficiency system. Figure 4 illustrates the identified weaknesses of current water efficiency system.

The empirical research findings emphasise a common challenge shared by both LEED certified and non-certified hotels regarding water efficiency. Within this context, it was revealed that both categories of hotels face distinct issues that impact

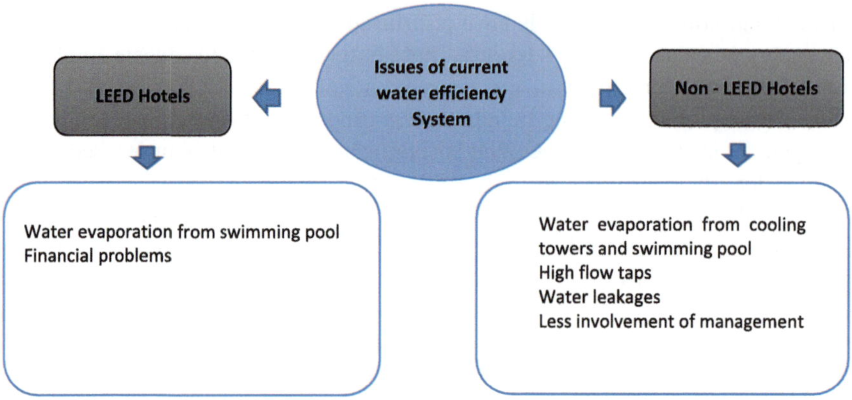

Fig. 4 Weaknesses of water efficiency system

their water management strategies. In one instance, a respondent from a LEED certified hotel noted the significant challenge posed by water evaporation from the swimming pool, particularly pronounced due to the hotel's location within a dry zone. This necessitated daily refilling of the pool, highlighting a tangible concern even within environmentally conscious establishments. Conversely, respondents from non-certified hotels identified many issues contributing to inefficiencies in their water systems. These encompassed water evaporation from cooling towers and pools, high-flow taps, water leakages and a lack of proactive management involvement. Notably, the utilisation of booster pumping systems for water distribution posed a unique challenge, rendering the detection of leaks difficult due to consistent pipeline pressure maintenance. In contrast, LEED certified hotels demonstrated a proactive approach to water efficiency. These establishments were selected for air cool chillers over water cool chillers to mitigate water evaporation in cooling towers, implemented BMS for continuous monitoring of water distribution systems, and embraced innovative, water-efficient fixtures to regulate consumption. Moreover, the management of LEED certified hotels exhibited a heightened commitment to water efficiency, driven by the imperative to uphold LEED certification standards. In contrast, non-certified hotels appeared to prioritise budgetary concerns over sustainability initiatives, thereby neglecting opportunities for future benefits. While it is evident that both LEED certified and non-certified hotels face water efficiency challenges, the severity and proactive measures taken to address these issues differ significantly. LEED certified hotels, by adhering to certification requirements and embracing sustainable practices, exhibit a more robust framework for water management. In contrast, non-certified hotels often struggle with a lack of proactive management involvement and insufficient investment in sustainable solutions, leading to a greater prevalence of water-related challenges.

4.5 Required Improvements of Water Efficiency System

To solve the problems of the current water efficiency system in the hotels, respondents proposed a few improvements which have been summarised in Fig. 5.

According to empirical findings, there is a pressing need for enhancements in both water efficiency fixtures and mechanisms within the hotel industry, particularly in Sri Lanka. Respondents have pointed out that compared with developed countries, Sri Lankan hotels currently lack innovative technologies to achieve water efficiency. Developed nations have made significant strides in reducing the water footprint associated with food production, indicating a clear direction for improvement within the hotel sector. Also, respondents from LEED certified hotels noted the difficulty in altering current designs due to architectural constraints and financial limitations. Hence, it is crucial to conduct a comprehensive cost–benefit analysis before implementing new technologies in existing buildings to ensure maximum benefits from these improvements. While LEED certified hotels have made notable progress in enhancing their water efficiency systems, non-certified hotels still lag behind in adopting innovative water efficiency technologies. These establishments can significantly enhance their water management practices by integrating advanced solutions such as sewage water treatment plants, rainwater harvesting systems and air-cooled chillers instead of traditional water-cooled ones. Implementing BMS to monitor water distribution, incorporating low-water-consuming plants for landscaping and adopting efficient water management policies and practices are also vital steps. It is worth noting that non-certified hotels can achieve substantial water consumption reductions by implementing water efficiency practices beyond merely upgrading fixtures. Measures such as water audits, robust water management policies and employee induction programmes focused on water-saving concepts can significantly contribute to overall water conservation efforts, without imposing excessive costs on hotels.

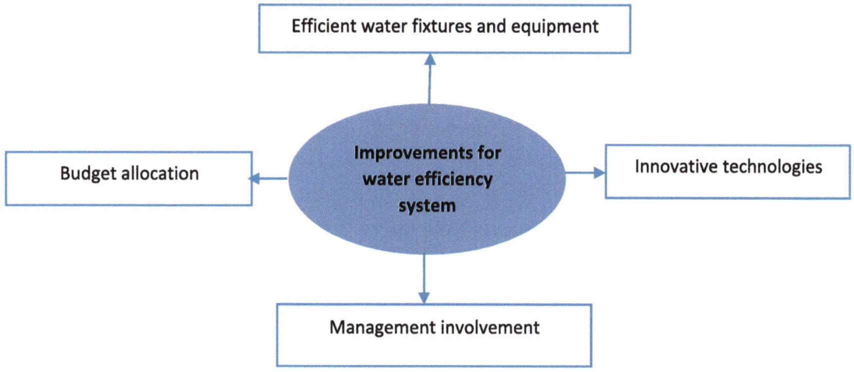

Fig. 5 Improvements in water efficiency system

4.6 General Merits and Demerits of LEED Certification

According to empirical findings, LEED certified hotels boast numerous technological improvements stemming from their adherence to LEED certification standards. These advancements serve as notable merits for LEED certified buildings. Respondents from such hotels have asserted that LEED serves as a powerful marketing tool within the hotel industry, attracting guests who actively seek accommodations in environmentally friendly establishments. LEED stands out as the world's most renowned green rating system, making it highly prestigious for any organisation to attain such certification. Guests inclined towards environmentally conscious choices tend to favour LEED certified hotels for their global recognition as eco-friendly destinations. Notably, LEED certification fosters a culture of water conservation within hotels, resulting in significantly reduced water consumption compared to non-certified counterparts. This emphasis on water efficiency incentivises hoteliers to implement systems aligned with LEED requirements. Furthermore, LEED certification correlates with increased revenue for hotels, often allowing for higher service charges. This financial boost positively impacts employee satisfaction within the hotel. Additionally, LEED guidelines lead to reduced operational costs, energy and water wastage, as well as diminished environmental pollution. Consequently, the overall value of LEED certified properties is augmented due to their efficient operations and associated benefits.

Most Sri Lankan hoteliers choose not to pursue LEED certification for their hotels, citing the difficulty of achieving it without significant renovations. Instead, they turn to alternatives such as Green SL, initiated by the Sri Lankan Green Building Council, which they find more suited to the local context. They believe that Green SL offers a more feasible path to sustainability in Sri Lanka. However, hoteliers who have obtained LEED certification argue that while Green SL may be well-suited to Sri Lanka, it lacks global recognition as a certification system. Given that Sri Lankan hotels primarily cater to foreign tourists, who are more familiar with LEED certification than Green SL, the perceived global recognition of LEED becomes a significant factor in their decision-making process.

While non-LEED hoteliers often emphasise the initial costs associated with the LEED certification process, proponents of LEED-certified hotels argue that decision-making should not solely focus on initial costs. Although the initial costs of LEED-certified buildings may be considerable, the overall life cycle costs could prove to be lower. Table 6 provides a summary of the advantages and disadvantages of LEED certification within the hotel industry.

According to the above table, it is evident that LEED certification offers numerous advantages that often outweigh the challenges associated with achieving it. LEED-certified buildings, particularly in the hotel industry, benefit from a wide range of technological advancements and operational efficiencies that not only enhance sustainability but also improve the overall value of the property. These merits extend beyond environmental considerations, positively influencing marketing, customer satisfaction, employee motivation and financial performance.

Table 6 General merits and demerits of LEED scertification

Merits of LEED certification	Use as a marketing tool
	Motivate to achieve water efficiency
	Motivate to achieve energy efficiency
	Reduce operational cost
	Enhance reputation
	Employee motivation
	Provide advanced technologies
	Enhance customer satisfaction
	Improve total building efficiency
	Support to achieve company goals
	Global recognition
	Waste reduction
	Reduce environmental pollution
	Increase the property value
	Tax benefits
	Long terms savings
Demerits of LEED certification	Difficulty in achieving requirements
	Higher initial cost
	Lack of focus on cultural aspects

4.7 *Merits of LEED Certification in Terms of Water Efficiency*

Empirical findings have shown that LEED-certified hotels prioritise the implementation of advanced water efficient fixtures and mechanisms. The theoretical findings explain that achieving LEED certification necessitates meeting several prerequisites. Respondents also explained that pursuing LEED certification prompts hoteliers to embrace sustainable practices, particularly in the realm of water efficiency and its associated technologies and mechanisms. Consequently, they have enhanced their water supply and distribution systems and are committed to their upkeep. The role of management in fostering water efficiency is deemed crucial for meeting targets. Respondents from LEED certified hotels indicated a substantial motivation among both employees and management to conserve water, largely driven by the aspiration to maintain the LEED certification.

Conversely, non-LEED buildings lack a similar motivational factor to align with. This highlights the direct influence of the LEED rating system on management and employee motivation. Thus, LEED certification catalyses inspiring both management and employees to enhance water efficiency within the hotel. The merits of LEED certification in terms of water efficiency are briefly summarised in Fig. 6.

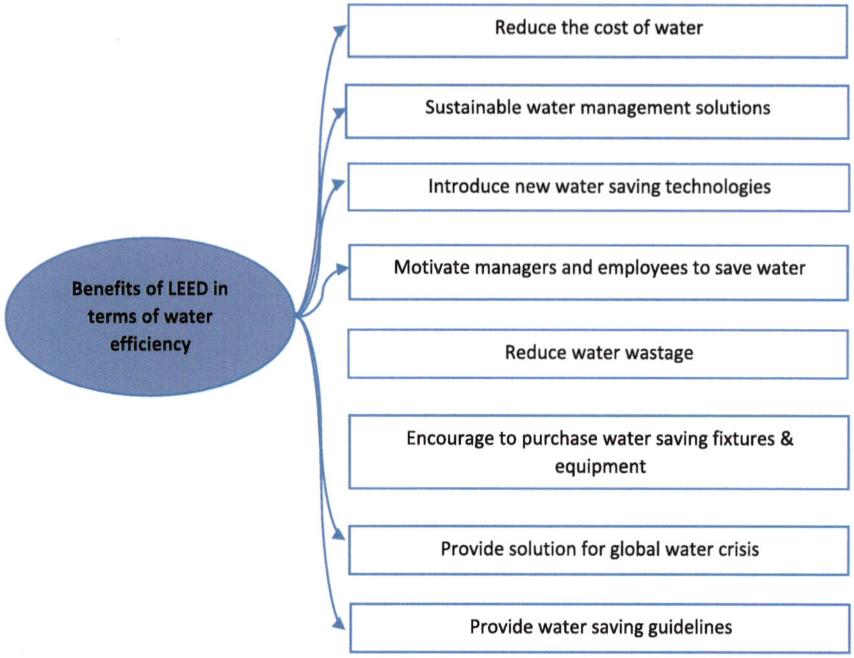

Fig. 6 Benefits of LEED in terms of water efficiency

4.8 A Framework to Compare Water Efficiency Practices in LEED Certified and Non-Certified Hotels in Sri Lanka

Figure 7 depicts a framework for comparing water efficiency practices utilised in LEED certified hotels and non-certified hotels in Sri Lanka. The framework emphasises the significance of achieving water efficiency and outlines the responsibilities associated with it at the top. The hotel industry is depicted as the central entity, divided into two sections to distinguish between LEED certified and non-certified hotels. To facilitate comparison, water efficient fixtures and equipment employed in both types of hotels are represented within blue-colour shapes, describing differences in water efficiency practices and technologies. In LEED certified hotels, these fixtures and equipment are further categorised according to LEED requirements, illustrated by black-colour boxes. At the bottom of the framework, the benefits of LEED certification for water efficiency are contrasted with the identified issues and shortcomings prevalent in non-certified hotels. Benefits specific to water efficiency in LEED certified hotels are highlighted in pink boxes, while general advantages of LEED certification for the hotel industry are represented in orange boxes. However, certain drawbacks identified through literature review and case study findings are indicated in grey boxes. The framework utilises dashed lines to illustrate connections between various concepts. It serves as a valuable resource for individuals in the hotel industry and students conducting research on water efficiency practices. Hotel personnel can readily identify different water efficient fixtures, technologies and practices, contributing to industry efficiency, while also gaining insights into the water efficiency requirements and benefits of LEED certification for the hotel sector.

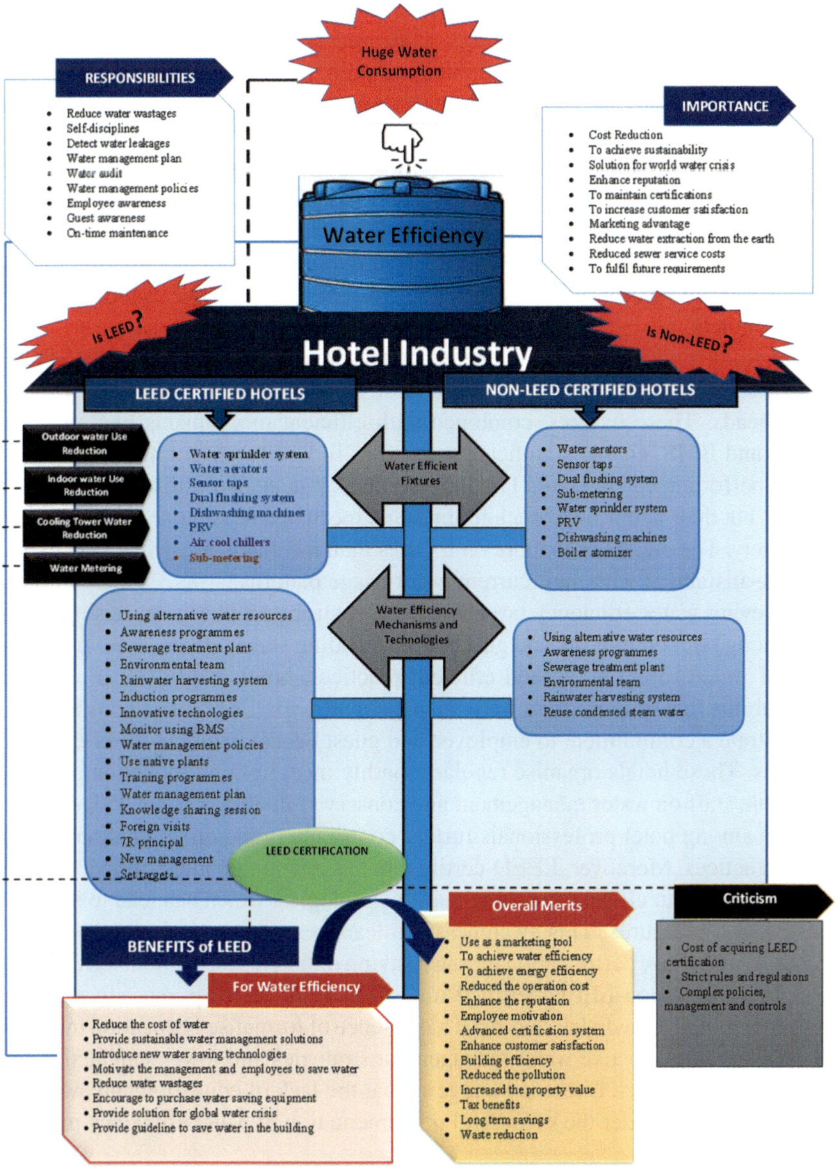

Fig. 7 Framework to compare water efficiency practices in LEED certified and Non-certified hotels

5 Discussion

In this study, the water efficiency practices used by the LEED certified hotels and non-certified hotels were compared and a framework was proposed to summarise the comparison between the LEED and non-LEED certified hotels. The results highlight the varying approaches adopted by the two types of hotels in terms of water resource utilisation, water efficiency technologies, awareness programmes and management strategies. LEED certified hotels exhibit a comprehensive approach to water resource utilisation by employing a mix of alternative water sources such as harvested rainwater, treated wastewater, tube well water and municipal water for different purposes within the hotel. These varied sources allow the hotels to reduce reliance on a single water source and enhance overall sustainability [2]. Non-certified hotels, on the other hand, primarily rely on tube well water as their main water source and use rainwater and treated wastewater for irrigation purposes. While they demonstrate some alternative water resource utilisation, their strategies are not as extensive as those of LEED certified hotels.

The study finds that LEED certified hotels invest in advanced water-efficient fixtures such as water aerators, sensor taps, dual flushing systems and low flow showerheads. These fixtures, combined with efficient mechanisms like air-cooled chillers and BMS, enable significant reductions in water consumption while maintaining performance standards [1]. Non-certified hotels also utilise water-efficient fixtures, but their adoption of such fixtures and mechanisms is more limited in scope and variety. The emphasis is more on fixtures than mechanisms, which may explain their dissatisfaction with their current water usage patterns.

Achieving water efficiency targets requires a strong emphasis on raising awareness among both employees and guests. By educating stakeholders about the importance of water conservation and efficient practices, hotels can foster a culture of sustainability that supports their water management goals [26]. LEED-certified hotels demonstrate a commitment to employee and guest education about water efficiency practices. These hotels organise regular monthly meetings and induction processes to educate staff on water management and conservation efforts. Knowledge-sharing sessions among hotel professionals further contribute to the dissemination of innovative practices. Moreover, LEED-certified hotels encourage international visits for staff to learn about global water efficiency technologies, which can lead to improved conservation measures. This awareness can lead to more mindful water usage and encourage proactive participation in water-saving initiatives throughout the establishment. In contrast, non-LEED-certified hotels lack a structured approach to awareness programmes and knowledge sharing. This absence of formalised education initiatives may lead to less effective water management strategies and lower satisfaction with their current practices. Literature suggests that the lack of education and awareness programmes can hinder the successful implementation of water efficiency measures [22].

In LEED-certified hotels, dedicated sustainability teams have been established for implementing water management plans and policies. The systematic planning provided by these teams is essential for the successful application of water efficiency measures. LEED-certified hotels conduct annual water audits to identify areas for improvement and ensure compliance with LEED requirements, resulting in effective water conservation efforts and efficient resource utilisation. In contrast, non-LEED-certified hotels often lack the same level of strategic planning and monitoring when it comes to water management. This absence of comprehensive water management plans and oversight can lead to less effective water conservation efforts and reduced efficiency in resource utilisation [7]. This discrepancy aligns with literature on water management practices in the hospitality industry. For instance, researchers have noted that hotels with strong environmental certifications, such as LEED, tend to have more structured and systematic approaches to resource management [24]. They implement formalised strategies to monitor and control water usage, resulting in improved sustainability and cost savings. On the other hand, non-LEED-certified hotels may face challenges due to the lack of robust water management systems. This can lead to higher water consumption and inefficient practices, which ultimately affect their bottom line and environmental footprint [16].

Accordingly, the study explains the effectiveness of LEED certification in driving comprehensive water efficiency practices within the hotel industry. Non-certified hotels have room for improvement in water efficiency practices, particularly in terms of adopting a broader range of water-efficient technologies and fostering awareness among stakeholders.

6 Conclusions

The study underlines the different practices regarding water efficiency observed between LEED-certified and non-LEED-certified hotels. While both groups of respondents recognise the paramount importance of water efficiency for reducing operational costs, complying with environmental standards and ensuring long-term conservation, they also acknowledge additional benefits such as improved marketing appeal and enhanced consumer satisfaction. Despite these shared perspectives, notable differences exist in the adoption of specific water-saving practices. LEED-certified hotels tend to demonstrate a more comprehensive approach, incorporating advanced technologies like air-cooled chillers and building management systems for efficient water monitoring. Moreover, LEED certified hotels prioritise the use of native plants for outdoor landscaping, implement training programmes for staff education and adhere to sustainability principles such as the 7R framework. Furthermore, LEED hotels typically have sustainability teams responsible for developing water management plans and policies. These teams conduct annual water audits to ensure compliance with LEED requirements and identify areas for improvement. In contrast, non-LEED-certified hotels often lack these sophisticated measures, resulting in limited recognition beyond the local level and a diminished ability to capitalise on marketing opportunities. The advantages afforded by LEED certification, including global recognition and access to cutting-edge techniques, further accentuate the gap between the two groups. However, in the context of globalisation, technological advancements continually evolve and progress. Therefore, it becomes imperative for hoteliers to stay up-to-date on the latest developments in water efficiency technology to effectively enhance their water conservation efforts. The findings of this research offer valuable insights for professionals within the hotel industry in Sri Lanka, aiding in the identification and enhancement of water efficiency practices. Utilise the research findings as a framework to identify and evaluate water management practices appropriate to the hotel industry. By leveraging the results as a guide, industry professionals can separate effective strategies for optimising water usage within their establishments. Moreover, highlighting the correlation between LEED certification and enhanced water management practices can incentivise hoteliers to pursue certification, thereby fostering a culture of sustainability and resource conservation within the industry. In Sri Lanka, the number of LEED certified buildings remains relatively low due to several challenges associated with obtaining LEED certification. Therefore, there is a future opportunity for research aimed at identifying the barriers to achieve LEED certification in Sri Lankan hotel industry. By conducting such research, valuable insights can be gained into the motivators and obstacles faced by stakeholders in pursuing LEED certification for their buildings.

References

1. Abdou AH et al. (2020) A description of green hotel practices and their role in achieving sustainability a description of green hotel practices and their role in achieving sustainable development. Sustainability 12:9624. https://doi.org/10.3390/su12229624
2. Antonova N, Mendoza-jim J (2023) Determinants of water consumption in hotels : new insights obtained through a case study. Water 15(17):3h049. https://doi.org/10.3390/w15173049
3. Australian Government (2018) Water efficiency. Available at: https://www.energy.gov.au/households/water-efficiency. Accessed 14 March 2019
4. Berger MA et al. (2016) Exploring the energy benefits of advanced water metering. Energy analysis and environmental impacts division energy technologies area (August), p. 54. https://www.researchgate.net/profile/Michael_Berger18/publication/311302089_Exploring_the_Energy_Benefits_of_Advanced_Water_Metering/links/584101c908aeda696811adde.pdf
5. Boretti A, Rosa L (2019) Reassessing the projections of the world water development report. npj Clean Water 2(1):1–6. https://doi.org/10.1038/s41545-019-0039-9
6. Clark CE (2021) Green building overview and issues. https://www.everycrsreport.com/files/2021-03-12_R46719_390fdda22dc77d93fd977617aa09cdd792ded6c3.pdf
7. Cosgrove WJ, Loucks DP (2015) Water resources research. Water Resource Res 51(6):4823–4839. https://doi.org/10.1002/2014WR016869.Received
8. Denagama T (2020) Comparative study of green building rating systems in Sri Lanka: the case of social and cultural awareness comparative study of green building rating systems in Sri Lanka: the case of social and cultural awareness. In: International Research Symposium—University of Vocational Technology. https://www.researchgate.net/publication/344598231%0AComparative
9. Food and Agricultural Organization (2023) Water scarcity in Asia-Pacific casts a dark cloud over World Food Day, where the message 'Water is life, water is food' takes on a special urgency, Food and Agricultural Organization. Available at: https://www.fao.org/asiapacific/news/detail-events/en/c/1653491/#:~:text=Waterdemandisonthe,ImpactsofClimateChange

10. Gosh N (2022) Water-scarce economies and scarcity values: can water futures trading combat water scarcity? Observer Research Foundation, New Delhi, India
11. Greer F et al. (2019) Energy & buildings energy and water efficiency in LEED: How well are LEED points linked to climate outcomes? Energy Buildings 195:161–167. https://doi.org/10.1016/j.enbuild.2019.05.010
12. Jhariya MK, Banerjee A, Meena RS (2021) Importance of natural resources conservation: moving toward the sustainable world. Natural Resources Conservation and Advances for Sustainability. Elsevier Inc. https://doi.org/10.1016/B978-0-12-822976-7.00027-2
13. Kanhaiya KSS (2023) Green buildings for a sustainable future: project management perspective, institute project management. Available at: https://projectmanagement.ie/blog/green-buildings-for-a-sustainable-future-project-management-perspective/
14. Khoshbakht M, Rasheed E (2022) Do green buildings have superior performance over non-certified buildings? occupants' perceptions of strengths. Buildings 12(9). https://doi.org/10.3390/buildings12091302s
15. Kularatne T et al. (2019) Do environmentally sustainable practices make hotels more efficient? A study of major hotels in Sri Lanka. Tourism Management 71:213–225. https://doi.org/10.1016/j.tourman.2018.09.009
16. Lin C, Shyu C, Li C (2023) The greener the hotel, the better operating efficiency it has? A sustainable tourism perspective. Sustainability 15(14):1–16. https://doi.org/10.3390/su151410927
17. Liu Y et al. (2019) Research on green project certification in China based on LEED and GBES. In: IOP Conference Series: Earth and Environmental Science, p 052036. https://doi.org/10.1088/1755-1315/233/5/052036
18. Macaskill S, Becken S, Coghlan A (2023) Engaging hotel guests to reduce energy and water consumption: a quantitative review of guest impact on resource use in tourist accommodation. Cleaner Responsible Consumption 11:100156. https://doi.org/10.1016/j.clrc.2023.100156
19. Marchi L, Antonini E (2021) Green Building Rating Systems (GBRSs). Encyclopedia 1(4):998–1009. https://doi.org/10.3390/encyclopedia1040076
20. Moore S (2015) Water and tourism. Water Policy. https://doi.org/10.1016/S1366-7017(01)00056-3
21. New Hampshire Environmental Service (2013) Water efficiency: business or industry water use and conservation audit. https://www.des.nh.gov/organization/commissioner/pip/factsheets/dwgb/documents/dwgb-26-16.pdf
22. Park J et al. (2021) 'Can Hotel Companies' Water Conservation Management and Waste Reduction Measures Influence Hotel Customers' Willingness to Pay More and Intention to Revisit?' Int J Environ Res Public Health 18(17):9054. https://doi.org/10.3390/ijerph18179054
23. Preston E (2019) 10 water saving devices you can install yourself. https://www.thespruce.com/water-saving-devices-install-yourself-4096495. Accessed 13 March 2019
24. Soria E, Parra D (2021) Environmental approach in the hotel industry: riding the wave of change. Sustainable Futures 3:100050. https://doi.org/10.1016/j.sftr.2021.100050
25. The World Counts (2014) Average daily water usage. Available at: http://www.theworldcounts.com/stories/average-daily-water-usage. Accessed: 6 March 2019
26. Torres-bagur M, Ribas A (2020) Understanding the key factors that influence efficient water-saving practices among tourists: a Mediterranean case study. Water 12(8):2083. https://doi.org/10.3390/w12082083
27. Tortella BD, Tirado D (2011) Hotel water consumption at a seasonal mass tourist destination: the case of the island of Mallorca. J Environ Manage 92(10):2568–2579. https://doi.org/10.1016/j.jenvman.2011.05.024
28. United Nations Sri Lanka (2023) Integrated water resources development: the way forward for Sri Lanka to tackle the Climate Crisis. Available at: https://srilanka.un.org/en/248878-integrated-water-resources-development-way-forward-sri-lanka-tackle-climate-crisis#:~:text=Adding to the issue%2C according,as "highly water stressed"
29. US Green Building Council (2024) LEED rating system, US Green Building Council. Available at: https://www.usgbc.org/leed

30. Vierra S (2023) Green building standards and certification systems, whole building design guide—National Institute of Building Sciences. Available at: https://www.wbdg.org/resources/green-building-standards-and-certification-systems
31. Waidyasekara KGAS (2015) Appraisal of water efficiency in green building rating systems', in *SLQS-QATAR Annual Get Together 2015*. Sri Lanka Quantity Surveyors. http://www.slqsqatar.org/appraisal-of-water-efficiency-in-green-building-rating-systems/
32. Waidyasekara KGAS, Silva MLD (2013) 'Comparative study of green building rating systems in terms of water efficiency and conservation', In: The Second World Construction Symposium 2013: Socio-Economic Sustainability in Construction [Preprint]. Available at: http://www.irbnet.de/daten/iconda/CIB_DC26706.pdf
33. Walsman, M. (2014) *It's not just cost savings: LEED certification boosts hotel revenue too, US Green Building Council*. Available at: https://www.usgbc.org/articles/it's-not-just-cost-savings-leed-certification-boosts-hotel-revenue-too (Accessed: 8 January 2019).
34. Wan Alwi SR et al. (2014) Benchmarking water utilisation using minimum water network technique
35. Wu P et al. (2017) A comprehensive analysis of the credits obtained by LEED 2009 certified green buildings. Renew Sustain Energy Rev 68:370–379. Available at: https://doi.org/10.1016/j.rser.2016.10.007
36. Zhang Yunlin et al. (2023) Importance and vulnerability of lakes and reservoirs supporting drinking water in China. Fundamental Research 3(2): 265–273. Available at: https://doi.org/10.1016/j.fmre.2022.01.035

Chapter 7
Drivers to Implement Circular Economy-Based Waste Management Practices in the Australian Construction Industry

Yuchen She, Nilupa Udawatta, Chunlu Liu, and Olubukola Tokede

Abstract The construction industry practices are criticised due to the high amount of waste generation and consumption of natural resources in Australia. Thus, it is necessary to transition from traditional linear to circular economy (CE)-based construction practices to overcome these challenges. This chapter aims to identify drivers for implementing CE-based waste management (WM) practices in Australian construction projects to mitigate the negative impacts of construction and demolition (C&D) waste generation. A qualitative research approach was used by conducting twenty preliminary interviews (Stage 1) and four case studies of commercial projects (Stage 2). The data was analysed using the content analysis method using NVivo software. The findings indicate that regulatory compliance, economic benefits, environmental sustainability, social responsibility, reputational benefits, stakeholder proactivity, time, cost and quality considerations are vital for implementing CE in managing C&D waste. Regulatory compliance requires collaboration among government, non-government organisations and industry stakeholders. Economic drivers include cost savings and new revenue streams, fostering environmental sustainability and innovation. Social responsibility promotes local growth and ethical advantages, while reputational benefits enhance brand values and commitment to sustainability. Stakeholder proactivity, driven by mutual interests, benefits both the environment and the community. Similarly, time, cost and quality considerations encompass the importance of balancing quality, cost and efficiency in implementing CE-related strategies.

Y. She · N. Udawatta (✉) · C. Liu · O. Tokede
School of Architecture and Built Environment, Deakin University, Geelong, Australia
e-mail: nilupa.udawatta@deakin.edu.au

Y. She
e-mail: yuchen.she@research.deakin.edu.au

C. Liu
e-mail: chunlu.liu@deakin.edu.au

O. Tokede
e-mail: olubukola.tokede@deakin.edu.au

T. Ramachandra et al. (eds.), *Concepts, Policies and Technology for Sustainable Buildings*, Advances in Sustainability Science and Technology,
https://doi.org/10.1007/978-981-96-3086-8_7

This balance advocates for a significant change in current construction practices by considering the interplay of policy, economy and social responsibility. These findings will be useful for construction professionals and policymakers to accelerate the implementation of CE-based C&D WM practices in construction projects. The findings will also contribute to advancing knowledge in the field of facilitating the implementation of CE.

Keywords Circular economy (CE) · Construction projects (CPs) · Construction and demolition (C&D) waste · Drivers · Waste management (WM)

1 Introduction

The Construction Industry (CI) practices are criticised due to the amount of construction and demolition (C&D) waste generation and natural resource consumption [50], which contribute to nearly 30% of the solid waste streams, which are called C&D waste, in many developed countries [27]. In Australia, C&D waste alone accounted for 38% of total waste generation in 2021, totalling 25.2 million tonnes [31]. Traditional waste management (WM) practices, which treat using the 3Rs (reduce, reuse and recycle) as a primary rule, are partially effective but have not curtailed the high landfill rates, with 4.9 million tonnes landfilled in 2021 [31]. This underscores a critical need for a more effective mechanism to prevent waste generation that prioritises resource efficiency and the implementation of sustainability [45], especially in Australia [31].

In addressing these challenges, the circular economy (CE) approach has gained wider attention from scholars and practitioners due to its focus on transitioning from traditional linear construction models to a “make-use-recreate” system [29]. This system aims to minimise waste and optimise the usage of resources [9]. Its primary principles follow a framework that is both restorative and regenerative by design, promoting a sustainable approach to production and consumption [9]. Obviously, the current construction industry, conducting the linear economic system, is considered one of the major industries with the highest potential to apply the CE concept [19], as it can promote the sustainable use of resources and address the limitations associated with traditional WM practices [18].

However, current CE practices need to fully exploit their potential in the construction sector [28]. Knowing the drivers of CE implementation in the CI can facilitate a more effective mechanism to prevent waste generation that prioritises resource efficiency and the implementation of sustainability [8]. Despite the growing interest in CE within the construction sector, a notable gap exists in understanding the specific drivers that facilitate the implementation of CE-based WM practices, particularly in Australian construction projects (CPs). Therefore, this chapter addresses the identified gap by exploring the drivers for CE-based C&D WM in Australian CPs. In particle, its findings contribute to the CI in Australia and other countries looking to facilitate the implementation of the CE in the C&D WM sector. Additionally, the findings contribute to the advancement of knowledge in the field of facilitating CE implementation.

2 Current Drivers to Implement Circular Economy-Based Construction and Demolition Waste Management

To effectively integrate CE into C&D WM practices, it is essential to understand the key drivers that can facilitate this transition. The existing literature identifies five main categories of drivers: regulatory compliance, economic benefits, environmental sustainability, social responsibility, reputational benefits and stakeholder proactivity, as shown in Table 1.

The above table shows that regulatory compliance is a critical driver for applying CE-based C&D WM in construction projects [16]. Some countries have already merged this concept into national strategic policy agendas [3] or have involved some concept elements in policies. For example, China directly applies the term "circular economy" in waste and resource policy [3] and treats the CE as a national political goal [14, 52], for example, the Circular Economy Promotion Law of the People's Republic of China in 2008 [25]. Since 2012, the German government has adopted the German Resource Efficiency Programme to track resource efficiency development (Federal Ministry for the Environment, Nature [11], this reflects parts of elements primarily focused on by the CE, that is, reducing and reusing products. Applying the partial components of the CE cannot necessarily be called CE actions, as they are just the initial components of a CE [23]. It is ineffective to transition towards the CE by focusing on an individual solution or blanket solutions since effective policies for achieving the CE rely on integrating numerous policy interventions [10]. In 2015,

Table 1 Drivers to apply CE-based C&D WM from the current literature

Driver	Explanation	References
Regulatory compliance	Pressure from government regulations and policies promoting waste prevention, recycling and resource recovery	[14, 16, 52]
Economic benefits	Cost savings from waste disposal, revenue from recovered materials and new business opportunities	[2, 15, 40, 41]
Environmental sustainability	Reducing environmental impacts, conserving natural resources and lowering the carbon footprint of the built environment	[21, 41]
Social responsibility	Enhancing local economic development, improving public health, and fostering social innovation and collaboration	[17, 30]
Reputational benefits	Enhancing brand value, attracting customers and partners and gaining a competitive advantage	[4, 34]
Stakeholder proactivity	Encouraging collaboration among stakeholders, sharing knowledge and effective practices and driving innovation	[13, 51]

the United Nations released a new universal 2030 agenda and its seventeen new sustainable development goals for calling all countries and stakeholders to transform the world into a sustainable and resilient path. After this agenda, the CE has been recognised as one of the possible ways for different countries and communities to support some of these sustainable development goals. In some countries, CE is framed and created as a bottom-up means to stimulate policies relevant to WM and the environment [14]. For example, in 2015, the European Union created a "Closing the Loop—An EU Action Plan for Circular Economy" package. The Netherlands published "A Circular Economy in the Netherlands by 2050" in 2016 [20]. In the same year, Scotland published "Making Things Last: A Circular Economy Strategy for Scotland" [39], and Finland also issued "Leading the Cycle—Finnish Road Map to a Circular Economy 2016–2025" in 2016 [12].

Under the urgent demands to achieve sustainable development worldwide and Australia's growing pressure from the recycling crisis after China's ban on foreign waste imports, the Australian Government [6] urges all industries to implement CE. In response to this call to action, some Australian state governments have released some specific CE policies or policies promoting CE in place to make the CE concept a top-to-bottom strategy prioritised by the government to deal with issues caused by waste, including C&D waste. In 2019, the [26] released a Circular Economy Policy Statement to positively affect the economy, society and environment. In the meantime, the [36] released a Waste Management and Resource Recovery Strategy to transition to the CE for waste, while the Tasmania government (2019) created a Draft Waste Action Plan on moving towards the CE for WM and resource recovery. Since 2019, the Western Australian government (2019, 2020) has created a Waste Avoidance and Resource Recovery Strategy Action Plan 2030, intending to move towards "a sustainable, low-waste, circular economy". In 2020, the [46] made a CE policy and action plan to fundamentally overhaul Victoria's current recycling system. However, for all of the above initiatives, the world has become around 9% circular [33], this seems to indicate that even with much work ahead, significant opportunities exist to transition to CE.

Another driver for adopting CE in C&D WM is the potential economic benefits derived from resource efficiency, waste reduction and recycling [15]. By applying CE, construction firms can cut down on costs associated with waste disposal and landfilling, as well as reduce their dependency on virgin materials [15]. For example, recycling C&D waste into new construction materials can provide a cost-effective alternative to traditional materials [41]. Additionally, the sale of recovered materials can generate revenue, further contributing to the economic viability of CE-based WM practices [2]. Moreover, the adoption of CE in C&D WM can lead to the creation of new business opportunities, such as waste brokerage, recycling and remanufacturing, thereby fostering innovation and entrepreneurship in the construction sector [40]. Companies that adopt CE may gain a competitive advantage by demonstrating their commitment to sustainability, which can enhance their reputation and brand value [34].

The third driver for applying CE in C&D WM is environmental sustainability. Construction and demolition WM practices based on CE contribute to reducing the environmental impacts of the CI by conserving natural resources, reducing waste generation and lowering the carbon footprint of the built environment [21]. By designing for disassembly, reuse and recycling, construction firms can minimise waste generation and extend the life cycle of materials, which can significantly reduce the overall environmental impact of CPs [43]. Furthermore, recycling C&D waste can help reduce the demand for virgin materials and decrease energy consumption associated with the extraction, production and transportation of these materials [41]. This, in turn, can contribute to mitigating the environmental impacts of resource depletion, habitat destruction and climate change [21].

Social responsibility is another important driver for the adoption of CE in C&D WM. By engaging in sustainable WM practices, construction firms can enhance local economic development, create employment opportunities and improve public health [17]. For instance, the development of waste recycling facilities and infrastructure can generate new jobs and stimulate economic growth in local communities [17, 30]. Moreover, responsible WM practices can contribute to improving public health by reducing the negative impacts of improper waste disposal, such as air and water pollution and the spread of diseases [24]. By adopting CE in C&D WM, construction firms can demonstrate their commitment to social responsibility and contribute to the overall well-being of society [35].

Reputational benefits serve as another driver for the adoption of CE in C&D WM. Construction firms that successfully implement CE can enhance their brand value and attract customers and partners who prioritise sustainability [4, 34]. By showcasing their commitment to environmentally responsible practices, these firms can gain a competitive advantage in the market and build trust among stakeholders, which can translate into long-term success [1]. In addition, stakeholder proactivity plays a crucial role in promoting the adoption of CE in C&D WM. Encouraging collaboration among various stakeholders, such as designers, contractors, waste managers and policymakers, can facilitate the sharing of knowledge, effective practices and innovation [13, 51]. In 2020, Planet Ark, a non-profit environmental organisation, established the Australian Circular Economy Hub to serve as an all-inclusive platform in Australia for businesses, governments and communities to enhance understanding, foster collaboration and accelerate the implementation of CE within the country (Planet [32]). By working together, stakeholders can identify opportunities for improving C&D WM practices and develop new strategies and technologies to support the transition towards CE in the CI [13, 51].

3 Methods

Addressing the critical issue of waste generation and resource consumption in the CI, this chapter used Australia as a case study. It adopted a qualitative research approach to identify the drivers for implementing CE-based WM practices in Australian CPs. The chapter collected data through preliminary interviews (PIs) and case studies (CSs).

In the first phase, PIs were used to explore industry professionals' insights on transitioning from traditional linear models to CE-based construction practices. Participants who had actively engaged in relevant sectors and had current knowledge of CE and C&D waste were involved in PIs. As Australia is relatively new to the concept of CE, these interviews were pivotal for firstly providing an in-depth understanding of the drivers of adopting CE in Australian CPs. The PIs were chosen for their ability to facilitate comprehensive discussions while allowing flexibility to explore emergent themes [7]. Twenty participants, including CE experts and professionals from the CI, were interviewed, with the sample size determined to ensure data saturation within the recommended range (i.e. 5–25 interviews) for qualitative studies [37]. Participants represented various stakeholder categories, including consultant-related, client-related, government-related and contractor-related stakeholders. From Participant 1 to 20, the range of experience among the participants in the CI varied significantly, from 3 to 49 years, showcasing a broad spectrum of insights and expertise. Similarly, their experience in C&D WM and CE extended from 2 to 40 years, ensuring a comprehensive array of perspectives on the topics of interest. This strategic composition of the sample facilitated a thorough examination of the study themes, rooted in deep professional experience and expertise.

In the second phase, four CSs were conducted to observe CE principles in action within real CPs. Projects that have effectively implemented CE strategies in C&D WM were involved. To ensure the validity of the case study findings, the data collection was conducted through a combination of semi-structured interviews and reviews of project documentation. These data sources help validate the information's accuracy and comprehensiveness. These CSs provided practical insights into the application of CE-based WM practices, allowing for exploring complex phenomena in their natural settings [49]. Four distinct CSs (A, B, C and D) from the Australian CI were selected, each highlighting different aspects of CE implementation at the project level. Case study A was a six-storey mixed-use commercial building executed from 2020 to 2022. It employed a Design and Construct (D&C) method with Early Contractor Involvement, aiming for high sustainability ratings, including a 6-Star Green Star and NABERS Energy Rating. Case study B was a mixed-use commercial project spanning from 2016 to 2022. Using D&C, it sought a 6-Star Green Star, a 6.5-Star NABERS Energy and a 4-Star Water Rating. Case study C was a commercial residential project conducted from 2020 to 2022, targeting a 5-Star Green Star Rating using D&C. Case study D was an eight-storey mixed-use commercial building completed between 2018 and 2020, aiming for a 5-Star Green Star Rating through D&C.

A content analysis technique was used to analyse PIs and CSs. This method was selected for its efficacy in systematically interpreting textual data and extracting meaningful patterns and themes, particularly relevant to CE principles and WM practices in construction [22]. Using a deductive approach, the analysis was structured to align with the context of CE in construction. NVivo software [5] facilitated an organised and detailed analysis, including cross-case comparisons to identify commonalities and differences, enriching the study's overall findings [22, 38]. The process of developing the coding scheme involved a rigorous review of the transcripts from both preliminary interviews and case studies. This preliminary examination was critical in ensuring that the coding scheme was not only comprehensive but also aligned with the nuanced complexities and multifaceted nature of CE and WM in CPs. Following this, an iterative process of coding, reviewing and refining was undertaken. This iterative approach facilitated the gradual refinement of the coding scheme, ensuring its capability to encapsulate rich insights and nuances from the collected data. The use of NVivo software in this process was invaluable, providing a systematic platform for coding data, developing categories and analysing coded data segments. This systematic approach to coding and analysis underscored the study's commitment to rigour and depth, enabling a nuanced exploration of CE and WM practices within CPs. Furthermore, the analysis transcended mere data organisation to include a dynamic interpretation of the findings. Through the adept use of NVivo's "Coding Matrices Queries", the study delved into the frequencies and patterns emerging from the data, thereby unveiling critical insights into the application and impact of CE in C&D WM. This not only facilitated a deeper understanding of the prevailing practices but also highlighted areas for potential improvement and innovation.

4 Analysis

The following two sections highlight the chapter findings from the PIs and CSs.

4.1 Drivers Identified from Preliminary Interviews

According to preliminary interview findings, drivers to apply CE to manage C&D waste can be outlined as shown in Table 2.

4.1.1 Regulatory Compliance

It was found that governmental policies are critical in managing the performance of key project stakeholders in implementing CE-based C&D WM. Thus, the Australian government needs to develop the related legislation for the CI to involve CE-based C&D WM. Then, the related building construction codes of practice or standards can be published to guide and lead stakeholders in appropriately implementing CE-based C&D WM in the workplace (Participants 2 and 15). Waste management regulations and policies need to be issued and unified as mandatory by the federal government at the regional and local levels. For example, separating appropriate waste materials is compulsory, and new levies for C&D waste are required (Participants 3, 11 and 17). In this way, all key stakeholders can be aware of their responsibilities and the effects of actions taken in implementing CE-based C&D WM (Participant 3).

To make WM policies visible, creating a national roadmap or route map is necessary to prepare for developing mandatory standards for implementing CE-based

Table 2 Drivers to apply CE to manage C&D waste from preliminary interviews

Category	Theme	Subtheme
Regulatory compliance	Government-related policy making (12/20)	A roadmap for developing mandatory standards for implementing CE-based strategies in the CI
		Developing legislation for construction involving CE-based C&D WM
		Standardising CE-based C&D WM to exhibit the practice direction
		Classifying and defining C&D waste materials need to be unified
		Developing regulations and policies to compulsorily apply CE-based C&D WM
		Policies on reducing newly built buildings and reusing building components/materials/existing buildings
		Developing policies necessitating detailed materials passports for all building materials or components in building projects
		Recertifying builders according to new and updated policies based on CE
		Involving CE implementation in the building permits
	Non-government criteria (8/20)	A roadmap for developing mandatory standards for implementing CE-based strategies in the CI
		Developing legislation for CE-based C&D WM
		Standardising CE-based C&D WM to exhibit the practice direction
		Classifying and defining C&D waste materials need to be unified
Economic benefits	Financial support (6/20)	Incentivising organisations/companies related to the CI better from a fiscal/tax perspective if they involve CE in their WM
		Providing concessional loans for CE buildings
		Providing relevant insurance for CE-based building construction, building products and building materials/components
	Support with rewards (3/20)	Rewarding project stakeholders
		Creating value for all stakeholders based on the CE application
	Infrastructure support (3/20)	Developing minimal right infrastructures to make CE-based C&D WM possible

Notes: Numbers at the end of each statement indicate the number of participants who agreed on that particular point strategies in the CI (Participants 8 and 9). It was highlighted that the classification and definition of C&D waste materials need to be unified by a comprehensive regulatory apparatus about the definition of "waste" and "second-hand raw materials" (Participant 17). It is also crucial to "… only use what we need …" (Participant 15) when transitioning towards CE-based C&D WM. To support WM policy, governments need to develop self-sufficiency policies on reducing newly built buildings and policies on reusing building components/materials and existing buildings (Participant 5). Also, a need to track and know materials flows in the built environment sector calls for policies that require detailed materials passports for all building materials or components in building projects (Participant 15). Last but not least, governments need to involve CE implementation in building permits (Participant 13) and recertify builders according to new and updated policies based on CE (Participants 1 and 9).

Non-government criteria are also vital for implementing CE-based C&D WM. CE-based C&D WM plans need to be mandatory for all CPs (Participants 8 and 9). Also, it needs to identify principles of CE application in the CE-based C&D WM plans to get all stakeholders involved (Participant 20). In the architectural specifications, clarifying and describing CE-related guidelines in detail (e.g. materials to be used and installation methods) is key to communicating those to builders effectively to implement them in the construction process (Participant 4). These strict guidelines from the start make CE implementation viable and provide clear steps for working with subcontractors (Participants 10 and 12). It is especially vital to have a specification of how to reuse and recycle those materials (Participant 12). Meanwhile, nothing can be done without acknowledging the risk and eliminating the risk in CI (Participant 6). Construction companies need a formal risk analysis and risk management plans in terms of involving CE in the C&D WM for their projects (Participant 6). Also, it is necessary to consider the whole life cycle of buildings in rating criteria when implementing CE-based C&D WM in CPs (Participants 5 and 9).

4.1.2 Economic Benefits

In terms of financial support, governments need to incentivise organisations/companies related to the CI better from a fiscal/tax perspective if they involve CE in their WM, for example, subsidies and more environmental, social and governance investments (Participants 2, 8 and 11). For example, Participant 9 mentioned that banks need to provide concessional loans for CE buildings, such as funding those who purchase CE buildings with CE certifications or credits and providing a loan with a reduced interest rate (Participant 9). Also, insurance companies need to provide appropriate insurance for CE-based building construction, building products and building materials/components to avoid possible risks in their implementation (Participant 5). Meanwhile, it is also essential to reward all project stakeholders with appropriate practices for CE-based C&D WM (Participant 10). Similarly, a value proposition needs to be made for each stakeholder at the very end, with an alignment of values to implement CE principles. Furthermore, for infrastructure support, governments need to develop at least a minimal number of suitable infrastructures to make CE-based C&D WM possible.

4.2 *Drivers Identified from Case Studies*

Figure 1 presents the drivers for implementing CE-based WM strategies to minimise C&D waste generation in selected CSs. These drivers are categorised under five broad themes: time, cost and quality considerations; economic benefits; reputational benefits; building rating tools and certifications; and environmental sustainability.

4.2.1 Economic Benefits

In the first subtheme, "financial benefits of diverting waste to recycling industries", an essential factor is the role of cost-savings in WM. For example, a project manager from Case Study A expressed a growing comprehension of the economic benefits of not discarding materials. Similarly, a contractor administrator and a senior design manager from Case Study B underscored the significant costs associated with construction waste disposal. These views are echoed by a waste management contractor from Case Study A, who emphasised that minimising landfill usage was in their best interest due to high landfill fees. Another aspect of this subtheme is getting credits back for recycling items. A construction project manager from Case

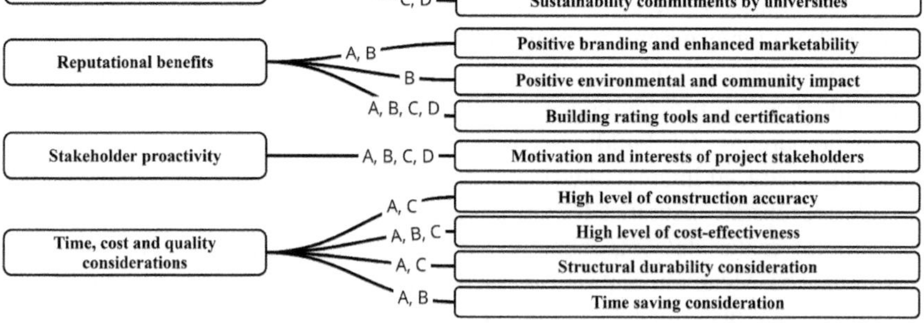

Fig. 1 Drivers to apply CE-related strategies to minimise C&D waste from case studies. *Note* A = Case study A; B = Case study B; C = Case study C; D = Case study D

Study D revealed an innovative scheme wherein credits are received for recycling items, which can reduce construction or labour costs. This system provides a direct economic incentive to recycle, highlighting its potential to reduce waste effectively. Additionally, getting subsidies and incentives from the government offers intriguing insights. A waste management contractor from Case Study A mentioned how some recyclers pay them, creating a financial benefit. They also underlined the influence of government subsidies, describing them as the primary driver for the construction industry to engage in recycling. This underscores the power of policy mechanisms to driving CE practices. Furthermore, construction companies can earn discounts for meeting recycling targets through proper waste management; this reasserts the economic advantage of these practices.

The second subtheme, "financial support of using recycled materials to promote CE-based C&D WM", emphasises the potential influence of government policy in promoting sustainable practices. A plastering contractor in Case Study A confirmed that they would utilise CE-related materials if financial support was available from the government.

4.2.2 Environmental Sustainability

Regarding subthemes "focusing on low carbon buildings" and "sustainable built environment consideration", stakeholders showcased a clear trend towards greener and low carbon construction methods. For example, the construction project manager outlined their commitment to a CE by eliminating all C&D waste to minimise their carbon footprint, a sentiment echoed by the project procurement director. Project Design Director C discussed the carbon sequestration properties of timber, demonstrating alignment with sustainability targets. The notion of a sustainable built environment is integral to these commitments. The Project Director of Sustainability, Environments and Logistics D mentioned using recycled materials and furniture reuse, indicating a consideration for the environment at all stages of the construction process.

4.2.3 Social Responsibility

In the subthemes "sustainability commitments by universities" and "sustainable considerations on well-being and living", universities play a pivotal role in promoting sustainability. The Project Director of Sustainability, Environments and Logistics D discussed the university's sustainability commitments and their implementation into the construction project, highlighting their influence in steering broader industry trends. Sustainable considerations for well-being and living conditions are also of great importance. Architect A and Client Representative A cited the sustainable operation of the building and the effective recovery of resources as key strategic goals. Developer Representative A highlighted the psychological benefits of sustainable design, noting how timber enhances workplace well-being by reducing stress and fostering a connection to nature.

4.2.4 Reputational Benefits

In the subtheme "positive branding and enhanced marketability", various stakeholders acknowledge the importance of meeting sustainability standards and showcasing innovative construction methods. For example, Sustainability Consultant A stated that the local government adopted CLT primarily to appear sustainable and meet GS design requirements, positively impacting their reputation. Developer Representative B echoed this sentiment, highlighting that their involvement in sustainable projects can enhance their market reputation as progressive and sustainable, potentially attracting increased investments and tenant satisfaction. Moreover, Senior Design Manager B noted that promoting the retention of 60% of the existing structure can serve as a valuable marketing tool for leasing the building.

The subtheme "positive environmental and community impact" underscores the importance of considering both the environmental and social aspects of CPs. End-user B elaborated on the benefits of retaining a building's core, creating a unique and elegant structure that the city can be proud of. They also acknowledged significant community environmental savings and emphasised the importance of considering broader environmental and community impacts.

When discussing "building rating tools and certifications", GS, National Australian Built Environment Rating System (NABERS), and Passive House certifications emerge as crucial methodologies. Green Star Design and the As-Built rating tool, employed by the Green Building Council of Australia, are consistently referred to throughout the excerpts. Client Representative A described the aim for a six-star GS rating as an "aspiration", emphasising the need to meet external standards. This sentiment is echoed by Client Representative C, who highlights their commitment to a five-star GS community rating. Similarly, Construction Project Manager A (builder side) acknowledges the mandatory nature of GS requirements for WM in their project. Several other stakeholders, including Construction Project Manager C (client side), Contract Administrator B, Developer Representative B, Quality Assurance Engineer B, Quantity Surveyor C, Senior Development Manager B and Site Engineer A, recognise GS as a key driver in their project's design and construction approach.

Furthermore, Client Representative A emphasised the importance of achieving a NABERS six-star waste rating for the project's operation phase. Contract Administrator B agreed with this, emphasising the need for the building to be highly efficient for end-users and noting that CE factors are considered throughout the project's life cycle. This theme was reiterated by Construction Project Manager A (client side) and Contract Administrator B, who both mention NABERS as part of their project's sustainability requirements.

While less frequently mentioned, Passive House certification is depicted as an aspirational target. Construction Project Manager C (client side) and Project Design

Director C nevertheless mentioned this certification as a desirable but ultimately unachieved sustainability certification due to budget constraints.

4.2.5 Stakeholder Proactivity

In terms of the "motivation of project stakeholders", the analysis of stakeholder commitment to sustainability unveils various motivations across various roles involved in the construction project. There is a distinct shift in builders' attitudes towards sustainable practices and waste minimisation. For instance, Construction Project Manager A (client side) noted a collective push from all parties with a growing awareness of the responsibility and the need to recycle construction waste as much as possible. This sentiment is corroborated by Project Design Director C, who highlighted the necessity of builder interest in sustainable practices, despite the unconventional nature of these processes. Waste Management Contractor A also indicated that the builder requires them to recycle as much as possible to meet their obligations. Clients and developers play a crucial role in driving sustainable practices. Client Representative C discussed their influence over design decisions, while Construction Project Manager D (client side) mentioned the client's policy of achieving carbon neutrality by 2025. Developer Representative B demonstrated a commitment to sustainability through detailed sustainability targets and the drive towards the CE. The motivation and interest of individuals working at the ground level, such as site engineers and project managers, are also essential. Site Engineer A expressed personal interest in GS and sustainable buildings, while Construction Project Manager A (client side) showed interest in understanding and improving knowledge of CE. End-users and waste management contractors also display a strong commitment to sustainability. End-user B reflects on the broader impact of sustainable methods on the community, and Waste Management Contractor A emphasises the financial and environmental benefits of avoiding landfills. Cooperation among stakeholders is another critical factor. Construction Project Manager C and Site Engineer A highlighted the collaborative effort required to achieve sustainability outcomes, underscoring the shared commitment of all parties.

4.2.6 Time, Cost and Quality Considerations

In the subtheme "high level of construction accuracy", Architect A emphasises the importance of precision, stating that CE can improve construction accuracy, minimising the need for additional work and waste.

The subtheme "high level of cost-effectiveness" highlights the significance of balancing cost and environmental concerns in construction. While many participants preferred sustainability, they also acknowledged that commercial value often takes precedence. For instance, Plastering Contractor A noted that although environmental considerations are important, cost remains the primary concern in business decisions. Contract Administrator B similarly pointed out that cost is crucial when managing construction waste. These viewpoints indicate that cost considerations are a significant driver for the adoption of CE in CPs.

"Structural durability consideration" is another essential subtheme, emphasising the importance of long-lasting, high-quality buildings. Project Director C, representing the client, explains their decision to use cross-laminated timber (CLT) and to expose services without adding extra finishes. This approach increased the building's durability and demonstrated the potential benefits of CE principles for structural longevity.

Lastly, the subtheme "timesaving consideration" underlines the importance of reducing construction time through CE implementation. Architect A stated that off-site prefabrication can significantly reduce on-site construction time. Similarly, Developer Representative B highlights that retaining the concrete core in a case study project saved nearly two years of construction time. These examples illustrate that CE can significantly decrease construction time, benefiting both the environment and project efficiency.

5 Discussion

This section illustrates drivers in implementing CE-based C&D WM across literature review (LR), preliminary interview analysis (PIA) and case study analysis (CSA), including regulatory compliance, economic benefits, environmental sustainability, social responsibility, reputational benefits, stakeholder proactivity, time, cost, and quality considerations, and building rating tools and certifications.

5.1 Regulatory Compliance

The LR has highlighted the importance of regulatory compliance as a critical driver for applying CE in C&D WM. Furthermore, the PIA revealed the two main types of policy-related supports for implementing CE-based C&D WM, that is, government-related policy making and non-government criteria. This section discusses the roles of these two types of policy-related supports in implementing CE in C&D WM.

The LR demonstrated that the CE concept has been applied through government policies at local, regional and national levels. Several countries have integrated the CE concept into their national strategic policy agendas or included some elements of the concept in their policies [3]. For example, China directly applies the term "circular economy" in its waste and resource policy and treats the CE as a national political goal [14, 52]. The PIA also revealed that governmental policies are significant in managing the performance of key project stakeholders in implementing CE-based C&D WM. The Australian government needs to develop legislation for the CI to involve CE-based C&D WM and issue-related building construction codes of practice or standards to guide stakeholders (PIA). This is in line with the literature's emphasis on the role of governments in facilitating the transition to a CE. However, the literature also pointed out that applying the components of the CE does not necessarily mean CE actions are being taken, as they are just the initial components of a greater CE [23]. This suggests that governments should focus on developing legislation and policies and monitoring and enforcing compliance with these regulations to ensure that CE is effectively implemented in C&D WM.

The PIA identified the importance of non-government criteria in implementing CE-based C&D WM. For instance, PIA noted that CE-based C&D WM plans should be mandatory for all CPs, and the principles of CE application in these plans should be clearly identified to ensure stakeholder buy-in. Moreover, PIA emphasised the need to clarify and describe CE-based guidelines in architectural specifications to ensure builders and subcontractors understand and follow these requirements. This aligns with the literature's argument that non-governmental organisations, such as the Green Building Council of Australia, should involve rating criteria regarding CE-based C&D WM in the whole life cycle of buildings in green building rating systems (PIA).

5.2 *Economic Benefits*

The economic benefits of adopting CE in C&D WM have been widely acknowledged in the literature. The PIA also highlights the importance of economic benefits in promoting the successful implementation of CE-based C&D WM. Similarly, the CSA demonstrates the significance of economic considerations in the decision-making process for CPs and the factors influencing stakeholders' decisions.

A strong consensus is found in the literature, PIA and CSA on the cost-saving potential of adopting CE in C&D WM. For instance, cost savings can be achieved by reducing waste disposal and landfilling costs and decreasing dependency on virgin materials. This is consistent with the CSA, where one project emphasised the importance of minimising waste, while another highlighted the significance of time savings in reducing costs. The PIA also suggests that governments need to incentivise organisations and companies related to the CI better from a fiscal and tax perspective if they involve CE in their WM. Additionally, recycling C&D waste into new construction materials can provide a cost-effective alternative to traditional materials [41]. This is supported by the CSA, where the financial benefits of diverting waste to recycling industries were discussed, such as receiving credits for recycling items and subsidies for using recycled materials.

Moreover, the sale of recovered materials can generate revenue, further contributing to the economic viability of CE-based WM practices [2]. This aligns with the PIA, where participants suggest rewarding project stakeholders and creating value for all stakeholders based on CE application, which is essential. The CSA also demonstrates that the government's financial support for using recycled materials can influence stakeholders' decisions, as mentioned by Plastering Contractor A. Furthermore, the adoption of CE in C&D WM can create new business opportunities, such as waste brokerage, recycling and remanufacturing, thereby fostering innovation and entrepreneurship in the construction sector [40]. Companies that adopt CE principles may gain a competitive advantage by demonstrating their commitment to sustainability, which can enhance their reputation and brand value [34].

5.3 Environmental Sustainability

In the context of environmental sustainability, adopting CE in C&D WM can significantly contribute to reducing the environmental impacts of the CI. Both the LR and CSA highlight the importance of conserving natural resources, reducing waste generation and lowering the carbon footprint of the built environment. Designing for disassembly, reuse and recycling can minimise waste generation and extend the life cycle of materials, which can considerably reduce the overall environmental impact of CPs [21]. The CSA demonstrates how focusing on low-carbon buildings aligns with these objectives, as it emphasises achieving carbon neutrality through eliminating C&D waste and minimising the carbon footprint.

Recycling C&D waste can help reduce the demand for virgin materials and decrease energy consumption associated with the extraction, production and transportation of these materials [41]. This can contribute to mitigating the environmental impacts of resource depletion, habitat destruction and climate change [21]. The CSA reveals that sustainable built environment considerations, including using recycled materials and furniture reuse, were already being considered in CPs even before the term "circular economy" gained widespread recognition.

5.4 Social Responsibility

In the discussion of social responsibility, this chapter has demonstrated the significant role that the adoption of CE plays in C&D WM. The LR highlighted the benefits of sustainable WM practices in enhancing local economic development, creating employment opportunities and improving public health [17]. In line with this, the CSA revealed the subtheme of "well-being and living consideration", emphasising the importance of sustainable construction in promoting better living conditions and well-being for occupants.

The findings of the LR and CSA collectively underline the connection between sustainability thinking and the desire to create spaces that promote well-being and quality of life. This connection is further exemplified by Design and Construction Project Manager A's description of the strategic importance of using timber in an office workplace for its biophilic design credentials, which can improve occupants' emotional state and reduce stress. This reinforces the notion that construction firms can demonstrate their commitment to social responsibility and contribute to the overall well-being of society by adopting CE in C&D WM [35].

A critical comparison of the LR and CSA reveals that both data sources support the argument that social responsibility is a vital driver for CE in C&D WM. The LR provided evidence from previous studies highlighting the potential benefits of responsible WM practices, such as developing waste recycling facilities and infrastructure [17, 30]. Meanwhile, the CSA illustrated the practical implementation of sustainability thinking in the built environment, focusing on enhancing occupants' well-being and living conditions.

However, it is essential to acknowledge that the CSA provides a more in-depth exploration of the connection between sustainability thinking and social responsibility. By examining specific examples from industry professionals, the CSA offers a more nuanced understanding of the practical implications of adopting CE in C&D WM.

5.5 Reputational Benefits

The LR highlighted the importance of reputational benefits as a driver for the adoption of CE in C&D WM. Similarly, the CSA emphasised the significance of reputational benefits, particularly in the subthemes "positive branding and enhanced marketability" and "positive environmental and community impact".

Comparing the literature and CSA, it is evident that construction firms that successfully implement CE principles can enhance their brand value and attract customers and partners who prioritise sustainability [4, 34]. This aligns with the subtheme "positive branding and enhanced marketability" from the CSA, where stakeholders acknowledged the importance of meeting sustainability standards and showcasing innovative construction methods. For example, the CSA notes that adopting CLT helped the local government appear sustainable and meet GS design requirements, positively impacting their reputation. Also, the CSA highlights that involvement in sustainable projects could enhance their reputation in the market as progressive and sustainable, potentially leading to increased investments and tenant satisfaction.

The subtheme "positive environmental and community impact" from the CSA complements the findings in the LR by emphasising the significance of considering both environmental and social aspects of CPs [15]. For instance, the CSA shows the benefits of retaining the core of a building, stating that it resulted in something unique and elegant for the city to be proud of, as well as significant savings in terms of the community environment.

In the subtheme "sustainability commitments (special for universities)", the importance of embedding sustainability commitments into all aspects of a project is emphasised, which further supports the relevance of reputational benefits as a driver for adopting the CE. The CSA mentions that their primary driver for input into the project was to embed the university's commitments to sustainability in all aspects. In the subtheme "building rating tools and certifications", GS, National Australian Built Environment Rating System (NABERS), and Passive House certifications emerge as crucial methodologies. This supports the argument that reputational benefits can be derived from meeting and surpassing sustainability standards [34].

5.6 *Stakeholder Proactivity*

This chapter has shown that stakeholder proactivity is pivotal in adopting CE in C&D WM. The LR and CSA provided comprehensive insights into various stakeholders' motivations, interests and willingness to adopt sustainable practices and undertake CE engagement.

The Australian Circular Economy Hub, established by Planet Ark, demonstrates a proactive initiative encouraging stakeholder collaboration to promote CE (Planet [32]). This finding aligns with the CSA, highlighting the importance of stakeholder proactivity in fostering a sustainable CI that benefits both the environment and the communities it serves. The case study subthemes reveal the proactive attitudes of builders, site engineers, project managers, end-users and WM contractors towards embracing CE and sustainable practices. The LR [13] and case study findings concur that collaboration among stakeholders is essential in identifying opportunities to improve WM practices and develop new strategies and technologies that support the transition towards CE in the CI. The CSA showcases the willingness of various stakeholders to cooperate in adopting sustainable practices, thereby contributing to the effective implementation of environmentally conscious projects. Moreover, the CSA demonstrates that clients, developers and investors significantly influence the direction of CPs and the prioritisation of sustainability. This resonates with the LR assertion that encouraging collaboration among stakeholders, such as designers, contractors and waste management policymakers, can facilitate knowledge sharing, effective practices and innovation [13, 51].

5.7 *Time, Cost and Quality Considerations*

The theme "time, cost and quality considerations", only identified in case studies, encompasses various subthemes, highlighting the importance of balancing quality, cost and efficiency in implementing CE-related strategies.

It was revealed that CE can improve construction accuracy, minimising the need for additional work and waste. At the same time, cost considerations are a major driver for the adoption of CE in CPs. CE also resulted in a building with increased durability, showcasing the potential benefits of CE principles in terms of structural longevity. CE can significantly decrease construction time, benefiting the environment and project efficiency.

6 Conclusions

This chapter critically examines the implementation of CE principles in C&D WM within Australian CPs. It identifies key drivers essential for integrating CE-based practices, which can significantly reduce the industry's environmental impact.

Regulatory compliance emerges as a pivotal driver, highlighting the necessity of both government and non-government policies in steering the CI towards sustainable practices. Legislation, standardisation and incorporating CE principles into building permits are crucial elements in this transition. Economic benefits also play a fundamental role in advocating for CE adoption. Cost savings, revenue generation and new business opportunities provide substantial incentives, fostering a more sustainable approach in the industry. This economic perspective aligns with the growing global emphasis on environmental sustainability, which is another key driver identified. The chapter underscores the industry's increasing awareness of the need to conserve natural resources, reduce waste generation and minimise carbon footprints, advancing towards sustainable built environments. Social responsibility is highlighted as a significant motivator, with sustainable construction practices enhancing local economic development, public health and community well-being. The reputational benefits associated with sustainable practices are also noted, where positive branding and the pursuit of sustainability-related accreditations enhance a firm's marketability and commitment to environmental and social responsibilities. Stakeholder proactivity is identified as a crucial facilitator for the effective implementation of CE in C&D WM. The proactive involvement of diverse stakeholders, from builders to waste contractors, reflects a shared commitment to sustainability and innovation. The time, cost and quality considerations encompass the importance of balancing quality, cost and efficiency in implementing CE-related strategies when implementing CE-based WM practices in CPs.

This chapter contributes valuable insights for industry professionals and policymakers, outlining a multifaceted approach to promoting CE-based WM practices. It underscores the need for a collaborative effort to transition from traditional models to a more sustainable, circular approach that benefits both environmentally, economically and socially. In essence, the chapter advocates for a paradigm shift in the CI towards sustainability, highlighting the interconnected nature of policy, economic incentives, environmental conservation and social responsibility. The chapter also contributes to advancing knowledge in the field of facilitating CE by providing insights into the drivers of implementation. The findings of this chapter, therefore, will be valuable for future research and practice in these fields. Qualitative methods have been predominantly adopted in this chapter. In the future, one could adopt quantitative or mixed methods to provide more statistically robust insights and triangulate the findings.

Acknowledgements There is no funding support for this chapter.

References

1. Abuzeinab A, Arif M, Qadri MA, Kulonda D (2019) Green business models in the construction sector: an analysis of outcomes and benefits. Constr Innov 18(1):20–42
2. Allam Z, Jones DS (2018) Towards a circular economy: a case study of waste conversion into housing units in Cotonou, Berlin. Urban Science 2(4):118
3. Antikainen R, Lazarevic D, Seppala J (2018) Circular economy: origins and future orientations. In: Lehmann H (ed) Factor X. Springer, pp 115–129
4. Awan U, Sroufe R (2022) Sustainability in the circular economy: insights and dynamics of designing circular business models. Appl Sci 12(3):1521
5. Bazeley P, Jackson K (2013) Qualitative data analysis with NVivo (2nd ed.). Sage
6. Commonwealth of Australia (2018) National waste policy 2018. Department of the Environment and Energy. https://www.environment.gov.au/system/files/resources/d523f4e9-d958-466b-9fd1-3b7d6283f006/files/national-waste-policy-2018.pdf
7. Creswell JW (2013) Qualitative inquiry and research design: choosing among five approaches (3rd ed.). Sage Publications
8. de Jesus A, Mendonça S (2018) Lost in transition? drivers and barriers in the eco-innovation road to the circular economy. Ecol Econ 145:75–89
9. Ellen MacArthur Foundation (2012) Towards the circular economy: economic and business rationale for an accelerated transition. https://www.ellenmacarthurfoundation.org/assets/downloads/publications/Ellen-MacArthur-Foundation-Towards-the-Circular-Economy-vol.1.pdf
10. Ellen MacArthur Foundation (2015) Delivering the circular economy: a toolkit for policymakers. https://www.ellenmacarthurfoundation.org/assets/downloads/publications/EllenMacArthurFoundation_PolicymakerToolkit.pdf
11. Federal Ministry for the Environment, Nature Conservation, Building and Nuclear Safety (2012) German resource efficiency programme. https://www.bmu.de/en/topics/economy-products-resources-tourism/resource-efficiency/overview-of-german-resource-efficiency-programme-progress/
12. Finnish Government (2016) Leading the cycle—Finnish road map to a circular economy 2016–2025. The Finnish Innovation Fund, Sitra
13. Gálvez-Martos J-L, Styles D, Schoenberger H, Zeschmar-Lahl B (2018) Construction and demolition waste best management practice in Europe. Resour Conserv Recycl 136:166–178
14. Ghisellini P, Cialani C, Ulgiati S (2016) A review on circular economy: the expected transition to a balanced interplay of environmental and economic systems. J Clean Prod 114:11–32
15. Ghisellini P, Ripa M, Ulgiati S (2018) Exploring environmental and economic costs and benefits of a circular economy approach to the construction and demolition sector. A literature review. J Clean Prod 178:618–643
16. Giorgi S, Lavagna M, Campioli A (2018) Guidelines for effective and sustainable recycling of construction and demolition waste. In: Benetto E, Gericke K, Guiton M (eds) Designing sustainable technologies, products and policies: from science to innovation. Springer International Publishing, pp 211–221
17. Górecki J, Núñez-Cacho P, Corpas-Iglesias FA, Molina V (2019) How to convince players in construction market? Strategies for effective implementation of circular economy in construction sector. Cogent Eng 6(1):1690760
18. Liu J, Wu P, Jiang Y, Wang X (2021) Explore potential barriers of applying circular economy in construction and demolition waste recycling. J Clean Prod 326:129400
19. López Ruiz LA, Roca Ramón X, Gassó Domingo S (2020) The circular economy in the construction and demolition waste sector—a review and an integrative model approach. J Cleaner Prod 248
20. Netherlands Government (2016) A circular economy in the Netherlands by 2050. https://www.government.nl/topics/circular-economy/documents/reports/2016/09/14/a-circular-economy-in-the-netherlands-by-2050
21. Hossain MU, Ng ST (2019) Influence of waste materials on buildings' life cycle environmental impacts: adopting resource recovery principle. Resour Conserv Recycl 142:10–23

22. Krippendorff K (2013) Content analysis: An introduction to its methodology (3rd ed.). Sage Publications
23. Levitzke PSMV (2020) The development of a circular economy in Australia. In: Circular economy: Global perspective (pp. 25–42). Springer
24. Lu W, Yuan H (2011) A framework for understanding waste management studies in construction. Waste Manage 31(6):1252–1260
25. National People's Congress (2008) Circular economy promotion law of the People's Republic of China. http://www.chinaenvironmentallaw.com/wp-content/uploads/2008/09/circular-economy-law-cn-en-final.pdf
26. New South Wales Government (2019) NSW circular economy policy statement—Too good to waste. https://www.epa.nsw.gov.au/-/media/epa/corporate-site/resources/recycling/19p1379-circular-economy-policy-final.pdf?la=en&hash=F80151EA9C2C3E27BA889D15D18041CDF7A4D25A
27. Organisation for Economic Co-operation and Development. (2015). Working party on resource productivity and waste: a policy study on the sustainable use of construction materials. http://www.oecd.org/officialdocuments/publicdisplaydocumentpdf/?cote=ENV/EPOC/WPRPW(2014)4/FINAL&docLanguage=En
28. Osobajo OA, Oke A, Omotayo T, Obi LI (2022) A systematic review of circular economy research in the construction industry. Smart Sustain Built Environ 11(1):39–64
29. Ossio F, Salinas C, Hernández H (2023) Circular economy in the built environment: a systematic literature review and definition of the circular construction concept. J Cleaner Prod 414
30. Papastamoulis V, London K, Feng Y, Zhang P, Crocker R, Patias P (2021) Conceptualising the circular economy potential of construction and demolition waste: an integrative literature review. Recycling 6(3):61
31. Pickin Wardle C, O'Farrell K, Stovell L, Nyunt P, Guazzo S, Lin Y, Caggiati-Shortell G, Chakma P, Edwards C, Lindley B, Latimer G, Downes J, Axiö I (2022) National waste report 2022. https://www.dcceew.gov.au/sites/default/files/documents/national-waste-report-2022.pdf
32. Planet Ark. (2023). Australian Circular Economy Hub. https://planetark.org/programs/australian-circular-economy-hub
33. Platform for Accelerating the Circular Economy (2019) The circularity gap report 2019. https://docs.wixstatic.com/ugd/ad6e59_ba1e4d16c64f44fa94fbd8708eae8e34.pdf
34. Prieto-Sandoval V, Jaca C, Ormazabal M (2018) Towards a consensus on the circular economy. Cogent Soc Sci 179:605–615
35. Purchase CK, Al Zulayq DM, O'Brien BT, Kowalewski MJ, Berenjian A, Tarighaleslami AH, Seifan M (2021) Circular economy of construction and demolition waste: a literature review on lessons, challenges, and benefits. Materials (Basel), 15(1)
36. Queensland Government. (2019). Waste management and resource recovery strategy. https://www.qld.gov.au/__data/assets/pdf_file/0028/103798/qld-waste-management-resource-recovery-strategy.pdf
37. Saunders MNK (2012) Choosing research participants. In: Qualitative organizational research: core methods and current challenges. SAGE Publications
38. Saldana J (2013) The coding manual for qualitative researchers (2nd ed.). Sage
39. Scottish Government (2016) Making things last: a circular economy strategy for Scotland. Scottish Government
40. Shooshtarian S, Caldera S, Maqsood T, Ryley T (2020) Using recycled construction and demolition waste products: a review of stakeholders' perceptions, decisions, and motivations. Recycling 5(4):31
41. Silva RV, de Brito J, Dhir RK (2017) Availability and processing of recycled aggregates within the construction and demolition supply chain: a review. J Clean Prod 143:598–614
42. Tasmanian Government (2019) Draft waste action plan. https://dpipwe.tas.gov.au/Documents/Draft%20Waste%20Action%20Plan.pdf
43. Udawatta N, Zuo J, Chiveralls K, Zillante G (2021) From green buildings to living buildings? Rating schemes and waste management practices in Australian educational buildings. Eng Constr Archit Manag 28(4):1278–1294

44. United Nations (UN) (2015) Transforming our world: the 2030 agenda for sustainable development. https://sustainabledevelopment.un.org/content/documents/21252030%20Agenda%20for%20Sustainable%20Development%20web.pdf
45. United Nations Environment Programme. (2020). Sustainable trade in resources: global material flows, circularity and trade. https://wedocs.unep.org/bitstream/handle/20.500.11822/34344/STR.pdf?sequence=1&isAllowed=y
46. Victorian State Government. (2020). Recycling Victoria: a new economy. The State of Victoria Department of Environment, Land, Water and Planning. https://www.vic.gov.au/sites/default/files/2020-02/Recycling%20Victoria%20A%20new%20economy.pdf
47. Western Australia Government. (2019). Waste avoidance and resource recovery strategy action plan 2020. https://www.wasteauthority.wa.gov.au/images/resources/files/Strategic_Direction_Waste_Avoidance_and_Resource_Recovery_Strategy_2030_Action_Plan.pdf
48. Western Australia Government. (2020). Action plan 2020–2021—Waste avoidance and resource recovery strategy 2030. https://www.wasteauthority.wa.gov.au/images/resources/files/2020-21_Action_Plan_03072020.pdf
49. Yin R (2009) Case study research: design and methods. Sage Publications
50. Zhang C, Hu M, Di Maio F, Sprecher B, Yang X, Tukker A (2022) An overview of the waste hierarchy framework for analyzing the circularity in construction and demolition waste management in Europe. Sci Total Environ 803
51. Zhao X (2021) Stakeholder-associated factors influencing construction and demolition waste management: a systematic review. Buildings 11(4):149
52. Zhu J, Fan C, Shi H, Shi L (2019) Efforts for a circular economy in China: A comprehensive review of policies. J Ind Ecol 23(1):110–118

Chapter 8
Effects of Building Defects on Construction and Demolition Waste Generation: Case Studies of Australian Residential Building Projects

Argaw Gurmu, Gayani Karunasena, C. Savindi Ranthika Perera, Nilupa Udawatta, and Salman Shooshtarian

Abstract The construction projects generate a significant amount of construction and demolition (C&D) waste, and part of this is due to the building defects encountered during and after the construction phase. To date, there has been no research conducted to quantitatively evaluate the impact of building defects on C&D waste generation in Australia. Thus, this research aims to quantify the waste generated by building defects to gain insights into its contribution to C&D waste generation in residential building projects in Australia. An in-depth evaluation of three case studies was conducted to analyse and quantify the waste arising from various building defects in residential building projects in Australia. A derivative of the site visit method was adopted to quantify the waste generated due to building defects in the selected case studies. This method requires investigators to visit the C&D sites for a realistic survey. The waste generated in the case studies was due to non-compliant bathroom wall tiles (CS1), incorrect roof-end capping (CS2) and flaking ceiling paint (CS3). The findings of this study revealed that non-compliant bathroom tiles have incurred a waste of 80.77 kg/m^2 of wall area and a cost of \$ 521.19/m^2. The defect associated with the roof capping has incurred a waste of 3 kg/m of roof capping and a cost of \$ 85.35/m, while the flaking of the ceiling has incurred a waste of 0.30 kg/m^2 of ceiling area and a cost of \$ 40.48/m^2. The findings underscore the impact of building

A. Gurmu (✉) · G. Karunasena · C. Savindi Ranthika Perera · N. Udawatta
School of Architecture and Built Environment, Deakin University, Geelong, Australia
e-mail: argaw.gurmu@deakin.edu.au

G. Karunasena
e-mail: gayani.karunasena@deakin.edu.au

N. Udawatta
e-mail: nilupa.udawatta@deakin.edu.au

S. Shooshtarian
School of Property, Construction and Project Management, RMIT University, Melbourne, Australia
e-mail: salman.shooshtarian@rmit.edu.au

T. Ramachandra et al. (eds.), *Concepts, Policies and Technology for Sustainable Buildings*, Advances in Sustainability Science and Technology,
https://doi.org/10.1007/978-981-96-3086-8_8

defects on waste generation, revealing critical insights into the contribution of waste generated by defects to total waste generation and the impact of such defects on the project cost. This research, thus, contributes valuable insights for builders, developers, materials suppliers and manufacturers, policymakers, and other stakeholders to adopt proactive measures aimed at minimising defects and waste generation and fostering sustainable construction practices. Furthermore, by quantitatively evaluating the impact of building defects on C&D waste generation, this research sheds light on the relationship between construction quality issues and waste production.

Keywords Construction waste · Building defects · Residential buildings · Waste quantification

1 Introduction

The construction industry consumes an extensive amount of natural resources and generates significant construction and demolition (C&D) waste [3, 56]. The study conducted by Zheng et al. [61] revealed that the construction industry produces about 35% of the total waste sent to landfills worldwide. Construction and demolition waste generation adversely affect both project costs and timelines and result in negative environmental impacts [45]. For instance, material waste brings about an extra 15% of construction project costs in the United Kingdom (UK), 20–30% in the Netherlands and 11% in Hong Kong [46]. Likewise, Saidu and Shakantu [50] found that an average of 4% of material waste contributes to project-cost overrun in Nigeria. As a result, C&D waste issues have been gaining significant attention from professionals and researchers worldwide [6, 38].

Previous scholars have identified many causes of waste generation in the construction industry [8, 9, 27, 28, 36]. While Aravindh et al. [4] broadly identified these as time and cost-related causes, Igwe et al. [27] categorised the causes as planning, construction method and communication-related, rework and poor quality-related, procurement-related, design and documentation-related and external factors. However, when studying construction waste, several researchers have focused primarily on material waste and identified it as 'construction materials that are lost in transit on or off-site, discarded without adding value to the project for which it was procured, including overproduction or leftover from a newly constructed facility [1]'. For instance, Arifullah et al. [5], Musarat et al. [44], Fitchett and Rambuwani [17], Mahamid [39], and Sweis et al. [56] in their studies specifically focused on causes of material waste. The identified causes included design and contract documents-related, project staff-related, material handling and storage-related, site management and supervision-related, and external factors and conditions-related causes.

Among the causes of waste, building defects in the construction industry have been a prominent topic for over two decades [23, 30]. A building defect may involve any issue that reduces the value of a building [2]. For instance, in the UK, a survey report showed that 69% of newly built houses were found to have more than five defects in 2018 [26]. Additionally, research conducted in Russia indicated that about 80% of construction projects had critical defects [7]. Previous studies conducted in Australia identified several defects in residential buildings [19, 21, 22, 24]. However, none of the studies have analysed the relationship between defects and C&D waste. Further, earlier studies did not attempt to quantify the volume of waste attributable to defects. Hence, to fulfil the aforementioned gap, this research is designed to enhance the understanding of the association between defects and C&D in the residential building sector. Additionally, the locations of defects in residential buildings were explored by previous studies. Accordingly, the most frequent defects in residential buildings in Australia are related to (i) ceramic tiles, (ii) paint and (iii) roofing [21]. Thus, this study focused on the analysis of waste related to defects in the three elements.

Previous research estimated that the remedy of building defects could take up between 4 and 7% of the contract value [31, 42, 48]. A study conducted by Johnston and Reid [30] estimated that at least one defect was detected in around 70% of buildings built between 2000 and 2018 and Australians have spent 10.5 billion dollars to manage building defects. A fraction of this figure is spent on managing defect waste. While previous research has separately addressed issues related to waste and defects in construction, none have made an effort to explore the interconnectedness of these two aspects in their investigations. Furthermore, previous research on the cost of rectification due to defects was based on the construction contract value, not on the actual measurement of wastes generated on construction sites. Also, little attention has been given to identifying the contribution of different types of defects to waste generation. Understanding the cost implications associated with each type of defect within construction projects provides invaluable insights for project planning and budgeting. For instance, by understanding the financial consequences of various defects, the stakeholders can prioritise resources towards preventing or rectifying the most costly or impactful defects. This proactive approach can help minimise overall project costs and optimise resource utilisation. Furthermore, understanding the potential cost implications of specific defects enables informed decision-making. Thus, this study aimed to quantify the waste generated by building defects in residential projects in Australia to gain insights into its contribution to C&D waste.

2 Literature Review

2.1 Building Defects and Their Cost Implications

Building defects represent a significant challenge in construction projects, encompassing a wide array of issues that deviate from the intended design, functionality or structural integrity of a building [18, 52, 58]. These defects can arise at any stage of the construction process, from design and planning to construction, occupancy and maintenance phases. The occurrence of building defects poses multifaceted implications, impacting the quality, safety, durability and overall performance of structures [7, 16, 32, 33].

Building defects fall into two main categories: structural and non-structural [62]. Major or structural defects hold significant implications as they compromise the safety, habitability or intended functionality of a building. Instances of structural defects within elements like earth retaining walls, columns, beams or flat slabs caused by flawed design, subpar materials or errors in construction are deemed major and can render a building unsafe or unfit for its intended purposes [2]. On the other hand, minor defects result from substandard workmanship or inadequate materials used during building erection [62]. These defects typically affect non-structural elements such as partition walls, doors or windows, often associated with issues like faulty workmanship, dampness in older structures or deficiencies in plasterwork [25]. The appropriate response to a building defect hinges on its severity and the potential impact it might exert on the building's safety, functionality or visual appeal.

Building defects in construction projects not only jeopardise buildings' structural integrity and safety but also incur substantial financial repercussions [30, 41]. The costs of rectifying defects extend beyond mere repairs, encompassing a wide spectrum of direct and indirect expenses. For instance, studying the defect costs incurred by a museum, a school, a university, an industrial building, a housing, a fire station and a shopping centre, Josephson and Hammarlund [31] found that the defect costs vary between 2.3 and 9.4% of the project cost. The results of this study further revealed that, on average, 32% of the defect costs were traced back to the early phases, specifically involving the client and the design aspects. Approximately 45% of the total defect costs were identified as stemming from on-site activities relating to site management, workers and subcontractors. It was also demonstrated that approximately 20% of the defect cost originated in materials or machines. Analysing data from a Victorian Government insurance organisation in Australia, Mills et al. [42] found on average, defect rectification accounts for about 4% of the contract price. In their research, Taggart et al. [57], using an action research approach, focused on the defects discovered in construction projects and found that contractors do not measure costs associated with defects, but they typically allow 1–1.25% for snagging in their tenders. Therefore, previous research has established that understanding the financial implications of building defects is crucial for stakeholders in the construction industry to assess the impacts of defects and implement effective cost management strategies. Previous research has overlooked a crucial aspect: accurately quantifying the specific costs linked to waste resulting from defects. Earlier studies have primarily focused on estimating the overall cost of defects without differentiating between those causing demolition and waste generation versus those that do not. This oversight has limited the understanding of the nuanced economic impacts of different defect types within construction projects. It is essential to recognise that certain defects can lead to immediate consequences like demolition or visible impacts, while others may not. However, earlier studies have not distinguished between the costs associated with defects that result in demolition and those that do not, which obscures a thorough understanding of the economic and operational impacts of defects within construction projects. Moreover, by analysing the proportion of defective work that leads to waste, we can gain deeper insights into how defects contribute to overall waste generation. This analysis is critical for developing targeted strategies to mitigate waste in construction processes and enhance overall efficiency and sustainability in the industry.

2.2 Construction and Demolition Waste Quantification and Its Adaptation for Defects

Accurately quantifying the waste can be challenging due to the complexity and variability of factors involved. Jain et al. [29] adopted a bottom-up material flow analysis approach to quantify the C&D waste generation rate in Indian cities. Lau et al. [34] used the layouts of the construction waste such as stockpiled (assumed to stay in the form of rectangular base pyramidal), gathered (assumed to stay in the form of rectangular prism) and scattered (the number of samples scattered around the site are counted, and multiplied by the average weight to get the total weight of the scattered waste) to quantify the construction waste generated in terms of weight. Based on an analytical review of fifty-seven papers on construction and demolition waste quantification, Wu et al. [59] identified six categories of C&D waste quantification methodologies, including site visit method, waste generation rate method, lifetime analysis method, classification system accumulation method, variables modelling method and other particular methods, as shown in Table 1.

3 Methodology

This research aimed to quantify the waste generated by building defects to gain insights into its contribution to the overall C&D waste. The methods identified in the literature (Table 1) focus on quantifying the waste of an entire construction project or for a larger area and thus were not appropriate to quantify the waste due to a defect. Yet, the basis of the waste quantification was calculating the mass of material waste [59], and thus, the principles and the concepts behind the indicated methods are adhered to in the study. Therefore, in this research, as the first step, materials wasted due to defects were identified. Then, the total mass per m^2 or per linear m was calculated using either the product's density or actual weight. This technique is named as a derivative of the site visit method in the context of this study.

A case study methodology was adopted to achieve the aim of this research. Yin [60, p. 13] defines the case study method as an '*empirical inquiry that investigates a contemporary phenomenon within its real-life context.*' On the other hand, case studies have the ability to answer 'why' questions as well as 'how' questions, thus

Table 1 Construction and demolition waste quantification methods

No	Methodology		Method
1	Site visit method	Direct measurement	Requires weighing the waste produced or measuring its volume on site. For stockpiled waste, a rectangular-based pyramid was assumed; for gathered waste, the layout shape was assumed cuboid. Three samples are randomly chosen from the total for scattered and stacked waste of similar size. The average weight of the selected samples was calculated
		Indirect measurement	The number of trucks for waste collection and the volume of each container are recorded. Based on this information, the total waste volume at a project level is derived
2	Generation rate calculation (GRC) method	Per-capita multiplier	The average C&D waste generation per person is calculated, and future C&D waste generation amounts are extrapolated using waste weight per person multiplied by the population
		Financial value extrapolation	The material waste on sites was weighed, and the total area of the buildings was collected to determine the generation rate (kg/m^2). Through reviewing the building permit data, the financial value per square metre is obtained (\$/m^2). Building financial value could be then established using wasted material per financial value (kg/\$)

(continued)

Table 1 (continued)

No	Methodology		Method
		Area-based calculation	Generation rate values can be derived from site visits, project consultancy, construction guides and literature reviews. Once the total construction or demolition area is identified, the total waste generation amount can be estimated by multiplying the generation rate by the total area
3	Lifetime analysis (LA) method	Building lifetime analysis	GRC could derive the quantity of demolition waste after knowing the generated volume per unit of building area and the material mass per unit of volume. Appropriate assumptions of building lifetime are required when conducting this method
		Material lifetime analysis	The consumption of construction materials and typical waste factors used for construction materials purchasing were used to estimate the mass of solid waste generated as a result of construction activities. The lifetime of the material is considered
4	Classification system accumulation (CSA) method		The waste amount is calculated by applying SV and GRC. Once the quantities of all items have been determined, the total amount of waste could be derived by accumulating all those items
5	Variables modelling (VM) method		Assumes that the total quantity of construction waste generated at a particular time on a construction site is the accumulation of waste quantities from each construction activity

Source Adapted from Wu et al. [59]

often used in both explanatory and exploratory research [51]. Case studies facilitate the collection of specific, measurable information pertinent to the study objectives and are thus often adopted in the construction industry to collect quantitative data [35, 37]. Case studies offer a comprehensive methodology to collect, analyse and interpret quantitative data, facilitating the extraction of meaningful insights within the study context. Accordingly, case studies were chosen as the most appropriate research approach to quantify the waste resulting from defects. Accordingly, this research selected three residential construction projects in Australia with defect rectification as case studies.

With the case study approach, multiple data collection techniques can be used individually or in combination. Saunders et al. [51] identify these data collection techniques as interviews, observation, questionnaires and documentary analysis. In this research, video and documentary analyses were conducted to obtain data within case studies. The following sections detail the case studies in this research and the defects considered. To maintain confidentiality and anonymity, the project details, such as the exact location of the building and project stakeholders, are not disclosed in this research.

3.1 Case Studies

3.1.1 Case Study 1

Case Study 1 (CS1) is a construction site with nine residential units in Victoria, Australia. During the lock-up inspection, it was found that the bathroom wall tiles were fixed on top of the chipboard, and the floor tiles were also non-compliant with the Australian Standards. It was also found that tilers have placed a substantial blob of stiff adhesive on each tile corner and in the centre (spot fixing). Australian Standard AS 3958.1–2007 states that 'spot-fixing where the tiles are fixed with four or five dabs of adhesive is not recommended and should be avoided at all times.' It further stipulates that the adhesive should be applied to the substrate and then spread using a towel, and the minimum contact coverage for wet areas must be 90%. Figure 1 demonstrates the bathroom tiling condition before a destructive test, and Fig. 2 shows the test result.

As a result of this defect, it was decided to demolish the bathroom tiles. All the wall tiles, plasterboard walls, waterproofing and floor screed were removed, generating a waste of materials. Figure 3 illustrates the waste generation from the defect.

3.1.2 Case Study 2

Case study 2 (CS2) is a two-storey house in Victoria, Australia. During the lock-up inspection, it was found that there was an incorrect fall on the roof-end capping, preventing water from going into the box gutter (Fig. 4) and the drainage system. The

Fig. 1 The bathroom before the destructive test. *Source* Site Inspections [54]

Fig. 2 Spot fixed tiles after the destructive test. *Source* Site Inspections [54]

Fig. 3 Waste generated from wall tile defects. *Source* Site Inspections [54]

Fig. 4 Incorrect fall of the capping. *Source* Site Inspections [55]

Fig. 5 Edges of the capping are cut-no allowance for the downturn. *Source* Site Inspections [55]

inspection also found no allowances for the downturn, as shown in Fig. 5. Therefore, it was advised to replace the capping with a new one, generating material waste (Fig. 6).

3.1.3 Case Study 3

Case study 3 (CS3) is a two-storey house in Victoria, Australia. During an inspection, it was identified that there were stains on the ceiling, which progressed to flaking, as demonstrated in Fig. 7. As a result, it was required to scrape it off, prepare the surface and repaint it.

Figure 8 illustrates the waste generation when the existing painting was peeled off due to the defect identified.

After identifying the defects and the waste resulting from defects, the process of quantifying the construction waste due to defects, in terms of weight and the cost impact, is presented in the following section.

Fig. 6 Waste generated from capping defect. *Source* Site Inspections [55]

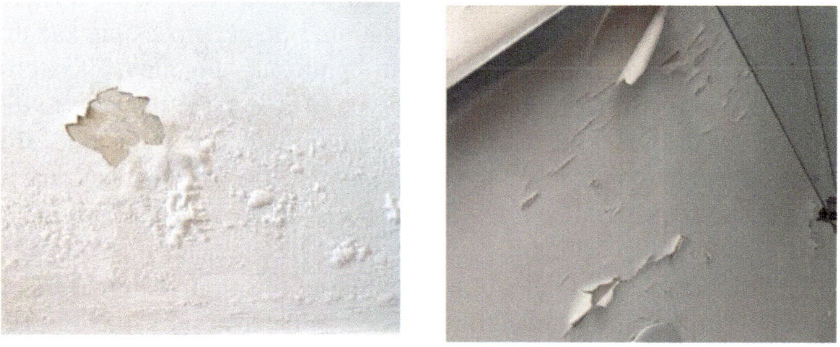

Fig. 7 Stains on the ceiling and flaking *Source Captured by authors*

Fig. 8 Waste generated from the painting defect. *Source* Captured by authors

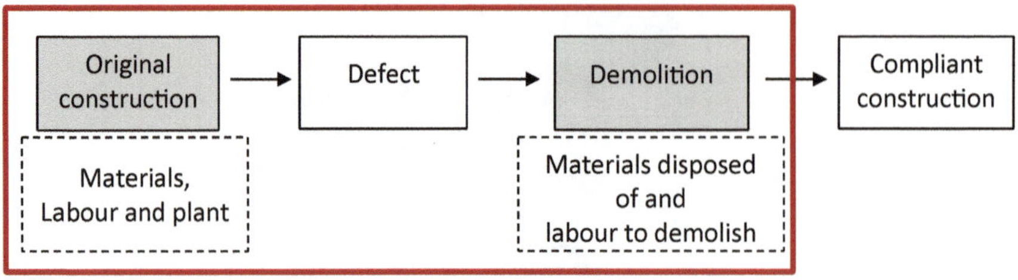

Fig. 9 Process of the cost of waste calculations

In addition to the mass of the waste, which is determined using a derivative of the site visit method, the cost of waste was calculated. For this research, the cost of waste is identified as the summation of the cost of wasted works, including material, labour and plant and the costs incurred to demolish the defective works (Fig. 9).

Therefore, as shown in Fig. 9, only the steps denoted by the thick red line were considered the waste cost. The costs associated with the handling of waste, which encompass tasks such as collection, transportation, storage, processing and disposal of waste, are excluded from the estimate due to significant variability. These costs can fluctuate based on regional regulations, the type of waste, the distance to disposal sites and the disposal methods used, among other factors. For example, stringent environmental regulations necessitate more expensive disposal methods, and remote sites incur higher transportation costs. Inconsistent data across various case studies can introduce substantial variability and potential inaccuracies in cost estimation, thereby undermining the reliability of the findings of this study. Similarly, estimates of productivity losses are excluded due to the complexity and variability of accurately measuring these losses. Productivity losses can vary greatly depending on factors such as the nature of defects and the operational efficiency of the workforce. For instance, frequent defects in critical components can cause substantial delays and increased labour costs, while minor issues might have minimal impact. The variability and context-dependency of productivity losses make them difficult to generalise across different projects. Moreover, reliable and consistent data on productivity loss and waste handling costs were not uniformly available, introducing further variability and potential inaccuracies in the analysis. By excluding these factors, this study maintains its integrity and avoids drawing misleading conclusions based on inconsistent data.

4 Data Analysis and Findings

This section synthesises and articulates the data analysis and the findings obtained from the study. For each case study/defect, the waste was quantified in terms of the weight of the waste generated due to the defect, and then the cost of the defect was calculated. The three cases analysed in this research included waste generated due to non-compliant bathroom wall tiles (CS1), incorrect roof-end capping (CS2) and flaking ceiling paint (CS3).

The waste quantification process included on-site assessments and waste audits to gather empirical data regarding the types of waste produced. The composition of the waste was analysed to identify its nature. Through sorting and categorising, a detailed understanding of their composition was obtained. To quantify, it was then required to weigh the waste materials. For this step, the original weights and densities of the materials were determined from the specifications provided by the material suppliers. Cost details were also obtained from the suppliers and Rawlinsons Construction Cost Guide (2023) to comprehend a cost analysis.

4.1 Findings of Case Study 1 (CS1)

4.1.1 Bathroom Wastes and Associated Costs

In CS1, non-compliant bathroom tiles were recognised, and thus, bathroom tiles were demolished. All the wall tiles, plasterboard walls, waterproofing and floor screed were removed, generating a waste of materials. Table 2 presents the identified material waste and the mass per m^2 for each disposed material (refer to Appendix A for the data sources).

Table 3 demonstrates the cost of defective bathroom tiles, labour costs and the costs incurred to demolish the defective bathroom tiles. The unit cost information was obtained from the Rawlinsons Cost Guide (2023).

Based on these findings, the defect identified in CS1 has resulted in a waste that weighs 80.77 kg/m^2 and has resulted in an additional cost of \$ 521.19/m^2 (\$ 459.29/m^2+\$ 61.90/m^2) to instal and demolish the defective bathroom tiling.

4.2 Findings of Case Study 2 (CS2)

4.2.1 Roofing Waste and Associated Costs

In CS2, because of the incorrect fall on the roof-end capping and no allowance for the downturn, the capping was replaced with a new one, generating material waste. Table 4 presents the identified material waste and the mass per m2 for each disposed material (refer to Appendix B for the data sources).

Table 5 demonstrates the cost of defective roof capping, labour costs and the costs incurred to demolish the defective roof capping. The unit cost information was obtained from the Rawlinsons Cost Guide (2023).

When CS2 is considered, the defect identified in CS2 has resulted in a waste that weighs 3 kg/m and has resulted in an additional cost of \$ 85.35/m (\$ 79.40/m + \$ 5.95/m) to instal and demolish the defective roofing caps.

Table 2 Calculation of the mass of the waste quantities for bathroom tiling

Material	Weight calculations	Mass per m^2
Wall tiles (200 mm × 200 mm plain tiles) *Source Bunnings Warehouse* [10]	Weight of 35 tiles = 16 kg Weight of one tile = 0.46 kg/tile Area of one tile = 0.04 m^2 Weight per m^2 = 0.46 kg/0.04 m^2	11.5 kg/m^2
Particle board (18 mm thick water resistant to timber wall framing) *Source Mister* [43]	Weight of one board = 37 kg Area of one plasterboard = 2.88 m^2 Weight per m^2 = 37 kg/2.88 m^2	12.84 kg/m^2
Floor tiles (200 mm × 200 mm plain tiles *Source Bunnings Warehouse* [11]	Weight of 35 tiles = 19 kg Weight of one tile = 0.54 kg/tile Area of one tile = 0.04 m^2 Weight per m^2 = 0.54 kg/0.04 m^2	13.5 kg/m^2
Waterproofing (polyethene film) *Source* [20]	Density = 930 kg/m^3 Thickness = 1mm Weight per m^2 = 930 kg/m^3 × 0.001m	0.93 kg/m^2
10 mm screed for floor tiles *Source Easy* [15]	Weight per m^2 (for 10mm thickness) = 20 kg/m^2	20 kg/m^2
Tile grout *Source Sika* [53]	Density = 2200 kg/m^3 Thickness = 10 mm Weight per m^2 = 2200 kg/m^3 × 0.01m	22 kg/m^2/m^2
Total mass per m^2		80.77 kg/m^2

4.3 Findings of Case Study 3 (CS3)

4.3.1 Ceiling Wastes and Associated Costs

In CS3, because of incorrect painting techniques, the ceiling painting was scraped off and repainted, generating material waste. Table 6 presents the identified material waste and the mass per m2 for each disposed material (refer to Appendix C for the data sources).

Table 7 demonstrates the cost of defective painting and labour costs and the costs incurred to demolish the defective work. The unit cost information was obtained from the Rawlinsons Cost Guide (2023).

Accordingly, the defect identified in CS3 has resulted in a waste that weighs 0.30 kg/m^2 and an additional cost of \$ 40.48/m^2 (\$ 24.59/m^2 + \$ 15.90/m^2) to instal and demolish the defective paint.

Table 3 Cost of defective bathroom tiling per m^2

The cost incurred to instal		
	Inclusions	Cost/m^2
Material cost		
Wall tiles (200 mm × 200 mm plain ceramic tiles)	Includes fixing with adhesive and pointing	\$124.00/m^2
Floor tiles (200 mm × 200 mm plain ceramic tiles)	Including bedding and pointing	\$125.00/m^2
Moisture-resistant plain particle board (18 mm thick)	Including fixing to timber framing	\$ 46.10/m^2
Waterproofing (polyethene film)	Including all laps and jointing	\$3 0.15/m^2
Total cost of defective bathroom tiling per m^2—(A)		*\$ 298.25/m^2*
Labour cost		
Particle board 10 mm thick fixed to timber frame in walls	0.24 h/m^2 × \$66/hr	\$ 15.84/m^2
Wall tile installation	1.10 h/m^2 × \$66/hr	\$72.60/m^2
Floor tile installation	1.10 h/m^2 × \$66/hr	\$72.60/m^2
Total labour cost to instal defective bathroom tiling per m^2—(B)		*\$ 161.04/m^2*
Total cost to instal defective works per m^2—(A + B)		*\$ 459.29/m^2*
The cost incurred to demolish		
		Cost/m^2
Demolition of ceramic floor tiles		\$ 21.80/m^2
Demolition of ceramic wall tiles		\$ 29.90/m^2
To strip the wall lining with particle board		\$ 10.20/m^2
Total costs to demolish defective bathroom tiling		*\$ 61.90/m^2*

Table 4 Calculation of the mass of the waste quantities for roofing capping

Material	Weight calculations	Mass per m
Barge capping of the steel roof *Source Bunnings Warehouse (2023c)*	Weight for 2.4 m = 1.8 kg Weight per m = 3.6kg/2.4 m	1.5 kg/m
Nylon anchor @ 500 mm c/c *Source Metal* [40]	Weight of 20 anchors = 10 kg Weight of one anchor = 10 kg/20 anchors = 0.5kg per anchor Anchors per m= (1/0.5) +1 = 3 anchors Weight per m= 3 × 0.5 kg per anchor	1.5 kg/m
Total mass per m		*3 kg/m*

Table 5 Cost of defective roofing capping per m

The cost incurred to instal			
		Inclusions	Cost/m
Material cost			
Barge capping of the steel roof		Includes fixing	$ 49.70/m
Total cost of defective roofing capping per m—(A)			*$ 49.70/m*
Labour cost			
Installation of the original capping	0.45 h/m × $66/hr		$ 29.70/m
Total labour cost to instal defective roofing capping per m—(B)			$ 29.70/m
Total cost to instal defective roofing capping per m—(A + B)			*$ 79.40/m*
The cost incurred to demolish			
			Cost/m
Demolition of the barge capping			$ 5.95/m
Total costs to demolish defective roofing capping			*$ 5.95/m*

Table 6 Calculation of the mass of the waste quantities for ceiling paint

Material	Weight calculations	Mass per m2
Ceiling painting—prep coat *Source Dulux (2023a)*	Weight for 1L = 1.42 kg Area covered by 1L = 14 m^2 Weight per m^2 = 1.42 kg /14 m^2	0.11 kg/m^2
Ceiling painting—top 2 coats *Source Dulux (2023b)*	Weight for 1L = 1.432 kg Area covered by 1L = 16 m^2 Weight per m^2 = 2 × 1.432 kg/16 m^2	0.19 kg/m^2
Total mass per m		*0.30 kg/m^2*

Table 7 Cost of defective paint per m^2

The cost incurred to instal		
	Inclusions	Cost/m^2
Material cost		
Painting (prime, one sealer undercoat and two topcoats)	Includes for preparation	$ 16.00/m^2
Total cost of defective painting per m^2—(A)		*$ 16.00/m^2*
Labour cost		
Painting	0.13 h/m × $66/hr	$ 8.58/m^2
Total labour cost to instal defective painting per m^2—(B)		*$ 8.58/m^2*
Total cost to instal defective works per m^2—(A + B)		*$ 24.58/m^2*
The cost incurred to demolish		
		Cost/m^2
Preparing existing plasterboard to receive new treatment		$ 15.90/m^2
Total costs to demolish defective works		*$ 15.90/m^2*

5 Discussion

This research focused on quantifying the waste arising from various building defects. This study holds paramount significance due to its pioneering exploration into quantifying waste generated by specific building defects within the construction industry. While previous research quantifies waste as a percentage of the total contract value [31, 42, 57], this research focused on analysing the most accurate waste volumes and associated costs attributed to distinct defects. Comparisons to benchmarks, policies or regulations are pivotal to demonstrating the significance of defects in waste generation within the C&D context. However, due to the limited number of studies that have specifically analysed the link between defects and C&D waste, there are currently no established benchmarks available for such comparisons. This absence of benchmarks highlights a critical gap in the existing literature, underscoring the need for further research in this area to establish comprehensive standards for evaluating the impact of defects on C&D waste generation. Addressing this gap would not only enhance the understanding of how defects contribute to waste but also support the development of targeted policies and practices aimed at reducing waste in construction projects.

While general mechanisms for waste quantification in construction and demolition exist, specific methodologies tailored to quantify waste attributable to building defects have not been extensively explored or applied. This study addresses this gap by focusing on quantifying the contribution of building defects to C&D waste generation, an area that has not been comprehensively studied previously. Hence, this research provides valuable insights into the distinct impacts of defects on waste generation in residential building projects in Australia. Therefore, the specific focus on building defects as a quantifiable source of waste represents a novel contribution of this research.

The impact of defects within the construction industry on the economy can be substantial, albeit challenging to quantify precisely due to various influencing factors. Building defects, when widespread or recurring, can exert cascading effects on the economy. After analysing 212 building audit reports in three states of Australia (New South Wales, Queensland and Victoria), [30] found that 85% of all the buildings analysed had at least one defect. The average number of defects identified per building was fourteen. Based on the results of this study, if there are 1000 m^2 of wall and floor tiles in bathrooms that are defective, it will have a significant national economic impact (approximately AUD 521,190). In addition to the expenses for repairs, replacements and legal disputes, building defects would drain financial resources within the construction sector. Also, since this defect would generate approximately 80.77 tonnes of waste, such defects can impact ecosystems, contributing to environmental degradation and necessitating additional resources for environmental remediation. This revelation of a high incidence of defects underscores the imperative for the construction industry to reevaluate current practices, embrace more stringent quality assurance protocols, and prioritise ongoing maintenance and quality checks to ensure safer, more durable, and higher-quality built environments.

Based on the findings, this study revealed compelling insights into the quantified waste and associated costs for each specific defect. The findings underscore not only the significant volume of waste generated due to this defect but also the substantial financial implications incurred in rectifying and managing the issue. While the waste volume in each case differed, the incurred costs still reflected a tangible financial burden attributable to similar defects. These insights emphasise the imperative for robust quality control measures, adherence to standards and meticulous attention to detail throughout the construction process to mitigate waste generation and minimise the substantial economic impacts of building defects.

6 Conclusions

Adopting three case studies, this paper aimed to quantify the waste arising from various building defects. The defects selected were non-compliant bathroom wall tiles (CS1), incorrect roof-end capping (CS2) and flaking ceiling paint (CS3). Construction and demolition waste generated from the identified defects in CS1, CS2 and CS3 approximately amounted to 80.77 kg/m^2, 3 kg/m and 0.30 kg/m^2, respectively. For the selected defects, the main components of construction waste were wood, concrete, insulation, tiling metal and painting. Other wastes, such as waste generated by the packaging of materials, were not included in the calculations. The defects' rectification costs were \$ 521.19/m^2, \$ 85.35/m and \$ 40.48/m^2 in CS1, CS2 and CS3. By shedding light on these previously unexplored aspects, this research fills a critical gap in existing literature, offering valuable insights into the direct correlation between building defects and their consequential waste production. Hence, this research contributes to the knowledge in the areas of building defects and construction waste management by providing a piece of empirical evidence regarding the association between defects and waste. Furthermore, this study has practical implications for policymakers, regulatory bodies and other stakeholders in the construction industry by providing concrete evidence regarding the financial and environmental impact of building defects. Based on this research's findings, corrective or preventive actions in reducing defects and wastes can be taken by the respective stakeholders. Additionally, the research contributes to theoretical understanding of construction quality management by highlighting the role of defects in waste generation. Further, the study's findings can inform theoretical discussions on effective strategies to reduce C&D waste in residential construction. By quantifying the impact of building defects on waste generation, theoretical frameworks can be developed to guide interventions that target defect prevention and waste minimization.

While the study provides valuable insights into quantifying waste resulting from specific building defects, it is important to acknowledge certain limitations that impact its broader applicability. Firstly, the data collection was confined to three case studies, potentially limiting the generalizability of the findings beyond these specific contexts. Additionally, the analysis does not account for costs associated with productivity loss and waste handling, which could affect the accuracy of the cost estimates. These identified limitations, however, do not diminish the significance of the research findings but rather highlight areas for future investigation and refinement. To enhance the robustness and applicability of future studies, researchers could expand the scope of inquiry to encompass a more diverse range of cases, thereby facilitating the generalisation of findings. Moreover, incorporating factors such as productivity loss and waste handling costs into subsequent analyses would contribute to a more comprehensive understanding of the economic implications of building defects. These avenues for further exploration offer promising directions for advancing research in this area and improving the effectiveness of waste management strategies in construction and maintenance practices.

Appendix A

Material	Source
Wall tiles (200 mm × 200 mm plain tiles)	https://www.bunnings.com.au/johnson-200-x-200mm-antique-white-ripple-ceramic-wall-tile-35-pack_p6661108
Particle board (18 mm thick water resistant to timber wall framing)	http://www.misterplywood.com.au/products/particleboard
Floor tiles (200 mm × 200 mm plain tiles)	https://www.bunnings.com.au/johnson-tiles-200-x-200mm-elements-coffee-matt-ceramic-floor-tile-35-pack_p6661775
Waterproofing (polyethene film)	https://www.globalplasticsheeting.com/our-blog-resource-library/bid/72325/density-of-polyethylene
10 mm screed for floor tiles	https://www.easymixsales.com.au/wp-content/uploads/2019/08/Easy-Mix-Traditional-Floor-Screed-TDS.pdf
Tile grout	https://aus.sika.com/dam/dms/au01/4/sikagrout_-212_hp.pdf

Appendix B

Material	Source
Barge capping of the steel roof	https://www.bunnings.com.au/insulspan-2-4m-white-flashing-barge-capping_p1142518

(continued)

(continued)

Material	Source
Nylon anchor @ 500 mm c/c	https://www.metalroofingonline.com.au/products/colorbond-parapet-capping-150mm/ https://www.bunnings.com.au/ramset-6-5-x-50mm-nylon-anchor-20-pack_p2260190

Appendix C

Material	Source
Ceiling painting—prep coat	https://www.dulux.com.au/paint/1step-prep/1step-prep-water-based-primer-sealer-and-undercoat/
Ceiling painting—top 2 coats	https://www.dulux.com.au/paint/ceiling/ceiling-white-plus-kitchen-and-bathroom/

References

1. Adewuyi TO, Odesola IA (2015) Factors affecting material waste on construction sites in Nigeria. J Eng Technol 6(1):82–99. Retrieved from https://jet.utem.edu.my/jet/article/view/237
2. Ahzahar N, Karim NA, Hassan SH, Eman J (2011) A study of contribution factors to building failures and defects in construction industry. Proc Eng 20:249–255. https://doi.org/10.1016/j.proeng.2011.11.162
3. Ajayi SO, Oyedele LO, Bilal M, Akinade OO, Alaka HA, Owolabi HA, Recycling (2015) Waste effectiveness of the construction industry: understanding the impediments and requisites for improvements. 102:101–112
4. Aravindh MD, Nakkeeran G, Krishnaraj L, Arivusudar N (2022) Evaluation and optimization of lean waste in construction industry. Asian J Civil Eng 23(5):741–752. https://doi.org/10.1007/s42107-022-00453-9
5. Arifullah SM, Jagarapu DCK, Jeelani SH (2023) Redoing of work on waste material in building construction projects. In: Poluraju P, Jeelani SH, Kumar JDC, Reddy MAK, Raja KH (Eds.) American Institute of Physics Conference, pp 1–8. AIP Publishing. https://doi.org/10.1063/5.0144436
6. Aslam MS, Huang B, Cui L (2020) Review of construction and demolition waste management in China and USA. J Environ Manage 264, 110445
7. Baiburin AK (2017) Errors, defects and safety control at construction stage. Procedia Eng 206:807–813. https://doi.org/10.1016/j.proeng.2017.10.555
8. Bajjou MS, Chafi A (2020) Identifying and managing critical waste factors for lean construction projects. Eng Manag J 32(1):2–13. https://doi.org/10.1080/10429247.2019.1656479
9. Bajjou MS, Chafi A (2018) Assessing the critical sources of wastes in the Moroccan construction industry: an empirical study. In: Proceedings of 3rd International Conference on Smart City Applications, 1–5. https://doi.org/10.1145/3286606.3286783

10. Bunnings Warehouse (2023a) Johnson 200200mm Antique White Ripple Ceramic Wall Tile—35 Pack. Retrieved from: https://www.bunnings.com.au/johnson-200-x-200mm-antique-white-ripple-ceramic-wall-tile-35-pack_p6661108. Accessed 10 Sept 2023
11. Bunnings Warehouse (2023b) Johnson Tiles 200200mm Elements Coffee Matt Ceramic Floor Tile—35 Pack. Retrieved from: https://www.bunnings.com.au/johnson-tiles-200-x-200mm-elements-coffee-matt-ceramic-floor-tile-35-pack_p6661775. Accessed 10 Sept 2023
12. Bunnings Warehouse (2023c) Insulspan 2.4m White Flashing Barge Capping. Retrieved from: https://www.bunnings.com.au/insulspan-2-4m-white-flashing-barge-capping_p1142518. Accessed 10 Sept 2023
13. Dulux (2023a) 1Step® Prep Water Based Primer, Sealer & Undercoat. Retrieved from: https://www.dulux.com.au/paint/1step-prep/1step-prep-water-based-primer-sealer-and-undercoat/. Accessed 6 Sept 2023
14. Dulux (2023b). *Ceiling White +PLUS Kitchen&Bathroom.* Retrieved from: https://www.dulux.com.au/paint/ceiling/ceiling-white-plus-kitchen-and-bathroom/ [accessed 6 September 2023]
15. Easy Mix (2019) Technical data sheet easy mix traditional floor screed. Retrieved from: https://www.easymixsales.com.au/wp-content/uploads/2019/08/Easy-Mix-Traditional-Floor-Screed-TDS.pdf. Accessed 29 June 2023
16. Faqih F, Zayed T (2021) Defect-based building condition assessment. Build Environ 191:1–16. https://doi.org/10.1016/j.buildenv.2020.107575
17. Fitchett A, Rambuwani P (2022) Waste control in South African construction projects. Trans Royal Soc South Africa 77(2):105–112. https://doi.org/10.1080/0035919X.2022.2045383
18. Forcada N, Macarulla M, Gangolells M, Casals M (2014) Assessment of construction defects in residential buildings in Spain. Build Res Inform 42(5):629–640. https://doi.org/10.1080/09613218.2014.922266
19. Georgiou J (2010) Verification of a building defect classification system for housing. Struct Surv 28(5):370–383
20. Global Plastic Sheeting (2012) Density of Polyethylene. Retrieved from: https://www.globalplasticsheeting.com/our-blog-resource-library/bid/72325/density-of-polyethylene. Accessed 20 February 2024
21. Gurmu A, Galluzzo A, Kite J (2021) Modelling customers' perception of the quality of services provided by builders: A case of Victoria, Australia. Constr Econom Build 21(1):100–124
22. Gurmu AT, Krezel A, Mahmood MN (2023) Analysis of the causes of defects in ground floor systems of residential buildings. Int J Constr Manag 23(2):268–275
23. Gurmu AT, Shooshtarian S, Muhammad NM (2022) Critical evaluation of building defects research: a scientometric analysis. J Perform Constr Facil 36(3):1–14. https://doi.org/10.1061/(ASCE)CF.1943-5509.0001727
24. Gurmu A, Hosseini MR, Arashpour M, Lioeng W (2023a) Development of building defects dashboards and stochastic models for multi-storey buildings in Victoria, Australia. Construct Innov
25. Hinks J, Cook G (2002) The technology of building defects. Routledge, UK
26. Hopkin T, Lu S-L, Sexton M, Rogers P (2019) Learning from defects in the UK housing sector using action research. Eng Constr Archit Manag 26(8):1608–1624. https://doi.org/10.1108/ECAM-04-2018-0146
27. Igwe C, Nasiri F, Hammad A (2022) An empirical study on non-physical waste factors in the construction industry. Eng Constr Archit Manag 29(10):4088–4106. https://doi.org/10.1108/ECAM-02-2021-0145
28. Ismail H, Yusof ZM (2016) Perceptions towards non-value-adding activities during the construction process. In: Proceedings of The 4th International Building Control Conference, 1–8. https://doi.org/10.1051/matecconf/20166600015
29. Jain S, Singhal S, Jain NK (2019) Construction and demolition waste generation in cities in India: an integrated approach. Int J Sustain Eng 12(5):333–340. https://doi.org/10.1080/19397038.2019.1612967
30. Johnston N, Reid S (2019) An examination of building defects in residential multi-owned properties. https://dro.deakin.edu.au/articles/report/An_examination_of_building_defects_in_residential_multi-owned_properties/20751541

31. Josephson PE, Hammarlund Y (1999) The causes and costs of defects in construction: a study of seven building projects. Autom Constr 8(6):681–687. https://doi.org/10.1016/S0926-5805(98)00114-9
32. Kliukas R, Kudzys A (2004) Probabilistic durability prediction of existing building elements. J Civ Eng Manag 10(2):107–112. https://doi.org/10.1080/13923730.2004.9636294
33. Knyziak P (2019) The impact of construction quality on the safety of prefabricated multi-family dwellings. Eng Fail Anal 100:37–48. https://doi.org/10.1016/j.engfailanal.2019.02.042
34. Lau HH, Whyte A, Law P (2008) Composition and characteristics of construction waste generated by residential housing project. Int J Environ Res 2(3):261–268
35. Li L, Chen K (2017) Quantitative assessment of carbon dioxide emissions in construction projects: a case study in Shenzhen. J Clean Prod 141:394–408. https://doi.org/10.1016/j.jclepro.2016.09.134
36. Liao L, Teo EAL, Chang R, Li L (2020) Investigating critical non-value adding activities and their resulting wastes in BIM-based project delivery. Sustainability (Switzerland) 12(1). https://doi.org/10.3390/SU12010355
37. Liu K, Leng J (2022) Quantitative research on embodied carbon emissions in the design stage: a case study from an educational building in China. J Asian Architect Build Eng 21(4):1182–1192. https://doi.org/10.1080/13467581.2022.2046003
38. Lu W, Yuan H (2011) A framework for understanding waste management studies in construction. Waste Manage 31(6):1252–1260. https://doi.org/10.1016/j.wasman.2011.01.018
39. Mahamid I (2022) Impact of rework on material waste in building construction projects. Int J Constr Manag 22(8):1500–1507. https://doi.org/10.1080/15623599.2020.1728607
40. Metal Roofing (2018) Parapet Capping 150mm—COLORBOND® or ZINCALUME®. Retrieved from: https://www.metalroofingonline.com.au/products/colorbond-parapet-capping-150mm/. Accessed 19 April 2024
41. Milion RN, Alves, T.d.C.L., Paliari, J.C., & Liboni, L.H.B. (2021) CBA-based evaluation method of the impact of defects in residential buildings: assessing risks towards making sustainable decisions on continuous improvement activities. Sustainability 13(12):1–20. https://doi.org/10.3390/su13126597
42. Mills A, Love PE, Williams P (2009) Defect costs in residential construction. J Constr Eng Manag 135(1):12–16. https://doi.org/10.1061/(ASCE)0733-9364(2009)135:1(12)
43. Mister Plywood (2023) Retrieved from: http://www.misterplywood.com.au/products/particleboard. Accessed 12 Sept 2023
44. Musarat MA, Alaloul WS, Hameed N, Dhinaharan R, Qureshi AH, Wahab MMA (2023) Efficient construction waste management: a solution through industrial revolution (IR) 4.0 evaluated by AHP. Sustainability 15(1):1–16. https://doi.org/10.3390/su15010274
45. Nagapan S, Rahman IA, Asmi A, Memon AH, Zin RM (2012) Identifying causes of construction waste - case of central region of peninsula Malaysia. Int J Integr Eng 4(2):22–28. https://penerbit.uthm.edu.my/ojs/index.php/ijie/article/view/190
46. Oko John A, Emmanuel Itodo D (2013) Professionals' views of material wastage on construction sites and cost overruns. Organiz Technol Manage Construct 5(1):747–757. https://doi.org/10.5592/otmcj.2013.1.11
47. Pamera S, Gurmu A (2020) Framework for building defects and their identification technologies: case studies of domestic buildings in Melbourne, Australia. In: Proc. 54th Int. Conf. of the Architectural Science Association, 2020, pp 1–10
48. Plebankiewicz E, Malara J (2020) Analysis of defects in residential buildings reported during the warranty period. Appl Sci 10:6123
49. Rawlinsons Quantity Surveyors (2023) Rawlinsons Construction Cost Guide 2023 (31st ed.). Rawlinsons Publishing
50. Saidu I, Shakantu W (2016) The contributions of construction material waste to project cost overruns in Abuja, Nigeria. Acta Struct: J Phys Develop Sci 23(1):99–113. https://hdl.handle.net/10520/EJC193764
51. Saunders M, Lewis P, Thornhill A (2009) Research methods for business students (5th ed.). Prentice Hall

52. Schultz CS, Jørgensen K, Bonke S, Rasmussen GMG (2015) Building defects in Danish construction: project characteristics influencing the occurrence of defects at handover. Architect Eng Design Manage 11(6):423–439. https://doi.org/10.1080/17452007.2014.990352
53. Sika Grout (2023) Product data sheet SikaGrout®-212 HP. Retrieved from: https://aus.sika.com/dam/dms/au01/4/sikagrout_-212_hp.pdf. Accessed 18 April 2024
54. Site Inspections (2023a) 2023's Most Costly Building Mistake. Retrieved from: https://www.youtube.com/watch?v=nnnncSxcU3s. Accessed 16 Aug 2023
55. Site Inspections (2023b) Reinspection Shock: New Home Still Non-compliant Despite Builder's Assurance!. Retrieved from: https://www.facebook.com/siteinspectionaustralia/videos/1862801227435675. Accessed 3 June 2023
56. Sweis GJ, Hiari A, Thneibat M, Hiyassat M, Abu-Khader, WS, Sweis RJ (2021) Understanding the causes of material wastage in the construction industry. Jordan J Civil Eng 15(2):180–192. https://jjce.just.edu.jo/issues/paper.php?p=5923.pdf
57. Taggart M, Koskela L, Rooke J (2014) The role of the supply chain in the elimination and reduction of construction rework and defects: an action research approach. Constr Manag Econ 32(7–8):829–842. https://doi.org/10.1080/01446193.2014.904965
58. Tayeh BA, Maqsoom A, Aisheh YIA, Almanassra M, Salahuddin H, Qureshi MI (2020) Factors affecting defects occurrence in the construction stage of residential buildings in Gaza Strip. SN Applied Sciences 2(2):167. https://doi.org/10.1007/s42452-020-1959-1
59. Wu Z, Yu ATW, Shen L, Liu G (2014) Quantifying construction and demolition waste: an analytical review. Waste Manage 34(9):1683–1692. https://doi.org/10.1016/j.wasman.2014.05.010
60. Yin RK (2003) Case study research: design and methods (3rd ed.). SAGE Publications, Inc.
61. Zheng L, Wu H, Zhang H, Duan H, Wang J, Jiang W, Song Q (2017) Characterizing the generation and flows of construction and demolition waste in China. Constr Build Mater 136:405–413
62. Zuraidi S, Rahman MAA, Akasah ZA (2016) Elements of building defect: a review of the recent literature. In: International annual symposium on sustainability science and management, Kuala Terengganu, Malaysia, vol 93, pp 9–11

Part II
Technologies on Energy Efficiency and Waste Management

Chapter 9
An Innovative Tool for Empowering Occupants for Energy Efficiency: A Perspective of New Zealand Office Buildings

Achini Shanika Weerasinghe, James Olabode Bamidele Rotimi, and Eziaku Onyeizu Rasheed

Abstract The building and construction sector is a significant global energy consumer and a major contributor to CO_2 emissions. This high energy consumption is heavily influenced by occupant behaviours within buildings. Existing literature has emphasised the need to adopt a holistic approach, incorporating indoor environmental quality (IEQ) parameters (i.e., thermal, indoor air quality, visual, acoustic) and social-psychological factors (i.e., attitudes, norms, perceived behavioural control, knowledge), rather than relying solely on objective models. This chapter is dedicated to creating an innovative tool known as the Occupant Energy Behaviour Tool, primarily focusing on its application in New Zealand. The findings presented in this chapter were based on research methodology that comprises two main components: (1) a comprehensive review of existing literature, aiming to identify prevalent occupant energy behaviours and the multifaceted drivers that shape them, both within New Zealand and on a global scale, and (2) a quantitative approach that investigates the perceptions of 393 office building occupants regarding their decision-making processes related to energy behaviour. The sample was chosen through stratified random sampling to represent diverse environments across New Zealand. The analysis includes Partial Least Squares Structural Equation Modelling (PLS-SEM) and decision tree analysis to interpret occupants' energy behaviours. The developed tool offers a comprehensive framework for monitoring and collecting data on occupant energy behaviours. It is vital for analysing and improving energy behaviours in New Zealand's office buildings adapting to local needs and applying broadly, enhancing

A. S. Weerasinghe (✉) · J. O. B. Rotimi · E. O. Rasheed
School of Built Environment, Massey University, Auckland, New Zealand
e-mail: achinisweerasinghe@gmail.com

J. O. B. Rotimi
e-mail: J.O.B.Rotimi@massey.ac.nz

E. O. Rasheed
e-mail: E.O.Rasheed@massey.ac.nz

T. Ramachandra et al. (eds.), *Concepts, Policies and Technology for Sustainable Buildings*, Advances in Sustainability Science and Technology,
https://doi.org/10.1007/978-981-96-3086-8_9

energy efficiency and sustainability. This tool integrates occupant behaviours into energy strategies and simulations, significantly boosting building performance.

Keywords Occupant energy behaviour tool · Environmental factors · Energy efficiency · Sustainability · Office buildings

1 Introduction

Rising greenhouse gas emissions are projected to bring substantial climate shifts, leading to increased environmental threats like global warming and severe weather occurrences [1, 2]. These changes persist irrespective of human activities, given the current levels of greenhouse gases [2, 3]. The building sector, including construction and operations, notably contributes to global energy consumption and carbon dioxide emissions, necessitating industry and research innovations for a sustainable, low-carbon future [2]. In line with achieving net-zero carbon buildings by 2050, as proposed by the UNFCCC in 2015, the building and construction sector plays a crucial role in global energy and emissions scenarios [1].

Energy demand in buildings is influenced by a myriad of factors, including climate, building characteristics and occupant behaviours and interactions [4, 5]. Occupant behaviour has been recognised as a critical determinant of building energy consumption, leading to significant discrepancies between predicted and actual energy usage [6, 7]. The actions of occupants, ranging from adjusting thermostats to opening windows, significantly influence the building's Energy Use Intensity (EUI), highlighting the need for accurate energy prediction and management strategies [8, 9].

Factors influencing occupant energy behaviour are categorised into various domains, including physical, environmental, psychological and social factors [8, 10, 11]. Furthermore, studies indicate that even buildings with similar physical attributes can exhibit varying energy consumption patterns based on occupants' comfort preferences, behaviours and lifestyles [12, 13]. These insights underscore the importance of considering a broad spectrum of influences, including social-psychological factors, in understanding and modelling occupant behaviour for building performance simulations (BPS) [14, 15].

Recognising the significance of these factors, recent research has increasingly focused on the social-psychological aspects influencing occupant energy behaviour, integrating theories from social sciences to provide a more comprehensive understanding of how and why occupants interact with building controls and systems [16, 17]. This comprehensive approach is vital in effectively modelling and managing energy consumption in buildings, particularly in contexts like New Zealand, where understanding the dynamic nature of occupant behaviour is crucial for addressing energy impacts. This chapter highlights the need for continued exploration and evaluation of occupant behaviours and their various drivers to optimise energy consumption in buildings globally. Further, by developing a tool that captures the complex interplay of these behaviours and their drivers, we aim to offer a unique contribution to the existing body of research and practice, particularly in New Zealand's efforts to achieve energy efficiency and sustainability in commercial buildings.

2 Literature Review

In buildings, occupants interact with various systems, exhibiting adaptive and non-adaptive energy behaviours [9, 18]. Adaptive behaviours include modifying windows, blinds, thermostats and lighting, using plug-in equipment like heaters and fans to tailor the indoor environment to personal preferences and adjusting clothing or moving around for comfort. Conversely, non-adaptive behaviours involve less interaction or adaptation to the environment, such as merely being present or using plug-in equipment without modifying the environment [9].

Empirical research has explored the potential for energy savings through these behaviours, noting significant variations in energy use intensity due to actions like window opening in commercial buildings [8, 19]. Detailed studies have delved into patterns of plug-in equipment use, revealing that most energy consumption occurs outside active office hours [20]. Estimates suggest that occupant behaviours could lead to energy savings of 5–30% in commercial settings through simple modifications of habits like thermostat adjustments or covering diffusers, indicating a substantial potential for emission reductions in the U.S. Hong and Lin [21], Dietz et al. [22] and Pothitou et al. [23]. However, actual energy usage often deviates from these estimates, underscoring the need for a deeper understanding of occupant behaviours for effective energy optimisation [24].

The factors influencing these behaviours are multifaceted. Physical discomfort often prompts occupants to adjust their environment or behaviours to achieve comfort. Studies have shown that comfort preferences, satisfaction levels and perceptions of the indoor environment vary widely, contributing to the complexity of occupant behaviours in buildings [25, 26]. These behaviours are influenced by a range of factors, categorised as internal (physical, psychological, social) and external (building properties, environment, time) [10, 27]. However, focusing solely on objective factors in energy research might overlook the comprehensive insights offered by subjective aspects of occupant behaviours. Incorporating social psychology knowledge into energy practices can enhance the understanding and prediction of these behaviours, leading to better energy modelling and simulation approaches [16, 28]. Figure 1 depicts the conceptual framework developed to understand the drivers of occupant behaviours.

Multiple studies affirm the profound influence of occupant behaviour on building energy use, highlighting its equal importance alongside technological advancements and physical building attributes [16, 30]. Significant discrepancies are often noted between expected and actual energy performance, largely due to standard occupant behaviour assumptions and the oversimplification of adaptive and non-adaptive behaviours [31, 32]. While the field of occupant behaviour research is

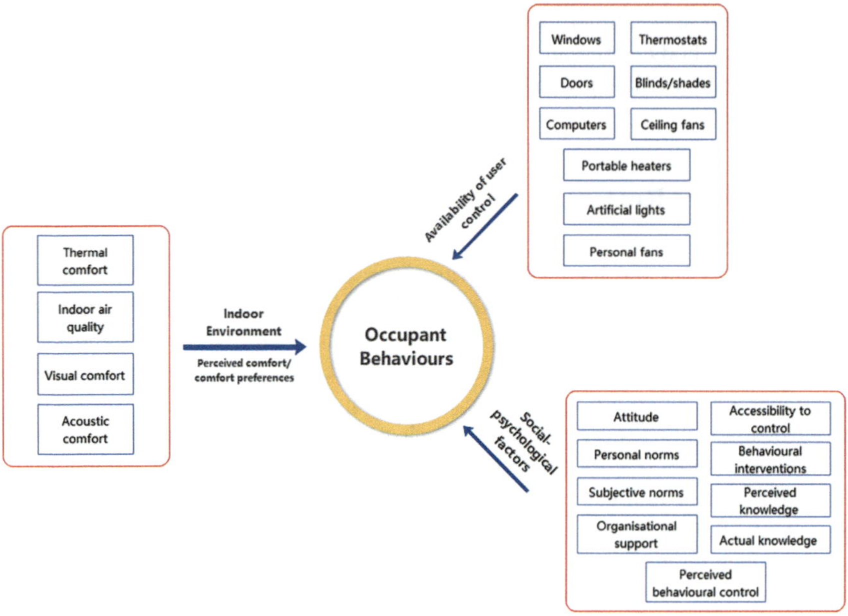

Fig. 1 Conceptualising the drivers of occupant behaviours [29]

growing, a small fraction delves into what drives changes in these behaviours, often focusing narrowly on environmental factors and overlooking other crucial influences like contextual factors, routines and habits [33, 34]. Therefore, multi-disciplinary approaches are encouraged, especially considering the varying influences of social, cultural and economic factors across different contexts to comprehensively understand these behaviours [35, 36].

In New Zealand, building energy consumption has been rising like other OECD nations. However, compared to its OECD peers, New Zealand needs to enhance its energy productivity and reduce GHG emissions, a significant portion of which comes from energy use in buildings [37–39]. Research highlights the importance of understanding and incorporating occupant perceptions and behaviours related to thermal, air quality and acoustic comfort in energy-saving strategies. Studies suggest that occupants' control over their environment is crucial, yet current building designs may not adequately reflect diverse occupant needs and expectations [40, 41]. In New Zealand, where commercial buildings, especially office buildings, constitute a considerable portion of the building stock, therefore understanding and optimising occupant behaviour is key for energy efficiency strategies as it can significantly impact energy savings [42, 43].

Moreover, energy codes and standards in New Zealand and other countries often simplify occupant-related aspects, indicating a need for more comprehensive and occupant-informed energy modelling. Despite the potential of occupant behaviour research, only a small fraction of studies specifically addresses this area in the New

Zealand context. Those that do, largely focus on thermal comfort in green versus conventional buildings and the role of multi-domain comfort preferences in occupant behaviour [44–46]. These studies call for a more detailed understanding of occupant behaviour, including lighting and fan use, to inform energy conservation strategies and occupant-centric building designs and operations.

A literature review of global and local research reveals a critical gap where current occupant behaviour models and tools, while crucial, still fall short due to a lack of comprehensive data. The dynamic nature of occupant behaviour adds complexity to building energy simulations, leading to differences between predicted and actual energy use. By enhancing these models with detailed subjective and objective data, the operational performance of buildings can be optimised, and future building designs can be improved.

Given the identified research gap, this chapter aims to develop the Occupant Energy Behaviour Tool tailored for New Zealand office buildings. This tool aims to provide a comprehensive framework for monitoring, collecting and interpreting data related to occupant energy behaviours, integrating both objective and subjective parameters including environmental, contextual, physiological and time-related factors alongside occupants' comfort preferences and social-psychological influences. By doing so, the tool seeks to enhance our understanding, modelling and optimisation of occupant energy behaviours, contributing to more effective energy management strategies, improved building performance simulations, and ultimately, promoting sustainability within the built environment.

3 Brief Description of the Case

This chapter encompasses an empirical case study focusing on the multifaceted influences on occupant behaviours in office environments within New Zealand. It integrates social-psychological determinants, indoor environmental quality, control measures and demographic characteristics to explore the complex relationships between these domains and energy-related behaviours. The chapter commenced with an extensive review of behavioural theories and adapted the MOA framework to conceptualise the research hypotheses, drawing from the foundational works of Abrahamse and Steg [47] for attitudes, subjective norms and perceived behavioural control (PBC) measures, and Zhang et al. [48] for personal norms. These theories provided the basis for understanding the interplay between individual behaviours and environmental factors.

3.1 Research Methods

A nationwide survey targeting a wide demographic of office occupants across New Zealand was conducted to evaluate the formulated hypotheses. The participant cohort was carefully selected to represent a broad cross-section of the office workforce, with a sample size of 289 respondents determined with a 95% confidence level and a 5% margin of error, considering the proportion of office workers in the selected population. This approach was guided by recent survey data and population proportions from the 2018 Census data [49–52]. Accordingly, a total of 294 valid responses were continued for analysis.

In parallel, a case study survey approach was employed on a specific subset of office occupants within a university setting. This segment targeted 1258 office occupants from three medium-rise buildings characterised by their environmental, structural, social and temporal attributes. 99 valid responses were yielded from the 266 primary occupants who were the designated office space users during the surveyed period. The selection of buildings and the detailed profiling of their characteristics were informed by the works of Chen et al. [5], Bavaresco et al. [53] and Ozcelik et al. [54], providing a comprehensive understanding of the unique behaviours and factors influencing a specific population within the larger context of New Zealand's workforce. The data collection and preparation phase included rigorous cleansing and checks for Common Method Bias (CMB), with statistical adjustments for respondent-related and measurement-related effects, as discussed in the methodologies of Baumgartner and Steenkamp [55], Jakobsen and Jensen [56], MacKenzie and Podsakoff [57], Richardson et al. [58], Podsakoff et al. [59] and Kock et al. [60]. The analysis employed Partial Least Squares Structural Equation Modelling (PLS-SEM) and decision tree analysis.

PLS-SEM explored the relationships among the identified constructs, adhering to the significance and relevance criteria outlined by Pearl [61], Schumacker and Lomax [62] and Sarstedt et al. [63]. The constructed SEM model was depicted in a path diagram, with circles illustrating latent factors and rectangles denoting indicators within the measurement model. According to Coltman et al. [64], measurement models are categorised as reflective and formative. The reflective measurement model describes Eq. (1) which establishes the connection between a latent variable and its observed indicators.

$$\begin{gathered} x = lY + e \\ \textbf{Eq. 1} = \text{Measurement model} \end{gathered} \tag{1}$$

where

x = observed indicator

Y = latent variable, the loading l is a

l = regression coefficient between x and Y (loading)

e = random measurement error.

According to Bollen and Bauldry [65], causal indicators include an error term which indicates that the indicators do not flawlessly measure its constructs. This error term accounts for all the additional causes not captured by the causal indicators. A measurement model utilising causal indicators can be elucidated through Eq. (2):

$$Y = \sum_{k=1}^{K} w_k \cdot x_k + z \tag{2}$$

Eq. 2 = Measurement model with causal indicators

where

w_k = the contribution of x_k ($k = 1\ldots, K$) to Y

z = error term associated with Y.

Bollen and Bauldry [65] state that the reliability of the SEM hinges on the validity of these causal assumptions. Therefore, in formative modelling using composite indicators, the error term is designated as zero ($z = 0$), leading to the adoption of Eq. (3):

$$Y = \sum_{k=1}^{K} w_k \cdot x_k \tag{3}$$

Eq. 3 = Measurement model using composite indicators

where

Y = linear combination of indicators x_k ($k = 1\ldots, K$) weighted by w_k = indicator weight.

For the current investigation, the minimum sample size for PLS-SEM was determined using the inverse square-root method, as described by Kock and Hadaya [66]. This method bases the sample size on the probability that the path coefficient exceeds a critical value at a given significance level. With an expected minimum path coefficient (βmin) between 0.15 and 0.20, the required sample size ranges from 155 to 274. The 294 responses considered for this chapter, meet this requirement, making the sample size adequate for SEM analysis.

$$n\min = \left[\frac{2.486}{\beta\min}\right]^2 \tag{4}$$

Eq. 4 = The minimum sample size

Decision tree analysis, employing the Classification and Regression Trees (CART) algorithm, provided a predictive understanding of occupant behaviours based on the multi-domain aspects surveyed, with the CART methodology [67–69]. CART employs historical data to create a decision tree that effectively handles categorical data and is resilient to outliers, as noted by Shao and Lunetta [70]. It modifies the independent variables numerically without altering the tree's structure. CART focuses on dividing data to achieve maximum homogeneity, aiming for uniform value distribution in each terminal node concerning the target attribute. The purity of these splits is determined using an impurity measure, which is essential for identifying predictor nodes with maximum uniformity [68]. The Gini index was utilised as the primary measure in the current analysis, a common choice in research within this domain [71]. The Gini index employs a squared probability formula for each classification of the dependent variable as part of its data-splitting methodology [71].

$$Gini\ index\,(V) = 1 - \sum_{i=1}^{n} (p_i)^2 \tag{5}$$

Eq. 5 = Gini index

V represents the Gini index

n iss the number of class labels

p_i is the proportion of ith class label.

4 Analysis

4.1 *Underlying Social-Psychological Effects on Occupant Behaviours*

Before building the SEM model, an examination of common method bias (CMB) was carried out through a single-factor analysis to determine its variance impact, as indicated by the results in Table 1 which showed 18.87% of the total variance attributable to the first set of factors. This level of CMB variance is considered negligible for affecting the outcomes as it is well below the 50% threshold. Consequently, the internal consistency of the 22 variables considered was evaluated using Cronbach's alpha, yielding a value of 0.754, surpassing the generally accepted benchmark of 0.70, thus ensuring reliability [72].

The analysis then proceeded to ascertain the reliability and validity of the social-psychological constructs through the PLS-SEM algorithm, forming the measurement model before hypothesis validation.

Table 1 Total variance explained [73]

Component	Initial eigenvalues			Extraction sums of squared loadings		
	Total	% of Variance	Cumulative %	Total	% of Variance	Cumulative %
1	4.151	18.868	18.868	4.151	18.868	18.868
2	2.431	11.052	29.920			

Extraction method: principal component analysis

This step is crucial to confirm the proposed model appropriately represents the sample data. The construct reliability was assessed through the Composite Reliability (CR) parameter, with values in the range of 0.60–0.70 deemed acceptable in exploratory studies [74]. As delineated in Fig. 2 and Table 2, all CR values for latent variables fell between 0.710 and 0.865, establishing the construct reliability, except for constructs with a singular indicator showing a perfect score.

Convergent validity was then evaluated to determine the extent to which the indicators effectively measure their corresponding construct, utilising standardised factor loadings and Average Variance Extracted (AVE) as metrics [72, 74]. The results depicted in Table 2 confirmed, that while most factor loadings exceeded the 0.7 mark all AVE values were above the 0.50 threshold, collectively affirming the construct reliability and convergent validity for each construct under investigation.

Further, the analysis assessed discriminant validity to measure the uniqueness of each construct and its distinction from other constructs in the model. This involved utilising the HTMT Ratio, Fornell-Larcker criterion and cross-loadings [75, 76].

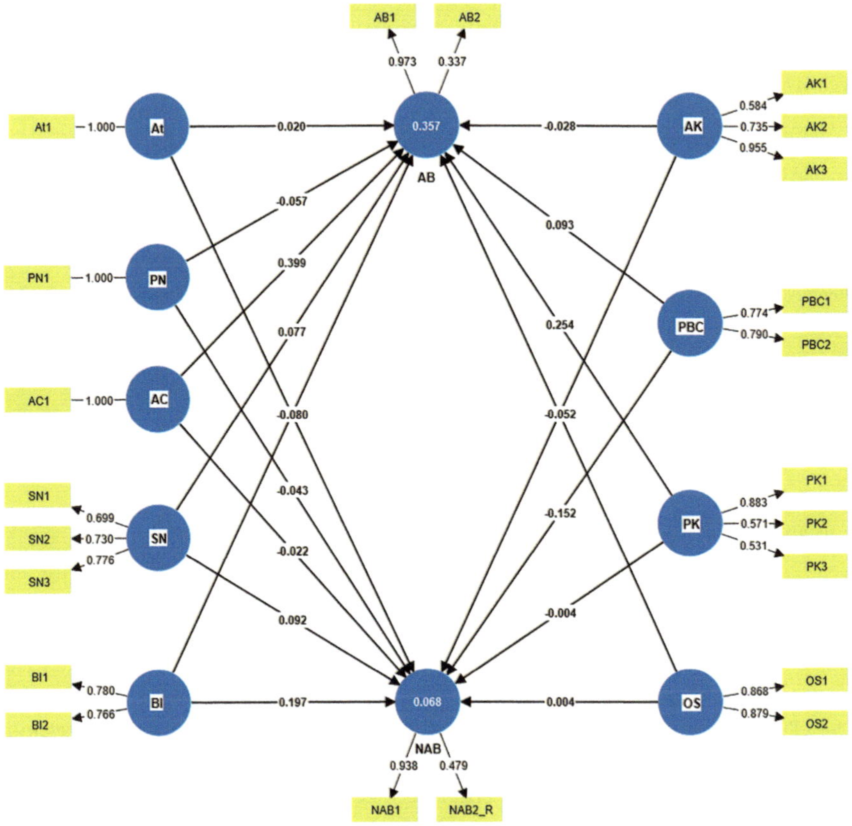

Fig. 2 Standardised estimates of the measurement model [73]

Table 2 Results of PLS-SEM algorithm analysis [73]

Constructs	Indicators	Statement	CR
Attitude—At	At1	Saving energy is important to me	1.000
Personal norms—PN	PN1	I feel responsible/obliged to save energy	1.000
Subjective norms—SN	SN1 SN2 SN3	My co-workers expect me to save energy and turn off electrical appliances at the workplace, and it is easy to share control over building systems with my co-workers	0.781
Organisational support—OS	OS1 OS2	My company encourages employees to save energy and rewards employees for saving energy	0.865
Behavioural interventions—BI	BI1 BI2	The building management team provides feedback on our energy use, and this is important for changing energy-driven behaviours	0.748
Accessibility to control—AC	AC1	I have personal control over most of the appliances in my workspace	1.000
Perceived behavioural control—PBC	PBC1 PBC2	Actions I take to save energy entirely within my control and those actions depend on my comfort preferences	0.758
Actual knowledge—AK	AK1 AK2 AK3	I am aware that reducing energy use will reduce costs and emissions, and improve organisation's image/reputation	0.853
Perceived knowledge—PK	PK1 PK2 PK3	I take actions to save energy such as close windows, turn off the lights, heaters, fans, computers, unplug appliances and adjusting clothing level	0.774
Non-adaptive behaviours—NAB	NAB1 NAB2	I often report discomforts related to IEQ or sometimes willing to tolerate it	0.710
Adaptive behaviours—AB	AB1 AB2	I frequently adjust building appliances for comfort and adapt myself to the workspace environment through clothing, beverages or movement	0.714

The HTMT ratios, as displayed in Table 3, were all below the maximum threshold of 0.90, and the square roots of the AVE values surpassed the inter-construct correlations as per the Fornell-Larcker criterion, except for the organisational support and behavioural intervention pairing. The cross-loadings further substantiated the

Table 3 Correlations and SQRT of AVE [73]

	AK	At	PK	PN	AC	PBC	SN	OS	BI	Sq.roots of AVE
AK	1.000	0.519	0.271	0.483	0.097	0.184	0.245	0.246	0.460	0.773
At		1.000	0.162	0.726	0.045	0.117	0.333	0.263	0.395	1.000
PK			1.000	0.195	0.265	0.461	0.31	0.101	0.288	0.680
PN				1.000	0.099	0.214	0.395	0.31	0.451	1.000
AC					1.000	0.689	0.372	0.086	0.056	1.000
PBC						1.000	0.736	0.459	0.593	0.782
SN							1.000	0.741	0.694	0.736
OS								1.000	1.246	0.873
BI									1.000	0.772

highest associations of indicators with their respective constructs. Despite high cross-loadings between attitude and personal norm indicators, the overall discriminant validity was upheld for most constructs.

Although the path coefficients for indirect and mediating relationships were mostly non-significant, the model's covariances and direct effects, as illustrated in Fig. 3, provided evidence for the causal relationships within the analysis. The validation of hypotheses through path analysis, as shown in Table 4, confirmed the significant positive relationships of various constructs with occupant energy-related behaviours, except for motivation, which displayed a negative association.

4.2 Occupants' Decision-Making of Their Behaviours

The aggregated data in Fig. 4 reflects occupant perceptions regarding the six IEQ parameters: temperature, air quality, natural and artificial light, and inside and outside noise.

The data shows a consensus of "about right" conditions across all parameters. Yet, this agreement is typically around the 50% mark, with the other half expressing varying degrees of discomfort across the extremes (e.g., too cold/hot, too stuffy/draughty, too dark/bright, too quiet/noisy).

Specifically, temperature responses indicate a seasonal influence, with 43% reporting the office as "too cold" or "cold" during winter, and 49% finding it "too hot" or "hot" in the summer. This suggests a need for building managers to consider occupant thermal comfort when calculating heating and cooling loads. While around 45% of respondents find the temperature "about right," indicating a fair level of thermal comfort satisfaction, air quality results show less extremity, with the majority, around 63%, finding it "about right," and fewer occupants perceiving it as "too stuffy" or "draughty."

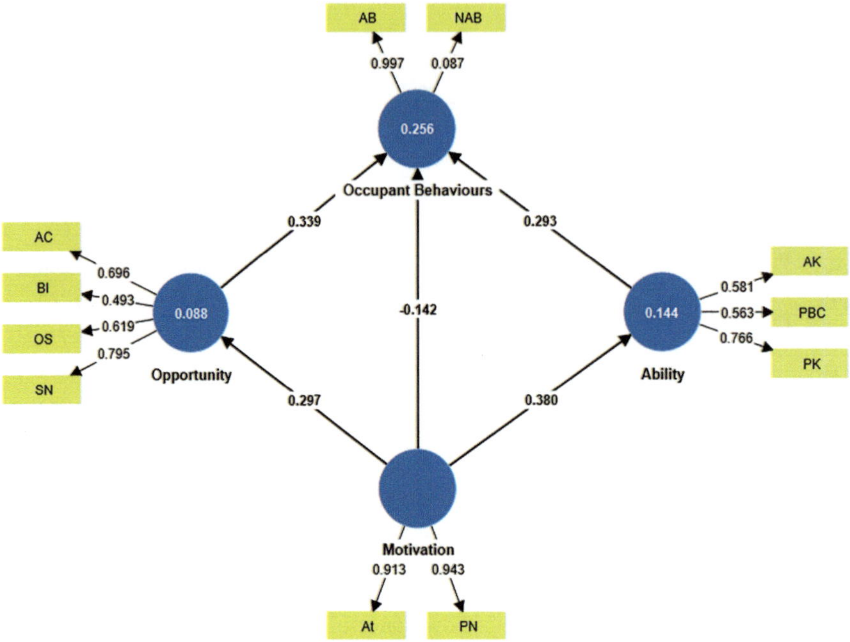

Fig. 3 Structural paths of direct and mediating effects [73]

Table 4 Path analysis and hypotheses validation [73]

Hypotheses	Regression paths	β
h1	At → Motivation	0.913
h2	PN → Motivation	0.943
h3	SN → Opportunity	0.795
h4	OS → Opportunity	0.619
h5	BI → Opportunity	0.493
h6	AC → Opportunity	0.696
h7	PBC → Ability	0.563
h8	PK → Ability	0.766
h9	AK → Ability	0.581
h11	Opportunity → Occupant energy-related behaviours	−0.142
h12	Ability → Occupant energy-related behaviours	0.339
h13	Motivation → Opportunity → Occupant energy-related behaviours	0.293
h1	At → Motivation	0.101
h2	PN → Motivation	0.111

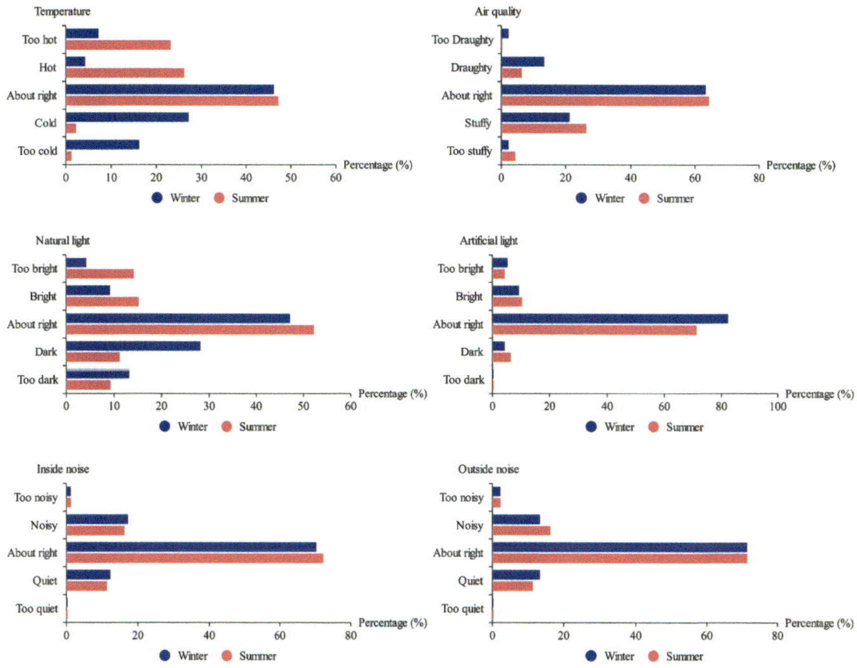

Fig. 4 Perceived comfort in temperature, air quality, lighting and noise [29]

Natural lighting discomfort is more pronounced than artificial lighting, with 40% finding it "too dark" in winter and 21% "too bright" in summer. In contrast, artificial lighting seems well-adjusted to meet year-round needs, with 80% satisfaction. Both inside and outside noise levels are generally considered "about right" by 71% of respondents, indicating a minor impact on the overall occupant comfort.

The distribution of individual control over various building systems and appliances is detailed in Fig. 5.

It reveals that a significant majority have full control over items like computers, doors, shades and blinds. However, less control is reported over portable heaters and thermostats, with only 38% and 17% respectively reporting full control. Ceiling fans, on the other hand, show the lowest control availability, with most occupants indicating no control.

These findings suggest a correlation between control availability and occupant perceptions of comfort, particularly linking thermal discomfort with limited control over heating and cooling systems. Conversely, the relatively high control over windows, doors and fans suggests a possible mitigating effect on perceptions of air quality discomfort. It is clear from these findings that user control plays a pivotal role in shaping occupants' perceptions and behaviours regarding their indoor environment.

Occupants' social-psychological expectations concerning energy behaviours are encapsulated using a five-point Likert scale, with nine factors considered: attitude,

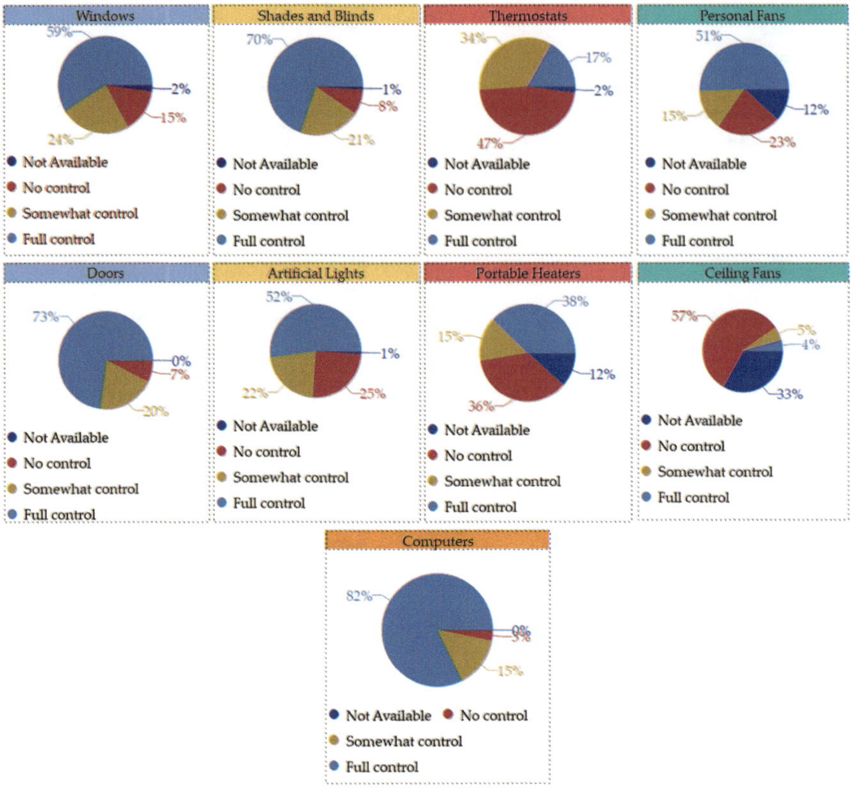

Fig. 5 Availability of user control in building systems and equipment [29]

personal norms, subjective norms, behavioural interventions, organisational support, accessibility to control, PBC, and actual and perceived knowledge. Table 5 presents the mean and standard deviation of these factors, highlighting higher mean scores for actual knowledge, accessibility to control and perceived knowledge, amongst others. In contrast, subjective norms, organisational support and behavioural interventions had mean scores below 3, indicating more negative perceptions towards these aspects.

The analysis explores the predictors and classification accuracy for behaviours such as window operation, lighting adjustment, thermostat, portable heater use and more, delineating the influential factors like control availability, environmental conditions and various social-psychological aspects.

Notably, the analysis highlights that accessibility to control is a predominant factor across behaviours, with other influential factors including actual knowledge, indoor environmental conditions, and personal and organisational aspects. This complex interplay suggests that while control availability is critical, other social-psychological and environmental factors substantially contribute to the occupants' behaviour.

The chapter analysis also considered behaviours such as adjusting clothing and moving through spaces, yet decision tree analysis for these behaviours was not

Table 5 Mean and SD of each social-psychological aspect relating to the occupant behaviours [29]

Factor	Mean	SD
AK	3.979	0.964
AC	3.953	0.850
PK	3.942	0.985
At	3.872	0.901
PN	3.563	0.984
PBC	3.160	1.168
SN	2.678	1.133
OS	1.932	1.121
BI	1.899	1.255

feasible due to insufficient data. However, the prevalence of these behaviours was noted, with a significant portion of occupants engaging in clothing adjustment and movement through spaces, pointing to the comprehensive nature of occupant behaviours and the variety of strategies employed for comfort and energy efficiency.

4.3 Development of the Occupant Energy Behaviour Tool

Drawing from the insights of the above analysis, this chapter formulated a tool for occupant behaviour, to efficiently gather and monitor data relating to occupant energy behaviours. The primary objective of the chapter was met by creating this tool, with its validation and suggestions for optimising energy performance in New Zealand office buildings. A Subject Matter Expert (SME) approach was utilised for gathering expert insights, contributing to the tool's precision. The validation process assessed the tool's comprehensiveness, clarity, coherence, relevance and practicality, identifying areas for enhancement, benefits and obstacles. The outcomes noted the necessity for more data and a classification system tailored to building features and stakeholder categories. It acknowledged the pros and cons of adopting a tool and offered strategies to address these challenges (Fig. 6).

5 Discussion

The chapter delves deep into how motivation, driven by attitudes and personal norms, is a critical driver of energy-efficient behaviours among occupants. High factor loadings for these components signal their strong influence on behavioural intent. Occupants with strong energy-saving attitudes and a personal sense of responsibility towards energy conservation are more likely to engage in behaviours that reduce energy use. The sophisticated relationship between attitudes and personal norms, and

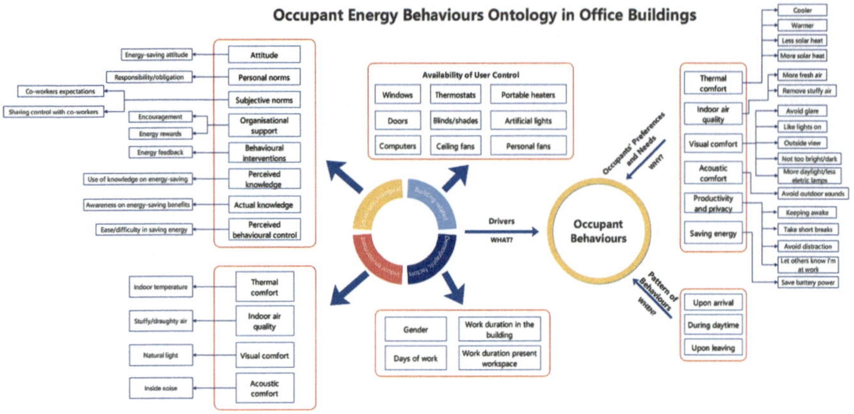

Fig. 6 Occupant energy behaviours ontology in office buildings [77]

their collective impact on adaptive and non-adaptive behaviours, suggests a comprehensive interplay of fundamental motivation factors affecting energy behaviours [61, 78, 79].

Opportunity, as shaped by subjective norms, organisational support, behavioural interventions and accessibility to control, provides the external context within which individuals operate. The significant correlations of these factors with adaptive and non-adaptive behaviours underscore the importance of a conducive organisational and social environment in facilitating or hindering energy-saving actions. When organisations provide clear support, guidelines and incentives for energy conservation, they create opportunities that individuals are more likely to exploit for energy-saving [16, 80, 81]. Incorporating insights from recent studies, such as those by Kosonen and Kim [82] who reviewed the advancement of behavioural energy interventions in commercial buildings, highlights the crucial role of occupant behaviour in achieving energy efficiency. These studies suggest understanding and influencing occupant behaviour through tailored interventions can significantly enhance energy savings. Moreover, research by Han and Zhang [83] on integrating energy-saving building systems with smart and low-cost sensing/control networks for sustainable environments provides practical examples of innovative practices. Their demonstration case study offers valuable insights into how technology and occupant behaviour can synergistically contribute to sustainable and healthy living environments.

The ability to act on energy-saving opportunities is contingent upon individual experiences with energy use, perceived ease or difficulty in implementing energy-saving measures (PBC), and the knowledge (both actual and perceived) individuals have about energy-saving strategies. This construct reflects the capability aspect of energy-saving behaviours, suggesting that even when motivation and opportunity are present the ability to carry out specific actions is crucial. Providing occupants with the necessary knowledge and control can empower them to make more energy-efficient choices [79].

By identifying this critical role of social-psychological factors in influencing occupant energy-saving behaviour, the building managers can enhance occupants' perceived efficacy and control, fostering a more favourable attitude toward energy conservation. The building occupants play a crucial role, and it is important to educate them about energy conservation and engage them in feedback and reward programmes to encourage energy-efficient practices. Integrating these social-psychological insights into building design and operation not only aids in meeting energy reduction targets but also supports broader policy goals toward net zero carbon and energy sustainability.

Occupants' perceptions of IEQ—specifically temperature, air quality and lighting—significantly influence their energy behaviours. Discomfort in these areas often prompts behaviours that lead to increased energy use, such as adjusting thermostats or lights. Seasonal variations particularly impact temperature and lighting comfort, suggesting the need for adaptive building designs and systems that can respond dynamically to changing external conditions and occupant needs [84–86]. The insights from Heydarian et al. [86] highlight the interconnectedness between occupant behaviour, IEQ satisfaction and energy use, urging building design, operations and management professionals to prioritise occupant-centric approaches in their sustainability strategies.

The ability of occupants to control their environment is a powerful mediator of comfort and, consequently, energy use. Enhanced control over heating, ventilation, and air conditioning (HVAC) systems, lighting and shading devices allows occupants to tailor their environment to their comfort needs, often leading to more efficient energy use. This finding highlights the importance of user-centric designs and technologies that provide occupants with intuitive, accessible controls over their environment [79, 87]. Furthermore, Deme Belafi et al. [88] underline the importance of a comprehensive approach that combines technological innovation with behavioural insights to optimise energy use in buildings. Accordingly, the developed tool enables seamless integration with existing building management systems to monitor, collect and analyse data on occupant energy behaviour. Such a system with functionalities, including real-time data analysis, predictive modelling based on historical data, and personalised feedback mechanisms, offers several benefits. These benefits encompass more effective equipment scheduling leading to reduced energy consumption and operational costs which building managers are often tasked with.

The decision tree analysis underscores the importance of control availability in influencing occupant behaviours aligning with existing research that points to the psychological and practical benefits of control in energy use. It also identifies other influential factors, including individual satisfaction with IEQ, attitudes toward energy conservation and the frequency of interaction with control systems. These findings complement the view that occupant behaviour is the result of a complex interplay of individual preferences, environmental conditions and available controls, all of which must be understood and catered to in building design and management [71, 88, 89]. Therefore, the chapter underscores the necessity of incorporating occupant energy behaviour data into energy modelling and simulations, traditionally overlooked in favour of building fabric and operational parameters.

Introducing an innovative tool designed to enhance energy efficiency in New Zealand office buildings, what the chapter presents aligns with the call by Franceschini and Neves [30] for a comprehensive approach to occupant behaviour modelling. This approach includes a wider array of behaviours and influencing factors and underscores the importance of model validation for credible and applicable research findings in building energy efficiency. The tool aids energy modellers in integrating occupant behaviours into both design and operational phases, enhancing the precision of energy consumption predictions, and advocating for the inclusion of occupant behaviour data in energy modelling and simulations, traditionally overshadowed by building fabric and operational parameters.

6 Conclusions

The chapter analysis develops an Occupant Energy Behaviour Tool following an in-depth examination of the social-psychological, environmental and control factors influencing occupant energy behaviours in New Zealand office buildings in Sect. 4. Its focus on occupant behaviour highlights the complex interplay between human actions and building energy dynamics, offering a path to more sustainable energy management practices.

The analysis reveals that attitudes, personal norms and perceived control are key to motivating energy-saving behaviours, while subjective norms, organisational support and accessibility to controls influence opportunities for such actions. These elements crucially predict both adaptive and non-adaptive energy behaviours among occupants. Despite discomfort with certain environmental factors, control over their surroundings is a vital mediator of occupant comfort and significantly affects energy behaviour, leading to more efficient and adaptive patterns.

The findings underscore the importance of understanding occupants' decision-making regarding energy use, offering valuable insights for building users, managers and energy modellers to improve energy performance and achieve net zero energy targets. Integrating these insights into building control technologies promotes an occupant-focused design and operational strategy.

Future research should focus on expanding the tool's applicability to diverse building types and occupant demographics, enhancing its predictive accuracy through machine learning algorithms, and exploring integration with IoT devices for real-time energy monitoring and control. Additionally, case studies demonstrating the tool's impact in actual building settings would further substantiate its value and encourage widespread adoption.

In conclusion, the Occupant Energy Behaviour Tool addresses a critical gap in sustainable building management and aligns with global efforts to achieve energy efficiency and environmental sustainability. Integrating social-psychological aspects with environmental and control factors offers valuable insights for developing more sustainable, efficient and user-friendly energy strategies in the built environment.

Acknowledgements We would like to extend our sincere gratitude to Massey University, New Zealand, for their financial support through the Doctoral Scholarship. This research would not have been possible without their generous assistance. Additionally, we would like to thank all the participants who contributed their time and effort to the data collection process. Your contributions were invaluable to this study.

References

1. IPCC (2019) Summary for Policymakers. In: Climate change and land: an IPCC special report on climate change, desertification, land degradation, sustainable land management, food security, and greenhouse gas fluxes in terrestrial ecosystems. https://doi.org/10.1017/9781009157988.001
2. UNEP (2019) Global environment outlook–GEO-6: healthy .planet, healthy people. U.P. House. https://wedocs.unep.org/20.500.11822/27539
3. IPCC (2014) Climate change 2014: synthesis report. Contributios of working groups I, II and III to the fifth assessment report of the IPCC. http://epic.awi.de/37530/1/IPCC_AR5_SYR_Final.pdf
4. Widén J, Wäckelgård E (2010) A high-resolution stochastic model of domestic activity patterns and electricity demand. Appl Energy 87(6):1880–1892. https://doi.org/10.1016/j.apenergy.2009.11.006
5. Chen CF, Xu X, Day JK (2017) Thermal comfort or money saving? Exploring intentions to conserve energy among low-income households in the United States. Energy Res & Soc Sci 26:61–71. https://doi.org/10.1016/J.ERSS.2017.01.009
6. Blight TS, Coley DA (2013) Sensitivity analysis of the effect of occupant behaviour on the energy consumption of passive house dwellings. Energy Build 66:183–192. https://doi.org/10.1016/j.enbuild.2013.06.030
7. Gaetani I, Hoes PJ, Hensen JLM (2016) Occupant behavior in building energy simulation: towards a fit-for-purpose modeling strategy. Energy Build 121:188–204. https://doi.org/10.1016/j.enbuild.2016.03.038
8. Fabi V, Andersen RV, Corgnati S, Olesen BW (2012) Occupants' window opening behaviour: a literature review of factors influencing occupant behaviour and models. Build Environ 58:188–198. https://doi.org/10.1016/j.buildenv.2012.07.009
9. Hong T, Yan D, D'Oca S, Chen CF (2017) Ten questions concerning occupant behavior in buildings: the big picture. Build Environ 114:518–530. https://doi.org/10.1016/j.buildenv.2016.12.006
10. O'Brien W, Gunay HB (2014) The contextual factors contributing to occupants' adaptive comfort behaviors in offices-a review and proposed modeling framework. Build Environ 77:77–87. https://doi.org/10.1016/j.buildenv.2014.03.024
11. Peng C, Yan D, Wu R, Wang C, Zhou X, Jiang Y (2012) Quantitative description and simulation of human behavior in residential buildings
12. Andersen RV, Toftum J, Andersen KK, Olesen BW (2009) Survey of occupant behaviour and control of indoor environment in Danish dwellings. Energy Build 41(1):11–16. https://doi.org/10.1016/j.enbuild.2008.07.004
13. Maier T, Krzaczek M, Tejchman J (2009) Comparison of physical performances of the ventilation systems in low-energy residential houses. Energy Build 41(3):337–353
14. Day JK, O'Brien W (2017) Oh behave! Survey stories and lessons learned from building occupants in high-performance buildings. Energy Res Soc Sci 31:11–20. https://doi.org/10.1016/j.erss.2017.05.037
15. Hess AK, Samuel R, Burger P (2018) Informing a social practice theory framework with social-psychological factors for analyzing routinized energy consumption: a multivariate analysis of three practices. Energy Res Soc Sci 46:183–193. https://doi.org/10.1016/J.ERSS.2018.06.012

16. D'Oca S, Chen CF, Hong T, Belafi Z (2017) Synthesizing building physics with social psychology: an interdisciplinary framework for context and occupant behavior in office buildings. Energy Res Soc Sci 34:240–251. https://doi.org/10.1016/j.erss.2017.08.002
17. Shi H, Fan J, Zhao D (2017) Predicting household PM2.5-reduction behavior in Chinese urban areas: an integrative model of theory of planned behavior and norm activation theory. J Clean Prod 145:64–73. https://doi.org/10.1016/j.jclepro.2016.12.169
18. Schweiker M, Carlucci S, Andersen RK, Dong B, O'Brien W (2018) Occupancy and occupants' actions. In: Wagner A, O'Brien W, Dong B (eds.) Exploring occupant behavior in buildings: methods and challenges, pp 7–38. Springer International Publishing. https://doi.org/10.1007/978-3-319-61464-9_2
19. Sun K, Hong T (2017) A simulation approach to estimate energy savings potential of occupant behavior measures. Energy Build 136:43–62. https://doi.org/10.1016/j.enbuild.2016.12.010
20. Gunay HB, O'Brien W, Beausoleil-Morrison I, Gilani S (2016) Modeling plug-in equipment load patterns in private office spaces. Energy Build 121:234–249. https://doi.org/10.1016/j.enbuild.2016.03.001
21. Hong T, Lin H-W (2013) Occupant behavior: impact on energy use of private offices. Lawrence Berkeley National Laboratory, https://escholarship.org/content/qt6jp5w8kn/qt6jp5w8kn.pdf
22. Dietz T, Gardner GT, Gilligan J, Stern PC, Vandenbergh MP (2009) Household actions can provide a behavioral wedge to rapidly reduce US carbon emissions. Proc Natl Acad Sci 106(44):18452–18456. https://doi.org/10.1073/pnas.0908738106
23. Pothitou M, Kolios AJ, Varga L, Gu S (2016) A framework for targeting household energy savings through habitual behavioural change. Int J Sustain Energ 35(7):686–700. https://doi.org/10.1080/14786451.2014.936867
24. Jia M, Srinivasan RS, Raheem AA (2017) From occupancy to occupant behavior: an analytical survey of data acquisition technologies, modeling methodologies and simulation coupling mechanisms for building energy efficiency. Renew Sustain Energy Rev 68:525–540. https://doi.org/10.1016/j.rser.2016.10.011
25. Schakib-Ekbatan K, Çakici FZ, Schweiker M, Wagner A (2015) Does the occupant behavior match the energy concept of the building?-Analysis of a German naturally ventilated office building. Build Environ 84:142–150. https://doi.org/10.1016/j.buildenv.2014.10.018
26. Abdulaali HS, Usman IMS, Hanafiah MM, Abdulhasan MJ, Hamzah MT, Nazal AA (2020) Impact of poor indoor environmental quality (Ieq) to inhabitants' health, wellbeing, and satisfaction. Int J Adv Sci Technol 29(4 Special Issue):1284–1296. https://www.scopus.com/inward/record.uri?eid=2-s2.0-85083007076&partnerID=40&md5=8fa629153d0b33ce09b98e69985229b4
27. Schweiker M (2010) Occupant behaviour and the related reference levels for heating and cooling. Tokyo City University
28. Yan D, Hong T, Dong B, Mahdavi A, D'Oca S, Gaetani I, Feng X (2017) IEA EBC Annex 66: definition and simulation of occupant behavior in buildings. Energy Build 156:258–270. https://doi.org/10.1016/j.enbuild.2017.09.084
29. Weerasinghe AS, Rasheed EO, Rotimi JOB (2023a) Occupants' decision-making of their energy behaviours in office environments: a case of New Zealand. Sustainability 15(3)
30. Franceschini PB, Neves LO (2022) A critical review on occupant behaviour modelling for building performance simulation of naturally ventilated school buildings and potential changes due to the COVID-19 pandemic. Energy Build 258:111831. https://doi.org/10.1016/j.enbuild.2022.111831
31. Guerra-Santin O, Silvester S (2017) Development of Dutch occupancy and heating profiles for building simulation. Build Res Inf 45(4):396–413. https://doi.org/10.1080/09613218.2016.1160563
32. Hong T, Taylor-Lange SC, D'Oca S, Yan D, Corgnati SP (2016) Advances in research and applications of energy-related occupant behavior in buildings. Energy Build 116:694–702. https://doi.org/10.1016/j.enbuild.2015.11.052
33. Haldi F, Robinson D (2011) The impact of occupants' behaviour on building energy demand. J Build Perform Simul 4(4):323–338

34. Stazi F, Naspi F, D'Orazio M (2017) A literature review on driving factors and contextual events influencing occupants' behaviours in buildings [Review]. Build Environ 118:40–66. https://doi.org/10.1016/j.buildenv.2017.03.021
35. Paone A, Bacher J-P (2018) The impact of building occupant behavior on energy efficiency and methods to influence it: a review of the state of the art. Energies 11(4):953. https://www.mdpi.com/1996-1073/11/4/953
36. Delzendeh E, Wu S, Lee A, Zhou Y (2017) The impact of occupants' behaviours on building energy analysis: a research review. Renew Sustain Energy Rev 80:1061–1071. https://doi.org/10.1016/j.rser.2017.05.264
37. MBIE (2017) Unlocking our energy productivity and renewable potential: New Zealand energy efficiency and conservation strategy 2017–2022. MBIE
38. EECA (2020) Programmes & funding. Energy efficiency and conservation authority retrieved 17 June from https://www.eeca.govt.nz/our-work/programmes-and-funding/
39. Conti J, Holtberg P, Diefenderfer J, LaRose A, Turnure JT, Westfall L (2016) International energy outlook 2016 with projections to 2040
40. Rasheed EO, Byrd H, Money B, Mbachu J, Egbelakin T (2017) Why are naturally ventilated office spaces not popular in New Zealand? Sustainability 9(6):902
41. Rasheed E, Rotimi J, Vilasini N, Weerasinghe A (2021) Proactive post-occupancy evaluation
42. Isaacs N, Saville-Smith K, Babylon M, Bishop R, Camilleri M, Donn M, Jowett J, Moore D, Roberti H (2010) Building energy end-Use study (BEES) year 3. BRANZ
43. Pivac N, Nižetić S, Zanki V (2018) Occupant behavior and thermal comfort field analysis in typical educational research institution a case study. Therm Sci 22:785–795. https://doi.org/10.2298/TSCI170915013P
44. Azizi NSM, Wilkinson S, Fassman E (2015) An analysis of occupants response to thermal discomfort in green and conventional buildings in New Zealand. Energy Build 104:191–198. https://doi.org/10.1016/j.enbuild.2015.07.012
45. Azizi NSM, Wilkinson S, Fassman E (2015) Do occupants in green buildings practice better energy saving behaviour in computer usage than occupants in conventional buildings? J Green Build 10(4):178–193. https://doi.org/10.3992/jgb.10.4.178
46. Weerasinghe AS, Rasheed E, Bamidele Rotimi JO (2020) Occupant energy behaviours–a review of indoor environmental quality (IEQ) and influential factors. In: Ali EA (ed) Ghaffarianhoseini, imaginable futures: design thinking, and the scientific method proceedings of the international conference of architectural science association
47. Abrahamse W, Steg L (2009) How do socio-demographic and psychological factors relate to households' direct and indirect energy use and savings? J Econ Psychol 30(5):711–720. https://doi.org/10.1016/J.JOEP.2009.05.006
48. Zhang Y, Wang Z, Zhou G (2013) Antecedents of employee electricity saving behavior in organizations: an empirical study based on norm activation model. Energy Policy 62:1120–1127. https://doi.org/10.1016/J.ENPOL.2013.07.036
49. Stats NZ (2022) 2018 Census place summaries. Retrieved March 4 from https://www.stats.govt.nz/tools/2018-census-place-summaries/
50. Calculator.net (2022) Sample size calculator. Retrieved March 6 from https://www.calculator.net/sample-size-calculator.html
51. Saunders M, Philip M, Thornhill LA (2019) Research methods for business students, 8th edn. Pearson
52. O'Kane P, Walton S, Ruwhiu D (2020) Remote working during COVID19. Department of Management, University of Otago
53. Bavaresco MV, D'Oca S, Ghisi E, Pisello AL (2020) Assessing underlying effects on the choices of adaptive behaviours in offices through an interdisciplinary framework. Build Environ 181:Article 107086. https://doi.org/10.1016/j.buildenv.2020.107086
54. Ozcelik G, Becerik-Gerber B, Chugh R (2019) Understanding human-building interactions under multimodal discomfort. Build Environ 151:280–290. https://doi.org/10.1016/j.buildenv.2018.12.046

55. Baumgartner H, Steenkamp J-BEM (2001) Response styles in marketing research: a cross-national investigation. J Mark Res 38(2):143–156
56. Jakobsen M, Jensen R (2015) Common method bias in public management studies. Int Public Manag J 18(1):3–30
57. MacKenzie SB, Podsakoff PM (2012) Common method bias in marketing: causes, mechanisms, and procedural remedies. J Retail 88(4):542–555
58. Richardson HA, Simmering MJ, Sturman MC (2009) A tale of three perspectives: examining post hoc statistical techniques for detection and correction of common method variance. Organ Res Methods 12(4):762–800
59. Podsakoff PM, MacKenzie SB, Podsakoff NP (2012) Sources of method bias in social science research and recommendations on how to control it. Annu Rev Psychol 63(1):539–569
60. Kock F, Berbekova A, Assaf AG (2021) Understanding and managing the threat of common method bias: detection, prevention and control. Tour Manage 86:104330
61. Pearl J (2009) Causal inference in statistics: an overview
62. Schumacker E, Lomax G (2016) A beginner's guide to structural equation modelling, 4th edn. Routledge, London
63. Sarstedt M, Ringle CM, Hair JF (2021) Partial least squares structural equation modeling. In: Handbook of market research, pp 587–632. Springer
64. Coltman T, Devinney TM, Midgley DF, Venaik S (2008) Formative versus reflective measurement models: two applications of formative measurement. J Bus Res 61(12):1250–1262
65. Bollen KA, Bauldry S (2011) Three Cs in measurement models: causal indicators, composite indicators, and covariates. Psychol Methods 16(3):265
66. Kock N, Hadaya P (2018) Minimum sample size estimation in PLS-SEM: the inverse square root and gamma-exponential methods. Inf Syst J 28(1):227–261. https://doi.org/10.1111/isj.12131
67. Lu H, Ma X (2020) Hybrid decision tree-based machine learning models for short-term water quality prediction. Chemosphere 249:126169. https://doi.org/10.1016/j.chemosphere.2020.126169
68. Ryu SH, Moon HJ (2016) Development of an occupancy prediction model using indoor environmental data based on machine learning techniques. Build Environ 107:1–9. https://doi.org/10.1016/j.buildenv.2016.06.039
69. Rizvi S, Rienties B, Khoja SA (2019) The role of demographics in online learning; a decision tree based approach. Comput Educ 137:32–47. https://doi.org/10.1016/j.compedu.2019.04.001
70. Shao Y, Lunetta RS (2012) Comparison of support vector machine, neural network, and CART algorithms for the land-cover classification using limited training data points. ISPRS J Photogramm Remote Sens 70:78–87
71. Bavaresco MV, Ghisi E, D'Oca S, Pisello AL (2021) Triggering occupant behaviour for energy sustainability: exploring subjective and comfort-related drivers in Brazilian offices. Energy Res Soc Sci 74:Article 101959. https://doi.org/10.1016/j.erss.2021.101959
72. Kline RB (2015) Principles and practice of structural equation modeling. Guilford Publications
73. Weerasinghe AS, Rotimi JOB, Rasheed EO (2023) Modelling of underlying social psychological effects on occupant energy-related behaviours. Build Environ 231:110055
74. Hair J, Alamer A (2022) Partial least squares structural equation modeling (PLS-SEM) in second language and education research: guidelines using an applied example. Res Methods Appl Linguist 1(3):100027. https://doi.org/10.1016/j.rmal.2022.100027
75. Fornell C, Larcker DF (1981) Evaluating structural equation models with unobservable variables and measurement error. J Mark Res 18(1):39–50
76. Henseler J, Ringle CM, Sarstedt M (2015) A new criterion for assessing discriminant validity in variance-based structural equation modeling. J Acad Mark Sci 43(1):115–135. https://doi.org/10.1007/s11747-014-0403-8
77. Weerasinghe A (2023) The impacts of occupant behaviours on energy consumption in New Zealand office buildings: a thesis submitted in partial fulfilment of the requirements for the degree of Doctor of Philosophy in Construction at Massey University, Albany, New Zealand. Doctoral dissertation, Massey University

78. Du J, Pan W (2021) Examining energy saving behaviors in student dormitories using an expanded theory of planned behavior. Habitat Int 107:Article 102308. https://doi.org/10.1016/j.habitatint.2020.102308
79. Li D, Xu X, Chen CF, Menassa C (2019) Understanding energy-saving behaviors in the American workplace: a unified theory of motivation, opportunity, and ability. Energy Res & Soc Sci 51:198–209. https://doi.org/10.1016/J.ERSS.2019.01.020
80. Mulville M, Jones K, Huebner G, Powell-Greig J (2016) Energy-saving occupant behaviours in offices: change strategies. Build Res & Inf 45(8):861–874. https://doi.org/10.1080/09613218.2016.1212299
81. Tang Z, Warkentin M, Wu L (2019) Understanding employees' energy saving behavior from the perspective of stimulus-organism-responses. Resour Conserv Recycl 140:216–223. https://doi.org/10.1016/J.RESCONREC.2018.09.030
82. Kosonen HK, Kim AA (2017) Advancement of behavioral energy interventions in commercial buildings. Facilities 35(5–6):367–382. https://doi.org/10.1108/F-04-2016-0044
83. Han K, Zhang J (2020) Energy-saving building system integration with a smart and low-cost sensing/control network for sustainable and healthy living environments: demonstration case study. Energy Build 214:109861. https://doi.org/10.1016/j.enbuild.2020.109861
84. Yun GY, Steemers K, Baker N (2008) Natural ventilation in practice: linking facade design, thermal performance, occupant perception and control. Build Res & Inf 36(6):608–624. https://doi.org/10.1080/09613210802417241
85. De Vecchi R, Candido C, de Dear R, Lamberts R (2017) Thermal comfort in office buildings: findings from a field study in mixed-mode and fully-air conditioning environments under humid subtropical conditions. Build Environ 123:672–683. https://doi.org/10.1016/j.buildenv.2017.07.029
86. Heydarian A, McIlvennie C, Arpan L, Yousefi S, Syndicus M, Schweiker M, Jazizadeh F, Rissetto R, Pisello AL, Piselli C, Berger C, Yan Z, Mahdavi A (2020) What drives our behaviors in buildings? A review on occupant interactions with building systems from the lens of behavioral theories. Build Environ 179:Article 106928. https://doi.org/10.1016/j.buildenv.2020.106928
87. Vellei M, Natarajan S, Biri B, Padget J, Walker I (2016) The effect of real-time context-aware feedback on occupants' heating behaviour and thermal adaptation. Energy Build 123:179–191. https://doi.org/10.1016/j.enbuild.2016.03.045
88. Deme Belafi Z, Naspi F, Arnesano M, Reith A, Revel GM (2018) Investigation on window opening and closing behavior in schools through measurements and surveys: a case study in Budapest. Build Environ 143:523–531. https://doi.org/10.1016/j.buildenv.2018.07.022
89. Park J, Choi CS (2019) Modeling occupant behavior of the manual control of windows in residential buildings. Indoor Air 29(2):242–251. https://doi.org/10.1111/ina.12522

Chapter 10
Navigating On-Grid and Off-Grid Solar Systems: Challenges and Strategies Within Sri Lankan Building Context

Diyana Sheharee Athidewa Ranasinghe, Anuradha Samarajeewa Waidyasekara, and Ganegodage Lakna Harindi

Abstract Renewable energy is increasingly recognised worldwide as crucial for sustainable development. In Sri Lanka, a developing nation, considerable efforts are put on attaining sustainable objectives, primarily solar energy. This chapter delves into economic aspects of two prevalent solar Photovoltaic (PV) systems: on-grid and off-grid installations in Sri Lankan context. The chapter provides a conceptual map outlining the financial challenges and strategies specific to each system, as well as those shared by both, drawing from insights gathered through twenty-four semi-structured interviews with users of both on-grid and off-grid solar systems and input from company and government authority experts. High capital costs and inadequate financing emerge as common challenges, addressed through strategies such as financing schemes, tax concessions and hybrid PV solar systems. Notably, the study underscores the pressing need for policy interventions and in-depth investigations to foster the profitability of solar PV systems, not only within Sri Lanka but also as part of the broader global shift towards renewable energy. By illuminating the economic complexities of solar energy deployment, this research significantly contributes to the global dialogue on renewable energy and provides a groundwork for future research endeavours and policy initiatives worldwide.

Keywords Challenges · Off-grid solar · On-grid solar · Strategies

D. S. A. Ranasinghe (✉) · A. S. Waidyasekara · G. L. Harindi
Department of Building Economics, University of Moratuwa, Moratuwa, Sri Lanka
e-mail: diyanar@uom.lk

A. S. Waidyasekara
e-mail: anuradha@uom.lk

G. L. Harindi
e-mail: L.H.Ganegodage@ljmu.ac.uk

T. Ramachandra et al. (eds.), *Concepts, Policies and Technology for Sustainable Buildings*, Advances in Sustainability Science and Technology,
https://doi.org/10.1007/978-981-96-3086-8_10

1 Introduction

The global consumption of resources to fulfil everyday electrical energy requirements implies the depletion of ores, minerals and petroleum [38]. As Kannan and Vakeesan [21] pointed out, the world's energy demand is growing fast because of the population explosion and technological advancements. Moreover, to meet the rising electricity demand, coal and gas power plant usage has increased by 2.5%, reflecting energy-related carbon emissions of 13 in 2018. Carbon dioxide releases many toxic gases, impacting climate changes and global temperatures [56]. Renukappa et al. [43] highlighted that even organisations are implementing carbon reduction strategies with the increasing concern about carbon emissions and climate change. There are four ways to implement energy-related carbon reduction strategies in organisational buildings [9]. They are passive building design, low embodied energy material for construction, energy-efficient domestic appliances and adopting building-integrated Renewable Energy (RE) technologies.

RE is considered one of the most important topics of discussion regarding sustainable consumption and environmental protection [18]. Aliprandi et al. [3] stated that electricity produced by fossil fuels raises atmospheric carbon dioxide levels, and, as a result, countries are focusing on replacing fossil fuels with RE technologies. Compared to conventional centralised electricity generation systems such as coal power plants, RE technologies allow customers to reduce their electricity costs and their effect on the environment [31]. Maradin [30] underscored that world energy consumption from RE sources is estimated to exceed 20% and continues to grow.

Biomass, wind, solar, hydroelectricity, geothermal energy and tidal power are familiar renewable sources [39]. Solar energy is rapidly gaining notoriety as an essential means of expanding RE resources [38]. According to Al-Ezzi and Ansari [2], employing sunlight to produce electrical energy has been demonstrated to be one of the most promising solutions to the world's energy crisis. Solar Photovoltaic (PV) technology has gained considerable attention worldwide, as it can help to reduce the adverse effects of CO_2 emissions [50]. Solar PV systems either supply electricity in their standalone mode or support the utility grid in their grid-tied mode [29]. Gawhade and Ojha [14] explained that an On-Grid Solar (ONGS) PV system is popular due to the abundance of solar light and advanced power electronics techniques. On the other hand, Off-Grid Solar (OFGS) systems are used exclusively in rural areas because installing the electrical grid is costly or technically complex [10]. Moreover, Xu et al. [54] underscored that the OFGS PV system is a significant step towards electrification in remote rural regions, and it is the most convenient and easy-to-instal technology.

Solar PV technology is a renewable way to produce energy and has many environmental benefits, such as lowering air and water emissions while reducing demand for traditional resources [16]. Given the vast potential and benefits of using solar PV technology, many challenges must be addressed in both global and local contexts [48]. Karakaya and Sriwannawit [22] have conducted a study on challenges under four categories: policy, management, economic and sociotechnical. However, Branker et al. [6] highlighted economic and financial challenges as the principal obstacles to applying solar PV systems. To use the PV systems for an entity, the financial benefits need to be determined [40].

2 Background of the Inquiry

2.1 *Energy Crisis and Renewable Energy*

According to Yoon et al. [55], the world's energy requirement will be multiplied by the year 2050 due to the economy's expansion and the global population. Rapidly growing energy consumption has raised concern over energy supply difficulties and extreme environmental impacts [41]. Khalil and Zaidi [25] pointed out that energy is a basic need of modern life, and the energy crisis is making destructive impacts and destroying the economy.

Thus, RE has become a vital choice to solve the energy crisis and environmental problems [26]. Supporting this, Gârdan et al. [18] explained that the RE concept is also generated by phenomena such as global warming and the accelerated degradation of the environment. Geothermal energy, wind energy, solar energy, tidal energy and biomass energy are among the RE sources, with solar energy being a notable approach for harvesting energy widely from sun radiation [47].

2.2 *Solar PV Systems*

According to Desai and Patel [12], generating electricity by photovoltaic arrays is gaining widespread attention and has been identified as a reliable technology for the coming years. Solar PV is one of the first clean energy technologies to be implemented globally [8]. Expressing a similar view, Zhang and He [57] stated that solar PV uses silicon-based material to convert sunlight directly into electricity and has been recognised as a primary solar power generating system. Further, categories of photovoltaic solar systems can be identified as stand-alone PV systems, grid-tied PV systems and hybrid solar power systems [5, 24].

2.2.1 On-Grid PV Systems

The ONGS PV technology is a power-generating PV system that integrates with the local utility grid [28]. It ensures maximum energy use from the utility grid and solar power system [23]. ONGS PV systems are widely practiced in many countries globally as the demand for alternative renewable energy against fossil fuels grows [35]. Orioli and Di Gangi [36] underscored that this is a significant technology as it integrates PV systems into the local grid and optimises the energy balance of

Table 1 Types of solar PV schemes. *Source* PUCSL [37]

Solar PV scheme	Description
Net metering	A billing mechanism that credits solar energy producers (Prosumers) for the electricity they add to the grid. However, no payment will be made for the excess electricity supplied over the consumption under this scheme
Net accounting	A programme that allows prosumers to export electricity produced through their solar PV systems to the national grid. The prosumer will be paid for the excess energy exported to the grid
Net plus	This involves getting paid for all the electricity generated using the premises' fixed solar panels

buildings, improving the system's infrastructure. Further, it lowers the functional costs and adds value to customers and utilities. Moreover, the grid-tied PV system's reliability is higher than that of the stand-alone PV system [17]. According to the Public Utilities Commission Sri Lanka [37], the ONGS rooftop market contains three solar PV schemes: net plus, net accounting and net metering as presented in Table 1.

2.2.2 Off-Grid Solar (Stand-Alone) PV Systems

Stand-alone PV systems operate independently from the local utility grid, designed and sized to cater to specific Direct Current (DC) and Alternating Current (AC) electrical loads [4]. This technology is usually a substitute for utility power, primarily used in places where the primary local utility grid is not connected [7]. According to Fara and Craciunescu [13], a stand-alone photovoltaic power system consists of a complete set of interconnected components for converting solar irradiance directly into electricity. The authors further highlighted that this system consists of a PV generator, charging controller, storage battery, solar inverter and system load. These systems have many applications in remote and isolated places to meet the ongoing demand for electricity [46]. Moreover, to ensure a stable energy supply, there should be compatibility between the stand-alone solar PV size systems and the electricity consumption in isolated sites.

2.2.3 Hybrid Solar PV Systems

Typically, 'hybrid' refers to two-generation energy sources like wind and solar [34]. It refers to a system that combines solar and batteries and can interact with the power grid in the solar world. According to Ghenai et al. [15], a hybrid solar system incorporates the best of ONGS and OFGS systems, with the system being either OFGS with utility backup power (grid backup) or grid-tied solar with additional battery storage. However, Sanni et al. [45] claimed that a hybrid solar PV system could be standalone, grid-connected or a combination of both. Ahmad et al. [1]

explained that a standalone system could function as a hybrid system by utilising battery storage to ensure a continuous power supply when no irradiance exists.

2.3 Economic Challenges in Solar PV Systems

Shukla et al. [48] stated that solar PV system adoption is often constrained by several challenges, which need to be overcome at the national and regional levels. Previous studies have identified various sociotechnical challenges under different categories related to PV technology [27]. Among these, economic challenges are the primary concern of energy consumers of solar PV systems [22], and they are the major constraint of solar PV implementation and future growth [52]. Table 2 presents a summary of economic barriers on both ONGS and OFGS systems.

Upon examining challenges in both ONGS and OFGS sectors, it becomes evident that the primary economic concern surrounding PV systems in the literature has been their capital-intensive nature.

Blessed with diverse geo-climatic conditions, Sri Lanka possesses a variety of RE resources. While some are already extensively used and developed to fulfil the country's present energy needs, solar energy emerges as a prominent renewable source. It offers the potential for integration into buildings through various technologies. Among these technologies, grid-tied PV technology has garnered attention across different sectors in Sri Lanka, offering the prospect of leveraging solar energy benefits. However, due to persistent power outages, the energy crisis in late 2019 compelled Sri Lankan communities to turn to OFGS systems for power supply. For instance, the findings of Ranasinghe et al. [42] emphasised that despite the popularity of solar PV systems in Sri Lanka, ONGS systems provide no advantages during power outages. Consequently, there was a growing demand for OFGS systems in Sri Lanka. Hemakumara et al. [19] shed light on Sri Lanka's initiation of a rooftop solar PV adoption programme to reduce imported fuels' socioeconomic and environmental impacts. Additionally, Silva [49] evaluated the current status of solar power generation in Sri Lanka, identifying opportunities and barriers to implementing solar energy programmes. Thanthirige et al. [51] compared the installation costs of ONGS, OFGS and hybrid solar PV systems to determine the most viable system to instal. The study by Ranasinghe et al. [42] provided a general discussion of the pros and cons within the OFGS sector, highlighting affordability challenges only for OFGS systems.

Despite this, a comprehensive comparative study addressing the economic challenges and required strategies for both OFGS and ONGS systems in the Sri Lankan building sector has not been conducted to the authors' knowledge. Consequently, this chapter aims to fill this gap by comparing OFGS and ONGS PV Systems in the Sri Lankan building sector, particularly focusing on financial considerations.

Table 2 Economic challenges of ONGS and OFGS systems

Challenges	Source of references					
ONGS systems	Lo et al. [27]	Karakaya and Sriwannawit [22]	Mondal and Islam [32]	Timilsina et al. [52]		
Capital intensive investment	✔	✔	✔	✔		
Lack of financing	·	✔	·	✔		
Long payback period	✔		✔	·		
High maintenance cost	✔	✔	·	·		
Low electricity cost of conventional energy sources	✔		·	·		
Low-cost effectiveness	✔	·	·	·		
High cost of solar PV modules	✔					
OFGS systems	Wattana and Aungyut [53]	Delgadillo et al. [11]	Ranasinghe et al. [42]	Mugisha et al. [33]	Samarakoon [44]	Hoeck et al. [20]
High price of battery	✔	✔	✔	✔	·	·
Capital intensive investment	✔	·	✔	·	✔	✔
No Return on Investment (ROI)	·	·	✔	·	·	·

3 Profile of the Respondents

Table 3 provides the profile of participants whose views were sought on both OFGS and ONGS PV systems. The interviewees comprised of fourteen individuals, from both ONGS and OFGS user categories, five experts from solar service provider companies and five experts from government authorities. This composition of participants allowed to gather diverse perspectives and insights, contributing to a robust analysis of the subject matter.

This study employed the purposive sampling technique to selectively identify users, focusing on individuals with awareness of solar PV systems. Specifically, users with at least two years of experience with ONGS systems were chosen. Assuming an extended usage period might provide valuable insights into their experiences with potential economic challenges, this criterion was set. Conversely, the snowball sampling method was employed to identify experts from companies and government authorities strategically. This method facilitated the selection of respondents with well-established expertise in solar PV systems.

Table 3 Profile of respondents

Respondent	Respondents' ID	Use of solar PV system	Designation
OFGS Users (OFU)	OFU1	Domestic	Engineer
	OFU2	Domestic	Lawyer
	OFU3	Domestic	Businessperson
OFGS and ONGS Users (BOU)	BOU1	Commercial	Electrical engineer
	BOU2	Domestic	Electrical engineer
	BOU3	Commercial	Electrical engineer
	BOU4	Domestic	Banker
ONGS Users (OU)	ONU1	Domestic	Banker
	ONU2	Domestic	Professor
	ONU3	Commercial	Businessperson
	ONU4	Commercial	Electrical engineer
	ONU5	Commercial	Electrical engineer
	ONU6	Commercial	Businessperson
	ONU7	Domestic	Doctor
Respondent	Respondents' ID	Experience on solar PV systems	Designation
Solar Service providers (SC)	SC1	OFGS & ONGS	Director
	SC2	OFGS & ONGS	Director
	SC3	OFGS & ONGS	Engineer
	SC4	OFGS & ONGS	Engineer
	SC5	OFGS & ONGS	Engineer
Government Authorities (GA)	GA1	OFGS & ONGS	Engineer
	GA2	OFGS & ONGS	Assistant director
	GA3	OFGS & ONGS	Project manager
	GA4	OFGS & ONGS	Assistant director
	GA5	OFGS & ONGS	Engineer

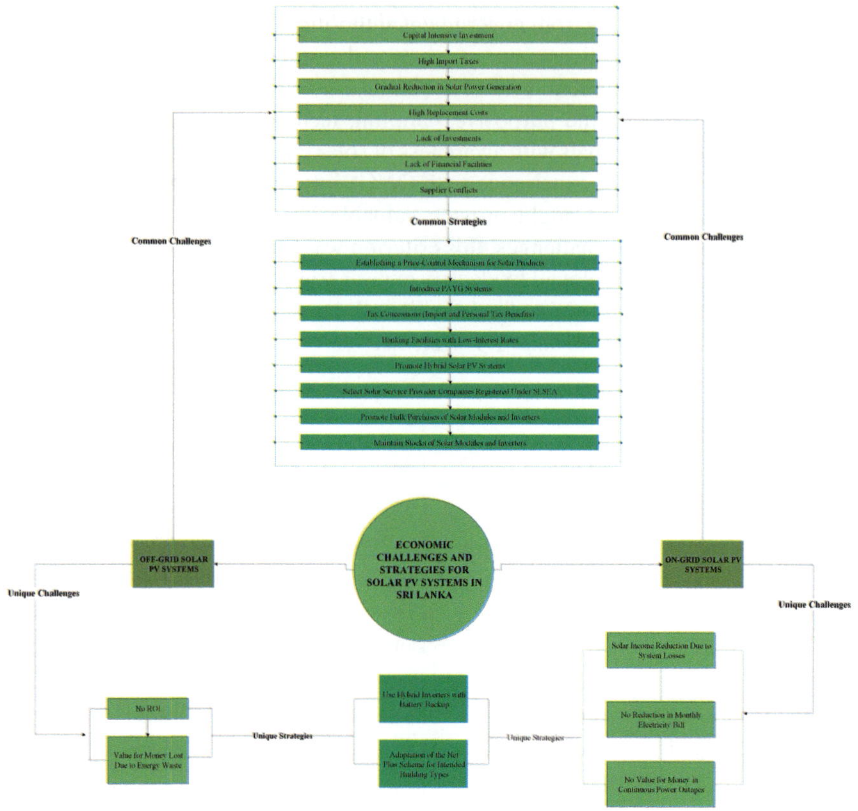

Fig. 1 Diagram of economic challenges and strategies for ONGS and OFGS systems

Collected views were analysed for the economic benefits, challenges and strategies of ONGS and OFGS systems separately analysed and then discussed collectively for both ONGS and OFGS systems to extract common and unique insights into economic challenges and associated strategies. A comparative analysis of the findings between both systems was done using a diagram (Fig. 1).

4 Analysis

This section delves into the exploration of economic barriers and strategies through a detailed analysis of qualitative data gathered from semi-structured interviews.

4.1 Economic Challenges of Off-Grid Solar Systems

All the participants cited the **absence of banking facilities** tailored to OFGS systems as a significant hindrance, unlike their ONGS systems. According to most government experts (90%), Sri Lankan solar companies also refrain from offering financing options for OFGS installations, further aggravating the financial constraints faced by users. Most users face compounded economic challenges due to the banking sector's hesitancy to provide loan facilities for OFGS systems. This is due to concerns about their ability to operate adequately for the specified period without any failure, as highlighted by a few company experts (20%).

Government support for solar systems is limited, making it harder for OFGS implementations to thrive, as acknowledged by the vast majority of respondents (95%). Even though there are experts in the solar industry, the **lack of investment and encouragement** for OGFS systems persists, especially with recent economic crises making matters worse. Additionally, 75% of respondents highlighted that tax breaks on certain solar equipment do not make up for the high taxes and import costs tied to OGFS systems, making them too expensive for many potential users.

On the other hand, most experts believed that OFGS systems have **no ROI**, which might discourage people from investing in these systems. According to SC4, "the substantial initial investment required, typically around 20 lakhs for an average household system, renders OFGS solutions financially unattainable for many". This exclusivity continues the socioeconomic divide, limiting OFGS and ONGS access to specific communities and exacerbating inequities in energy access in Sri Lanka as per the participants.

Capital-intensive investment, coupled with operational challenges such as the need for expensive earthing connections and import regulations, further compound the economic barriers associated with OFGS systems, as pointed out by all the participants. Escalating battery prices, aggravated by economic downturns, pose additional hurdles to system affordability and sustainability. The majority of GA experts (90%) pointed out that users opting for budget installations, such as using car batteries, often incur unforeseen maintenance costs due to inferior product quality.

Economic crises in Sri Lanka contributed to **delays in customs clearance and worsen import-related challenges**, further impeding OFGS system implementation, as per the perspective of all the SC experts. **Energy wastage**, a consequence of system inefficiencies and fluctuating household demand, undermines the economic viability of OFGS solutions compared to their ONGS counterparts, is another barrier highlighted by some of the participants.

4.2 Proposed Strategies to Overcome Economic Challenges of Off-Grid Solar Systems

Addressing the financial barriers inherent in the solar sector necessitates an intricate approach, as proposed by respondents in our study.

Absence of Banking Facilities and Capital-Intensive Investment: GA2 proposed implementing OFGS systems for low-income families through 'Samurdhi schemes,' leveraging existing financing facilities for poor communities. Half of the respondents endorsed interest-free loan programmes and the lifting of import bans on components to enhance affordability further. GA experts recommended implementing price control mechanisms to regulate high prices imposed by solar companies. OFU3 emphasised the necessity of government intervention in price control to prevent exploitation of customers during crises.

High Taxes: To make OFGS systems more financially accessible, respondents suggested making them fully or partially tax-deductible, following the example of other countries. The effectiveness of Pay As You Go (PAYG) systems in promoting affordability is highlighted by BOU4 and GA1, drawing parallels with successful implementations in African countries.

Limited Government Investment Initiatives: Reducing Prices and Promoting Local Manufacturing was opted as an effective strategy by a minority of respondents (20%). Further, they advocated for reducing prices by importing raw materials and investing in local manufacturing of solar components. This strategy aims to enhance financial affordability by minimising dependency on imported goods. BOU4 and BOU2 underscored the importance of establishing financial accessibility through concessions such as loan facilities, investment options or subsidies. Further, the majority of users (80%) advocated for reducing reliance on imported goods through government investments in local manufacturing capabilities. SC4 highlighted the long-term strategy of assembling or manufacturing essential system components domestically, mitigating the impact of currency fluctuations. This approach aligns with the goal of establishing technological independence and reducing dependency on foreign markets.

No ROI and Energy Wastage: Respondents from government and industry (40%) recommended hybrid solar systems as a solution to concerns about energy wastage and ROI. GA2 emphasised the flexibility of hybrid systems in mitigating energy crises, while SC4 stressed the need for a legal framework to expand their usage nationwide. OFU1 underscored the economic advantages of hybrid systems, especially during energy crises, while BOU2 emphasised the importance of safety mechanisms in promoting their reliability and durability.

4.3 Economic Challenges of Grid-Tied PV Technology

The economic journey of implementing ONGS PV systems is associated with various challenges, according to the respondents in this study. Similar to OFGS systems, all the participants (100%) agreed that the **capital-intensive investment** poses a major obstacle for grid-tied PV setups. ONU3, who runs a business with such a system, mentioned how this investment holds back system expansion. GA3 also emphasised the significant initial costs for boosting grid-tied capacity. Additionally, ONU4 shed light on the high transportation expenses incurred, especially for businesses located far from Colombo in Sri Lanka. **Market dynamics** added another layer of economic complexity, with the majority of the respondents (90%) noting ongoing price fluctuations in solar components. GA5 highlighted how import costs and taxes drive these fluctuations, aggravated by the absence of bulk ordering practices in Sri Lanka, as noted by SC2. GA1 further explained how irregular purchasing patterns lead to higher costs for clients.

A common thread among respondents was the absence of adequate financial support mechanisms for renewable energy ventures in Sri Lanka. ONU1 pointed out the lack of tailored loan schemes, underscoring the financial burden borne by users and organisations alike. This opinion was confirmed by GA1, who emphasised the absence of subsidies at the initial stage as a critical hurdle for solar adoption. The demand for subsidies and financing schemes as a means to alleviate capital-intensive investments was reiterated by ONU5 and ONU3. In addition to financial challenges, **delays and supplier conflicts** during the implementation stage have posed economic burdens for 20% of users. ONU6 and ONU5 cited delays from solar companies as contributing to increased indirect costs and operational setbacks. Supplier conflicts also emerged as a notable challenge, as highlighted by ONU7's experience. Operationally, **the replacement cost of system components** emerged as a significant economic concern, particularly due to the high import costs associated with spare parts as per 80% of participants. SC1 emphasised the logistical challenges of importing components, which incur additional courier costs and taxes. This sentiment was confirmed by SC2 and SC4, who underscored the expense of replacing inverters, solar modules and batteries.

Further economic challenges were expressed by a minority portion (28%) of respondents, including the **lack of substantial electricity bill reductions** and the gradual decline in solar power generation. GA4 highlighted various losses associated with PV systems, such as cable losses and environmental factors, which contribute to additional costs for users. Moreover, the **prevalence of power cuts** during energy crises has influenced user preferences, with many opting for OFGS systems as a reliable alternative. Domestic and industrial users alike have faced the economic implications of power outages, often resorting to additional investments in alternative energy sources.

In summary, this section of the chapter reveals numerous economic challenges spanning the initial, implementation and operational stages of ONGS PV technology, underscoring the economic complexities inherent in sustainable energy adoption.

4.4 *Proposed Strategies to Overcome Economic Challenges of Grid Tied Systems*

The economic challenges associated with ONGS PV systems necessitate innovative strategies for sustainable implementation and operation. Respondents in this study have identified several key strategies aimed at mitigating these challenges and enhancing the economic viability of ONGS PV systems in Sri Lanka.

Capital-Intensive Investment: A primary concern among ONGS PV users is the capital-intensive nature of system implementation. According to respondents, there are three primary solar PV schemes in Sri Lanka to sell generated solar units to recover the system's initial cost: Net accounting, Net metering and Net plus. To address the high initial cost challenge, all the respondents (100%) advocated for the adoption of the Net Plus scheme, a unique approach that allows users to sell generated solar units to recover the system's initial cost. Operating under this scheme, users benefit from payments by the Ceylon Electricity Board (CEB) for both consumed and generated energy, without a direct correlation between on-site generation and consumption. This approach ensures a comparatively high profit margin, making it an attractive option for domestic and industrial users alike. As explained by all the users, currently, CEB purchases units produced from solar PV systems at a rate of Rs. 22.00 for the first seven years, and from the eighth year to the twentieth year, will purchase at a rate of Rs. 15.50. Therefore, all the respondents noted that they expect to recover the system's initial cost and generate a profit through the Net Plus scheme. Government authority experts also confirmed that the Net Plus scheme enables long-term cost recovery and profitability, providing a sustainable solution to the capital-intensive nature of solar PV investments.

Supplier Conflicts: Supplier conflicts present additional economic burdens for solar PV users. To mitigate these challenges, respondents (75%) recommended selecting solar companies registered as energy service companies under the Sri Lanka Sustainable Energy Authority (SLSEA). By choosing registered companies, users can mitigate financial risks associated with supplier delays and conflicts, ensuring smoother implementation and operation stages. GA1 highlighted that "*selecting a reputed and government-organized service provider for installing systems will reduce conflict and associated costs for these system users.*"

High Replacement Costs: The high replacement cost of system components is another economic challenge pointed out by all ONGS PV users. As a strategy to mitigate this challenge, most of the respondents (55%) suggested promoting bulk purchases of solar modules and inverters to reduce replacement costs. According to solar service company experts, by encouraging bulk purchasing and maintaining stocks locally, users can minimise the impact of high component costs and ensure timely replacements, thereby enhancing the economic sustainability of ONGS PV systems.

In summary, this study identifies several suggested strategies to overcome the economic constraints of ONGS PV systems in Sri Lanka. Foremost among these strategies is the adoption of the Net Plus scheme, which offers a viable pathway to cost recovery and profitability for solar PV investments. Additionally, strategies such as selecting registered energy service companies and promoting bulk purchases contribute to mitigating supplier conflicts and reducing replacement costs, thereby enhancing the overall economics of ONGS PV systems.

5 Discussion

The discussion section centres on the economic barriers encountered by both OFGS and ONGS PV systems, along with the proposed strategies outlined in the preceding sections. Drawing from the analysis of both ONGS and OFGS unique and common challenges, the authors have developed a comprehensive overview represented in Fig. 1. This visual aid illustrates the shared and distinctive challenges and strategies within both fields, providing a clear roadmap for understanding the economic dynamics of solar PV systems.

5.1 Economic Challenges Between OFGS and ONGS PV Systems

All the participants identified the significant obstacle of capital-intensive investment for both systems. This financial strain is primarily attributed to reliance on imports, associated taxation and initial implementation costs imposed by solar companies. Moreover, this aligns with the literature findings which confirmed capital-intensive investment as the major economic challenge globally. Addressing these high capital costs emerges as a critical need for successful solar PV system implementations in various countries.

In terms of financing, all OFGS users mentioned the absence of tailored banking facilities, compared to just over half (51%) of ONGS users. This discrepancy underscores a significant barrier in promoting sustainable building practices in Sri Lanka. Notably, while conflicts with solar companies emerged as a significant cost barrier for the some (20%) of ONGS users, OFGS users did not explicitly mention this. The difference lies in the simpler installation process for OFGS systems, often carried out independently due to their standalone nature.

While only a minority of ONGS users (28%) noted no reductions in monthly electricity bills, most users observed a decline in bills, whereas OFGS users do not benefit from such reductions. Conversely, ONGS users expressed frustration over the lack of value during power outages, as excess energy goes to waste. Furthermore, concerns over solar income reduction due to system losses and gradual power reductions over time were highlighted by 40% of ONGS users, reflecting ongoing economic challenges common in many countries with ONGS solar PV systems.

High replacement costs pointed out as a common concern, particularly for ONGS users, while OFGS users may face similar issues due to variations in product quality. However, the primary hurdle for OFGS users lies in import taxes, aggravated by the system's relative novelty in Sri Lanka before the energy crisis. Conversely, ONGS users, having used their systems for over two years, have fewer concerns about tax issues and currency fluctuations.

In conclusion, while both OFGS and ONGS systems share common economic challenges such as capital-intensive investments and difficulties in securing financial support, they also face unique hurdles. ONGS systems contend with conflicts with solar companies and a gradual decrease in solar power production, reflecting longer-term operational concerns. Conversely, OFGS systems grapple with broader issues like a lack of investment and vulnerability to currency fluctuations. This comparative analysis underscores the need for comprehensive strategies tailored to address both shared and unique barriers, facilitating the widespread and sustainable adoption of solar energy solutions in the Sri Lankan building environment, as well as in similar situations worldwide.

5.2 Proposed Strategies Between OFGS and ONGS PV Systems

The strategies proposed for mitigating economic barriers in OFGS systems and ONGS PV systems reveal both commonalities and distinctions in their approaches. In OFGS systems, there is a significant emphasis on establishing price control mechanisms, introducing PAYG systems, and providing tax concessions and banking facilities. These strategies underscore a comprehensive financial support framework to overcome economic obstacles, relevant not only in Sri Lanka but in other countries facing similar economic barriers to solar adoption. The ONGS PV systems, on the other hand, focus on distinct measures tailored to their centralised nature. Strategies such as adapting existing schemes like the Net Plus Scheme and collaborating with registered solar companies under the SLSEA point towards a more regulatory and industry-focused approach.

However, some strategies bridge both ONGS and OFGS contexts. Bulk purchase promotion, maintaining stocks of solar modules and inverters, and using hybrid inverters with battery backups are examples of such strategies. These measures address common challenges such as reducing replacement costs and ensuring system reliability across different solar energy systems.

When considering specific strategies like adapting the Net Plus Scheme, its relevance extends beyond ONGS systems to include OFGS scenarios. Quality assurance and safe installations are universal concerns that warrant consideration irrespective of grid connections. Therefore, both types of users can benefit economically from having registered reputable solar companies handle their installations.

Similarly, the proposal to establish price controls, initially suggested by OFGS users, also finds merit as a strategy in ONGS systems. A price-controlling mechanism effectively mitigates high replacement costs, making it a valuable economic strategy applicable across diverse solar energy systems. Furthermore, the introduction of a PAYG system, initially proposed by OFGS users, emerges as a unifying strategy for enhancing sustainability in Sri Lankan buildings. By partnering with registered companies, the government can extend installment services to both ONGS and OFGS users, fostering affordability and accessibility.

The promotion of bulk purchases, as suggested earlier by ONGS users, also stands as a strategic initiative applicable to OFGS users. This approach mitigates delays associated with high demand or seasonal fluctuations and minimises vulnerability to currency rate fluctuations, presenting a viable strategy for both sectors. Moreover, adopting hybrid inverters, with advantages such as grid interaction and battery backup during power outages, emerges as a pivotal strategy for Sri Lanka's evolving power sector. Relevant authorities must prioritize this strategy to harness its dual benefits and ensure the sustainability of the solar PV sector.

In conclusion, the strategies proposed for both ONGS and OFGS PV systems demonstrate a blend of unique approaches and shared principles. Tailoring strategies to the specific needs of each context while recognising and implementing common strategies can collectively uplift the solar PV sector, fostering a sustainable and resilient energy future not only for Sri Lanka but for developing and developed regions worldwide.

6 Conclusions

This chapter has highlighted common economic challenges such as high initial investment costs and inadequate financial support. Additionally, it has identified specific barriers unique to each system and proposed strategies to mitigate them. An important finding of this analysis is the applicability of many strategies proposed for both ONGS and OFGS systems, including the integration of hybrid solar PV systems, tax concessions, financial support mechanisms and the introduction of PAYG systems.

Furthermore, this chapter has significant implications for both theory and practice, emphasising the necessity of analysing essential policies and regulations for hybrid solar PV systems. It reveals how users of ONGS systems often overlook the potential for power crises, leading to ineffective investments during outages, while users of OFGS systems face challenges due to insufficient considerations regarding their long-term economic sustainability and durability. Thus, the primary lesson for ensuring sustainability in building projects lies in anticipating power crises and economic difficulties before system installations.

This research extends beyond Sri Lanka's borders, highlighting the importance of thoroughly identifying and implementing RE systems for long-term success, particularly in regions facing economic challenges like Sri Lanka. International policymakers and authorities should consider adopting similar economic strategies, including providing financial aid, facilitating access to facilities and offering tax concessions to encourage renewable energy adoption. Additionally, reducing dependence on imported technologies, such as solar PVs, is crucial for long-term resilience.

Financial crises are inevitable globally from time to time, thus continued economic analysis, case studies for hybrid solar PV systems and studies on the dynamics of the solar energy industry remain vital. Accordingly, this chapter is a contribution to the economic discussion of solar PV systems worldwide, based on examples from the Sri Lankan building context. As explained, the insights from the proposed strategies provide valuable guidance to policymakers, researchers and industry professionals worldwide. As nations drive towards achieving a sustainable future, utilising renewable energy is a must. These similar economic obstacles from the Sri Lankan context can be overcome and mitigated through the findings of this chapter, thereby facilitating the successful adoption of solar PV systems in different geographies.

References

1. Ahmad NI, Ali Z, Kadir MZAA, Osman M, Zaini NH, Roslan MH (2021) Impacts of lightning-induced overvoltage on a hybrid solar PV–battery energy storage system. Appl Sci (Switz) 11(8). https://doi.org/10.3390/app11083633
2. Al-Ezzi, AS, Ansari MNM (2022) Photovoltaic solar cells: a review. Appl Syst Innov 5(4). https://doi.org/10.3390/asi5040067
3. Aliprandi F, Stoppato A, Mirandola A (2016) Estimating CO_2 emissions reduction from renewable energy use in Italy. Renew Energy 96:220–232. https://doi.org/10.1016/j.renene.2016.04.022
4. Bhatia S (2014) Solar photovoltaic systems. Adv Renew Energy Syst 144–157. https://doi.org/10.1016/b978-1-78242-269-3.50005-x
5. Bimenyimana S, Asemota GNO, Ihirwe PJ (2018) Optimization comparison of stand-alone and grid-tied solar PV systems in Rwanda. OALib 05(05):1–18. https://doi.org/10.4236/oalib.1104603
6. Branker K, Pathak MJM, Pearce JM (2011) A review of solar photovoltaic levelized cost of electricity. Renew Sustain Energy Rev 15(9):4470–4482. https://doi.org/10.1016/j.rser.2011.07.104
7. Chaar L (2007) Solar power conversion. Power Electron Handb 661–672. https://doi.org/10.1016/b978-012088479-7/50044-4
8. Chaurey A, Kandpal TC (2010) Assessment and evaluation of PV based decentralized rural electrification: an overview. Renew Sustain Energy Rev 14(8):2266–2278. https://doi.org/10.1016/j.rser.2010.04.005
9. Chel A, Kaushik G (2018) Renewable energy technologies for sustainable development of energy efficient building. Alex Eng J 57(2):655–669. https://doi.org/10.1016/j.aej.2017.02.0
10. Cuenca A, Oña C, Suquillo I, Miniguano H (2023) Design methodology of off-grid PV solar powered systems for rural areas in Ecuador. Revista Técnica Energía 20(1). https://doi.org/10.37116/revistaenergia.v20.n1.2023.537
11. Delgadillo J, Monterrey G, Aguilar J, Casano R (2022) Photovoltaic technology employment in Peru. A literature review. Eng Solid Mech 10(4):387–398
12. Desai HP, Patel HK (2007) Maximum power point algorithm in PV generation: an overview. Proc Int Conf Power Electron Drive Syst 624–630. https://doi.org/10.1109/PEDS.2007.4487766

13. Fara L, Craciunescu D (2017) Output analysis of stand-alone PV systems: modeling, simulation and control. Energy Procedia 112:595–605. https://doi.org/10.1016/j.egypro.2017.03.1125
14. Gawhade P, Ojha A (2021) Recent advances in synchronization techniques for grid-tied PV system: a review. Energy Rep 7. https://doi.org/10.1016/j.egyr.2021.09.006
15. Ghenai C, Merabet A, Salameh T, Pigem EC (2018) Grid-tied and stand-alone hybrid solar power system for desalination plant. Desalination 435. https://doi.org/10.1016/j.desal.2017.10.044
16. Grover S (2007) Energy, economic, and environmental benefits of the solar America initiative. https://doi.org/10.2172/914650
17. Gunerhan H, Hepbasli A, Giresunlu U (2008) Environmental impacts from the solar energy systems. Energy Sources Part A Recover Util Environ Eff 31(2):131–138. https://doi.org/10.1080/15567030701512733
18. Gârdan IP, Micu A, Paştiu CA, Micu, AE, Gârdan DA (2023) Consumers' attitude towards renewable energy in the context of the energy crisis. Energies 16(2). https://doi.org/10.3390/en16020676
19. Hemakumara GPTS, Athukorala SU, Yapa LGDS (2023) "Battle for solar energy" program: an environment-friendly sustainable approach for carbon neutrality in Sri Lanka. In: Sustainability and social marketing issues in Asia. https://doi.org/10.1108/978-1-80071-845-620231002
20. Hoeck I, Steurer E, Dolunay O, Ileka H (2022) Challenges for off-grid electrification in rural areas. Assessment of the situation in Namibia using the examples of Gam and Tsumkwe. Energy Ecol Environ 7(5). https://doi.org/10.1007/s40974-021-00214-5
21. Kannan N, Vakeesan D (2016) Solar energy for future world-a review. Renew Sustain Energy Rev 62. https://doi.org/10.1016/j.rser.2016.05.022
22. Karakaya E, Sriwannawit P (2015) Barriers to the adoption of photovoltaic systems: the state of the art. Renew Sustain Energy Rev 49:60–66. https://doi.org/10.1016/j.rser.2015.04.058
23. Karki P, Adhikary B, Sherpa K (2012) Comparative study of grid-tied photovoltaic (PV) system in Kathmandu and Berlin using PVsyst. In: 2012 IEEE third international conference on sustainable energy technologies (ICSET). https://doi.org/10.1109/icset.2012.6357397
24. Karthikeyan V, Rajasekar S, Das V, Karuppanan P, Singh AK (2017) Grid-connected and off-grid solar photovoltaic system. Green Energy Technol 0(9783319501963). https://doi.org/10.1007/978-3-319-50197-0_5
25. Khalil HB, Zaidi SJH (2014) Energy crisis and potential of solar energy in Pakistan. Renew Sustain Energy Rev 31. https://doi.org/10.1016/j.rser.2013.11.023
26. Li J, Liu P, Li Z (2022) Optimal design of a hybrid renewable energy system with grid connection and comparison of techno-economic performances with an off-grid system: a case study of West China. Comput Chem Eng 159. https://doi.org/10.1016/j.compchemeng.2022.107657
27. Lo K, Mah D, Wang G, Leung M, Lo A, Hills P (2018) Barriers to adopting solar photovoltaic systems in Hong Kong. Energy & Environ 29(5):649–663. https://doi.org/10.1177/0958305x18757402
28. Madeti SR, Singh S (2017) Online modular level fault detection algorithm for grid-tied and off-grid PV systems. Sol Energy 157:349–364. https://doi.org/10.1016/j.solener.2017.08.047
29. Mahela O, Shaik A (2017) Comprehensive overview of grid interfaced solar photovoltaic systems. Renew Sustain Energy Rev 68:316–332. https://doi.org/10.1016/j.rser.2016.09.096
30. Maradin, D. (2021). Advantages and disadvantages of renewable energy sources utilization. International Journal of Energy Economics and Policy, 11(3). https://doi.org/10.32479/ijeep.11027
31. McCabe A, Pojani D, Van Groenou AB (2018) The application of renewable energy to social housing: a systematic review. Energy Policy 114:549–557. https://doi.org/10.1016/j.enpol.2017.12.031
32. Mondal MAH, Islam AS (2011) Potential and viability of grid-connected solar PV system in Bangladesh. Renew Energy 36:1869–1874. https://doi.org/10.1016/j.renene.2010.11.033
33. Mugisha J, Ratemo MA, Keza BBC, Kahveci H (2021) Assessing the opportunities and challenges facing the development of off-grid solar systems in Eastern Africa: the cases of Kenya, Ethiopia, and Rwanda. Energy Policy 150. https://doi.org/10.1016/j.enpol.2020.112131

34. Newkirk M (2015) What is a hybrid solar system. Clean energy reviews. https://www.cleanenergyreviews.info/blog/2014/8/14/what-is-hybrid-solar
35. Nwaigwe K, Mutabilwa P, Dintwa E (2019) An overview of solar power (PV systems) integration into electricity grids. Mater Sci Energy Technol 2(3):629–633. https://doi.org/10.1016/j.mset.2019.07.002
36. Orioli A, Di Gangi A (2013) Load mismatch of grid-connected photovoltaic systems: review of the effects and analysis in an urban context. Renew Sustain Energy Rev 21:13–28. https://doi.org/10.1016/j.rser.2012.12.035
37. PUCSL (2021) Roof top solar power panel installation. https://columbiacollege-ca.libguides.com/apa/websites
38. Pangaribuan T, Manihuruk J, Hasiholan Pardosi C, Hutauruk J (2023) Analisis Perancangan Panel Surya Portabel Dengan Autotracking. J Electr Syst Control Eng 6(2). https://doi.org/10.31289/jesce.v6i2.8913
39. Patel A, Swathika OVG, Subramaniam U, Babu TS, Tripathi A, Nag S, Karthick A, Muhibbullah M (2022) A practical approach for predicting power in a small-scale off-grid photovoltaic system using machine learning algorithms. Int J Photoenergy 2022. https://doi.org/10.1155/2022/9194537
40. Phipps W, Sivadas D (2017) Economic evaluation of a grid-connected PV system at Otago Polytechnic. In: Second international conference on electrical, computer and communication technologies (ICECCT), pp 1–6. Coimbatore
41. Pérez-Lombard L, Ortiz J, Pout C (2008) A review on buildings energy consumption information. Energy Build 40(3):394–398. https://doi.org/10.1016/j.enbuild.2007.03.007
42. Ranasinghe DSA, Waidyasekara KGAS, Weerasooriya D (2023) Merits and demerits of off-grid solar systems: key stakeholders' perspectives. World Constr Symp 1. https://doi.org/10.31705/WCS.2023.63
43. Renukappa S, Akintoye A, Egbu C, Goulding J (2013) Carbon emission reduction strategies in the UK industrial sectors: an empirical study. Int J Clim Chang Strat Manag 5(3):304–323. https://doi.org/10.1108/ijccsm-02-2012-0010
44. Samarakoon S (2020) The troubled path to ending darkness: energy injustice encounters in Malawi's off-grid solar market. Energy Res Soc Sci 69. https://doi.org/10.1016/j.erss.2020.101712
45. Sanni SO, Oricha JY, Oyewole TO, Bawonda FI (2021) Analysis of backup power supply for unreliable grid using hybrid solar PV/diesel/biogas system. Energy 227. https://doi.org/10.1016/j.energy.2021.120506
46. Semaoui S, Arab A, Bacha S, Azoui B (2013) Optimal sizing of a stand-alone photovoltaic system with energy management in isolated areas. Energy Procedia 36:358–368. https://doi.org/10.1016/j.egypro.2013.07.041
47. Sharma VK, Singh R, Gehlot A, Buddhi D, Braccio S, Priyadarshi N, Khan B (2022) Imperative role of photovoltaic and concentrating solar power technologies towards renewable energy generation. Int J Photoenergy 2022. https://doi.org/10.1155/2022/3852484
48. Shukla A, Sudhakar K, Baredar P, Mamat R (2018) Solar PV and BIPV system: barrier, challenges and policy recommendation in India. Renewable And Sustain Energy Rev 82:3314–3322. https://doi.org/10.1016/j.rser.2017.10.013
49. Silva KPSD (2020) Opportunities and challenges of solar energy application in energy sector of Sri Lanka. Bull Korea Photovolt Soc 6(1)
50. Tanveer A, Zeng S, Irfan M, Peng R (2021) Do perceived risk, perception of self-efficacy, and openness to technology matter for solar PV adoption? An application of the extended theory of planned behavior. Energies 14(16). https://doi.org/10.3390/en14165008
51. Thanthirige TRM, Galappaththi UIK, Kadabadu JCK, Busawon K, Wedage TDG (2022) Life cycle analysis of solar photovoltaic and coal-based electricity generation: case study refers to Sri Lanka's domestic energy consumption. In: 2022 7th International conference on environment friendly energies and applications: renewable and sustainable energy systems, hybrid transportation systems, energy transition, and energy security, EFEA 2022-proceedings. https://doi.org/10.1109/EFEA56675.2022.10063822

52. Timilsina G, Kurdgelashvili L, Narbel P (2012) Solar energy: markets, economics and policies. Renew Sustain Energy Rev 16(1):449–465. https://doi.org/10.1016/j.rser.2011.08.009
53. Wattana B, Aungyut P (2022) Impacts of solar electricity generation on the thai electricity industry. Int J Renew Energy Dev 11(1). https://doi.org/10.14710/ijred.2022.41059
54. Xu L, Wang Y, Solangi YA, Zameer H, Shah SAA (2019) Off-grid solar PV power generation system in Sindh, Pakistan: a techno-economic feasibility analysis. Processes 7(5). https://doi.org/10.3390/pr7050308
55. Yoon J, Sun Y, Rogers JA (2010) Flexible solar cells made of nanowires/microwires. Semicond Nanomater Flex Technol 159–196. https://doi.org/10.1016/b978-1-4377-7823-6.00006-4
56. Zeren F, Akkuş HT (2020) The relationship between renewable energy consumption and trade openness: new evidence from emerging economies. Renew Energy 147:322–329. https://doi.org/10.1016/j.renene.2019.09.006
57. Zhang S, He Y (2013) Analysis on the development and policy of solar PV power in China. Renew Sustain Energy Rev 21:393–401. https://doi.org/10.1016/j.rser.2013.01.002

Chapter 11
Towards Sustainable Offsite Construction: Examining Embodied Energy in Shipping Container Buildings

Peter Nicholas and Buddhini Ginigaddara

Abstract This research explores the state of the construction industry in terms of its actions towards carbon emission reduction, current trends, and values relating to emissions and embodied energy in shipping containers. Shipping containers represent one type of offsite construction that is widely used for the construction of new buildings using recycled material. Given that shipping containers are considered environmentally friendly due to the use of recycled material, this research aims to investigate their carbon emissions and explore whether they are a sustainable solution. In-depth research into trends involving the conversion of shipping containers to buildings is conducted to see if there are any adverse side effects of the practice. Secondary data on emissions and embodied energy were reviewed to determine how this trend holds up to sustainability, as it is commonly associated with related published research. In comparison, traditional buildings and shipping container buildings had a comparative embodied energy requirement of 74,532.50 megajoules and 99,772.44 megajoules, respectively. The research concludes that well-known offsite construction methods, such as shipping containers, should not be readily assumed to be sustainable alternatives to traditional construction methods. The research suggests the importance of reviewing the embodied energy of buildings that utilise shipping containers, rather than assuming their sustainability, thus highlighting significant practical implications for the construction sector.

Keywords Sustainability · Carbon emissions · Embodied energy · Shipping containers · Offsite construction

P. Nicholas · B. Ginigaddara (✉)
School of Architecture and Built Environment, University of Newcastle, Callaghan, NSW, Australia
e-mail: Buddhini.GinigaddaraAppuhamilage@newcastle.edu.au

P. Nicholas
e-mail: Peter.Nicolas@uon.edu.au

T. Ramachandra et al. (eds.), *Concepts, Policies and Technology for Sustainable Buildings*, Advances in Sustainability Science and Technology,
https://doi.org/10.1007/978-981-96-3086-8_11

1 Introduction

The use of shipping containers as an alternative to construction materials with reduced construction time, cost, and carbon emissions has been in existence for several decades. Shipping containers have been used to build schools, onsite construction offices [1], sports stadiums [23], and portable hospitals [13], especially due to their versatility as a modular structure and fast onsite assembly. Under the offsite construction typology by Ginigaddara et al. [15], shipping containers are recognised as a volumetric building element that represents a complete building on its own or can be used as a module, which is a volumetric portion of a building. On account of their minimal waste generation, less material usage, being in line with circular economy principles, and based on life cycle cost analysis, many researchers have demonstrated the sustainability of shipping containers in the construction of buildings [7, 10, 13, 23]. Dara et al. [10] used the Athena Impact Estimator to conduct a life cycle cost analysis of shipping containers against light wood-designed buildings in Canada and found that over a 50-year life cycle, shipping containers had a closer to 10% reduction in cost.

According to Abrasheva et al. [1], the construction industry assumes the sustainability of shipping container homes based on recycling percentages of processed materials and has not accounted for the large amount of factors that could offset their apparent sustainability such as labour cost and energy costs to acquire these materials. Teng and Pan [33] confirmed how input parameters, assumptions, scenarios, and analytical model uncertainty of carbon estimating techniques of prefabricated high rise buildings lead to misinterpreted results. Interestingly, the supply chain that involves a myriad of labour, transportation, and other materials to install shipping containers as functional building elements has been overlooked in previous life cycle cost analysis. For example, a recent extensive study by Dara et al. [10] indicated how the life cycle cost of a typical recycled modular shipping container provides 3% energy efficiency compared to a traditional building unless the shipping container was upcycled using innovative materials and energy sources. This implies that even after upcycling, the energy efficiency of container buildings is lower compared to that of a traditionally constructed building.

With increased adoption of offsite construction and modular buildings, the construction sector is moving towards a more sustainable future [20, 33]. Offsite construction options such as passive houses and cross-laminated timber-based buildings have proven to be sustainable throughout their building life cycle [3]. Although shipping containers are recognised as a predominant type of offsite construction that helps to create volumetric building elements such as modules and complete buildings [15], their carbon emissions and sustainable features have not been sufficiently justified for mass-scale application. Therefore, prior to selecting shipping containers as a sustainable offsite construction solution, their carbon emissions need to be thoroughly examined. Failure to scrutinise this data could lead to a significant setback for the industry and the planet, further straining the economy, social well-being, and the environment.

This book chapter presents an investigation into contemporary practices in the construction industry scrutinising apparent 'facts' of sustainability of shipping container buildings, to determine whether they indeed help the industry prosper in a more sustainability focused world. The research question poised in this study is "what is the carbon 'cost' of building materials and more specifically the embodied energy of a home in comparison to one composed of shipping containers?". The aim of this research is to scale the sustainability of shipping container buildings compared to traditional ones to gauge the industry's progress towards a more sustainable future. All inclusive, are shipping container buildings a low carbon alternative to a traditional constructed building? If not, where do they fall on the carbon scale in relation to a traditional building?

The three-fold objectives of this research study are to (i) investigate the construction industry's practices and gauge its current trajectory towards sustainability goals, (ii) investigate the practice of using shipping containers in construction and determine whether it is benefiting the construction sector's sustainability goals, harming them, or keeping the 'status quo', and (iii) calculate the carbon emissions of shipping container buildings compared to a traditionally constructed building. Due to the geographical location and familiarity of the authors to the Australian construction industry, the research scope is limited to the Australian construction sector. Additionally, the research scope was constrained to evaluate embodied energy-related calculations under environmental sustainability assessment. The next sections of this chapter present the literature review, research methodology, results, and discussion, followed by conclusions of the study.

2 Literature Review

2.1 Australian Construction Industry's Trajectory Towards Reducing Greenhouse Emissions

Although Australia is home to 0.33% of the world's population with a declining phase of CO_2 emissions, it is among the first ten nations of the highest per capita greenhouse gas emissions worldwide [14]. This statement is backed up by the Organisation for Economic Co-operation and Development (OECD) which states the average output as $11.5tCO_2e$/capita [27]. The construction sector is highlighted as a key area where mitigation in carbon emissions should be achieved as it contributes greatly, although in many cases indirectly, due to its trading of materials with large quantities of embedded carbon, massive supply chains, and monumental use of electricity [22]. In line with the Paris Agreement [35], Australia has a Net Zero Plan with a target to reach nationwide net zero carbon emissions across all industries by 2050 [12]. According to Yu et al. [36], the massive scale of the national construction industry's carbon output means that if the construction industry were to commit to the emission reduction target of net zero by 2050, it could contribute 28% towards national emission reduction by 2030.

The Australian construction industry includes companies which are engaged wholly in the construction of residential and non-residential buildings, civil engineering projects, and roadways and bridge projects. Currently, the construction sector could make a greater effort, as the main driving factor behind most construction projects is compliance with building codes governed by the Australian Building Codes Board under the National Construction Code (NCC). However, these codes often fail to address construction practices that result in excessive carbon emissions [18]. Emissions attributed to construction within Australia primarily stem from operating costs and often do not account for the embodied carbon present within materials unless regulated by an external framework. Apart from these governmental bodies, other sector-specific environmental regulatory agencies offer voluntary frameworks such as the Greenstar accreditation framework or the Living Building Challenge [24]. Greenstar is a framework used internationally, developed to assist builders in constructing buildings which reach net zero emissions or minimum emission reduction targets [16]. The framework is more readily adapted opposed to the Living Building Challenge due to its semi-integrated nature in the NCC making it much easier to employ nationally and on a large scale. By undertaking construction using this framework, the builder will have to account for operating emissions, embodied carbon in building materials and comply with standards in every stage of construction [34]. Below are the objectives of the Greenstar framework applicable to the Australian construction sector:

- Management
- Indoor environment quality
- Energy
- Transport
- Water
- Materials
- Land use and ecology
- Emissions
- Innovation.

Overall, direct emissions from the construction, services, and transport industries saw a 15.5% reduction in emissions totalling 9.3 Mt CO_2-e in 2019–2020 [5]. Tumbag et al. [34] found how most greenhouse rated buildings in Australia were performing well in the areas of "greenhouse gas emissions", "indoor pollutants—engineered wood products", "peak electricity demand reduction", "potable water", and "sustainable transport" while whole life cycle assessments need further attention. The emissions of the construction sector are not where they should be, progress is being made rather slowly; however, the industry has seen the positive impacts of adopting greener functioning and the trend is beginning to grow challenging [18].

The construction industry is wide and varied with many activities occurring concurrently. Whether directly or indirectly, the sector contributes greatly to the national emissions of Australia. The carbon emissions of the three sectors construction, services, and transport industry fluctuate over the past 30 years and are currently sitting at 30% higher than their 1990 level [11].

Upon further examination to isolate construction industry carbon emissions, it became evident that, contrary to findings in contemporary literature, direct emissions from the construction sector have not exhibited a downturn in recent years. Instead, there has been an increase in direct emissions since 2015 [11]. In addition to these direct emission calculations, indirect emissions related to construction activities should also be assessed as they may provide a better understanding of variations in emissions data. After a long period of steady but slow growth from the late 1990s to the early 2010s, new peaks in indirect emissions have been reached. Although a slight downward trend has been recorded in the last couple of years, no true trend of decline is predicted as of yet [11].

As of 2023, the trajectory of the construction industry in terms of carbon emissions is looking quite negative. The issues the industry faced considering the COVID-19 pandemic have lasting effects where sustainability appears to not be the major concern, instead it is taking a 'back seat' to getting the industry running smoothly, while resolving the housing crisis. This is important to note, as if further research were to prove the sustainability claims of shipping container homes false, it would only exacerbate the existing issues.

2.2 Carbon Emissions Relating to Mineral Extraction and Refinement of Construction Materials

The efficiency in production of construction materials is one leading factor for emissions in the construction industry. It has been estimated that if certain production methods of building materials were improved, they would still reach a proverbial ceiling of how much reduction could be achieved with predictions for steel being 24% more efficient and 13% for concrete and related products [9]. The solution to this issue would be to prioritise the use of recycled materials [29]. However, in the case of material such as concrete, this can only constitute around 30–40% of the total volume of the new batch [9]. Sizirici et al. [30] studied carbon emissions present solely in the production of concrete and steel all inclusive of acquisition, processing, and transport. The study identified emissions associated with various stages of steel manufacturing, pertinent to shipping containers, specifying the type of fuel used and the corresponding CO_2 emissions per tonne of material produced [30].

The embodied energy value is an important measure in this research as it helps to accurately answer the research objectives. According to the Australian Government [4], the embodied energy is the amount of energy expended to extract, process, and form the component. Typically, the more processed a material is, the higher the embodied carbon value. This often takes the form of the material cost and the assembly cost for the structure to be built, and it is quantified in Mj/kg or megajoules per kilogram [4].

2.3 Structural and Architectural Makeup of Traditional Buildings and Shipping Container Buildings

Based on studies assessing the environmental impact of construction, the top-ranking concerns include environment and soil pollution, excessive resource consumption, and effects on biodiversity and waste generation while significant social implications encompass traffic congestion, disease spread, vulnerability to natural disasters, and reduction of crop productivity [32]. On a grand scale, the construction industry is estimated to contribute to 13% of the world's Gross Domestic Product (GDP), however, it also contributes to an estimated 36% of the global energy consumption and up to 39% of energy-related CO_2 emissions [22]. To curb these impacts on the environment, many have sought to build alternatives working towards a more sustainable industry [2]. This leads to compromises to be made in the large-scale adoption of new technology, which is seldom adopted on a large scale initially, with the industry finding it more sustainable financially and environmentally to instead repurpose old products for new application [29].

When discussing the reuse of materials, one might assume the likes of timber and brick; however other large-scale products have also found in favour with the industry such as the mass-produced shipping container [7]. These shipping containers provide rigid and strong specifications which make them widely applicable to form the basis of many structures. The sustainability of this form of construction is said to be quite high, based on the inherent merits of 100% recyclable steel when compared to traditional materials such as concrete [1]. Madkour [25] stated that the container home can be fitted with "as many green features" as a traditional building while being more sustainable for the environment with its recyclable factors. The major factor which seems to be present in the application of container homes is their ability to be moved, or in other words they can be picked up and transported globally should the need arise [7, 10]. Islam et al. [19] stated that container homes can be made using up to 75% recycled materials, and Bongiorno et al. [7] further justified their application of passive house concepts into container buildings to redefine construction by stating their commitment to "environmentally responsible design". The logical progression from previous research suggests that re-applying or using recycled materials would benefit the industry, especially concerning the principles of circular economy [23]. However, evaluating the carbon footprint of shipping container building through the study of labour and material cost implications is yet to be conducted.

A wide array of materials and designs in construction can be employed based on specifications designated by an architect or engineer and guided by local, state, and national governing legislation. The basic general structural and architectural elements of a traditional building are as follows:

- Foundations—They are typically formed from concrete and can be constructed in various ways. Footing pads are common in concrete foundations and consist of an isolated reinforced concrete section with piers. Another similar concrete footing is a strip footing which is a continuous stretch of reinforced concrete. This can be used for piles if the spacing is close enough; otherwise they are more commonly used for underfloor support for a reinforced concrete slab on top. Screw pile footings, in contrast to other footing types, are driven into the earth. They provide foundational support by either transferring load to solid earth types below the topsoil or by being supported by frictional force from the surrounding earth providing sufficient lateral support. These screw piles are typically formed in steel.
- Framing—The frame of a low-rise building is mostly constructed from softwood timber such as pine or structural laminated veneer lumber (LVL), with specific components necessitating the application of certain hardwoods such as oak timber. The framing is the largest step in the construction of a building with it encompassing the substructure for the floor, load-bearing or non-load bearing external or internal walls, and structure for the roof and supports for all cut-outs in the structure for any windows, doors, or services. The framing will rest on top of the foundation and provide the load transfer of weight through the foundation into the earth.
- Lining/cladding—The lining and cladding of a building is the next highly important step of construction where the material to enclose the frame is installed. The choice of materials here varies based on application, weather exposure, temperature variances, and moisture levels. The interior linings may include plasterboard, fibre cement sheeting, plywood, and fibreboard. The external linings will most likely include fibre cement sheeting, weatherboard cladding, and proprietary cladding systems (such as clip in board systems), and cement render.

An overview of the construction process of a traditionally constructed building is paramount, as it will assist in creating a comprehensive understanding of the potential benefits of shipping container buildings. This can now be applied to a shipping container to draw comparisons. The potential benefits which can be initially observed in the use of shipping containers in construction are their basic structure which can be observed in Fig. 1.

This structural makeup of a container negates the use of many of the abovementioned building elements in a traditionally constructed building. The container will still need adequate foundations to be constructed; however, it will supply its own form of framing with underfloor supports, wall cladding, and roofing. Therefore, the outward benefit of these containers will be that much of the structure of a building need not be constructed. Interior framing, insulation, lining, and proper weatherproofing is still required; however, this narrows the field of assessment. Knowing this information, the question then becomes, does the omittance of these construction elements amount to less embedded carbon and carbon emissions than a traditional building?

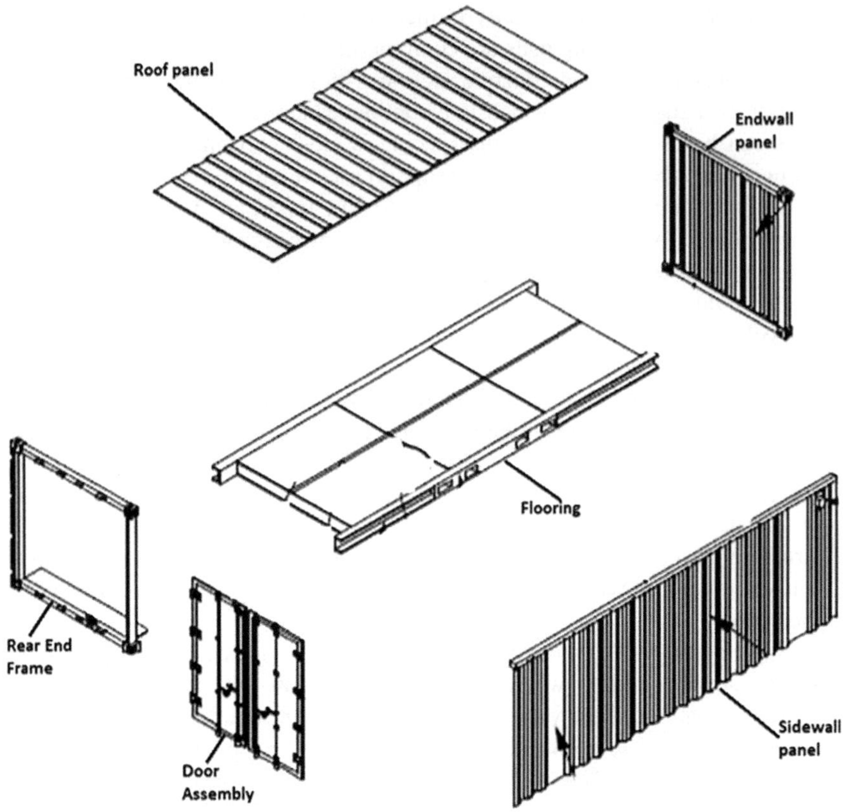

Fig. 1 Shipping container parts. [28] *Source* Adopted from Pacific Marine & Industrial

3 Research Methodology

3.1 Methodological Approach

Huang et al. [17] conducted a global country comparison quantitatively analysing carbon emissions in the construction industry, accounting for both carbon output and embedded carbon. The study provided more accurate results for embedded carbon and emissions, which are often assumed for imported products such as light fittings and sanitary fittings that are manufactured and imported from other countries [17]. Beginning with a similar quantitative assessment, Karlsson et al. [21] conducted a case study-based assessment into a Swedish infrastructure project which involved computer aided design (CAD) modelling-based total emissions of a typical segment of road. The concept is that total emissions from road infrastructure projects can be estimated based on a typical road segment, providing an accurate carbon emission output before project commencement to help manage the country's carbon emissions [21].

The research method for analysing the current state of carbon emissions of shipping container buildings and sustainability in the construction sector builds upon methodological approaches adopted in previous similar research. Huang et al. [17] provided verified carbon emission values, which served as the methodological foundation for assessing the current state of emissions in the construction industry and categorising emissions on a scale from sustainable to unsustainable. Furthermore, the approach seen in Karlsson et al. [21] was also adopted where case study analysis and secondary data sources were employed to provide a basis for CAD modelling to better understand and estimate the embedded carbon and emissions related to construction of container buildings.

3.2 Execution of Research Methods

Online data relating to carbon emissions was gathered from reputable online sources such as the Department of Climate Change, Energy, the Environment and Water and the Australian federal Government website which are in relation to the construction industry in Australia [4, 11]. Afterwards, potential sources/suppliers of shipping container buildings in Australia were explored, and those buildings were used as a basis for 3D CAD models/drawings to be drafted. These CAD models/drawings projected container buildings and traditional buildings of the same dimensions to accurately survey the quantities of building materials which are essential to the construction of these structures. Similar to the method followed by Islam et al. [19], a bill of quantities was created comparing the material cost of constructing a traditional building compared to a container building and presenting its findings in the form of comparisons. These comparisons thereafter stood to prove whether container buildings are in fact more sustainable in the sense that they have less embedded carbon and amount to less greenhouse gas production than the traditionally built structures.

3.3 Justification of Data

The data required for this research is diverse, yet as there is no single source available, multiple sources were reviewed. Many researchers have previously commented on the same issue where the non-existence of a universal standard for carbon emissions lead to many divergent calculations [10, 19]. The use of a single source would be ideal, however, given the nature of the research, this is unfeasible. Multiple sources must be compiled for the carbon data as no single source provides all the data required. This includes the embodied carbon data for construction materials, the emissions data related to transport, refinement/production, and total emissions data. For this reason, the sources to be used for each of them are as follows:

1. Total emissions—Department of Climate Change, Energy, the Environment and Water [11]
2. Embodied Energy—Your Home [4, 5]
3. Pacific Marine and Industrial [28].

The first source that was used for the collection of emission data was a government source, and hence the embodied energy values were taken as it is for calculations. The second source is an Australian government website that allows public access to lists and tables of building materials which define the amount of embedded carbon in materials. The third source was used to find the relevant data and information of shipping containers for drawings and bills of quantities for comparison between traditionally constructed buildings and the shipping container building.

4 Results and Discussion

4.1 Embodied Energy in Relevant Building Elements

Australian Government's embodied energy information published by the Australian government in 2020 (Table 1) was used for research data analysis. It is noted that only A1–A3 emissions were evaluated as only materials used in the produce stage were considered in this hypothetical scenario review.

LVL is a high-strength engineered wood product that is used for structural elements including beams and rafters. The lowest embodied energy of the selected materials is present in MDF which is also an engineered wood product that has a smooth surface and uniform density to be particularly used for moulding, shaping, cutting, and routing the surfaces. Compared to other wood-related materials present in Table 1, plywood has the highest embodied energy of 42.9 Mj/kg. It is a composite material made from thin layers of wood veneer bonded together with adhesive to create a strong and versatile building material. Hardwood and softwood are classifications of wood based on the type of trees they come from: hardwoods are derived

Table 1 Embodied energy in relevant materials

Material	Embodied energy Mj/kg
Laminated veneer lumber (LVL)	34.3
Medium density fibreboard (MDF)	22.0
Plywood	42.9
Softwood	19.0
Hardwood	26.9
Corrugated steel sheet	79.6
Corten steel (weathering steel)	79.6

Source Australian Government [4]

from broad-leaved trees, while softwoods come from cone-bearing, evergreen trees. Understandably, hardwood has a higher embodied energy compared to softwood. Corrugated steel refers to steel sheets that have been formed into ridges and grooves, creating a structural panel with increased strength and rigidity, commonly used in roofing and siding applications for buildings.

Data set published by the Australian Government [4] does not specify the embodied energy of Corten or weathering steel. The embodied energy of weathering steel is known as an unclassified property of the product and, therefore, no exact measurements can be made. The authors came across an unverified source, which denotes an embodied energy of 23.0 Mj/kg for weathering steel [26]. Given that weathering steel is an alloy composed of various metals such as chromium, nickel, copper, silicon, and phosphorus, an assumption was made that its embodied energy rating should be relatively high. For calculation purposes, it was assigned the same embodied energy value as corrugated steel sheets. This is because the components of a shipping container, to which the value will be applied, are similarly fabricated into an arbitrary shape.

The next step in the embodied energy calculations and generation of bills of quantities is to establish the materials required for constructing both a shipping container building and a traditionally built building of comparable size. To achieve this, the methodology used in Karlsson et al. [21] was adopted as CAD models serve as the basis for a better understanding of the structures.

As discussed previously, the structure of the shipping container eliminates the need for several building elements. Therefore, this research focuses solely on the elements rendered obsolete by the application of a shipping container compared to a traditional building. The container negates the use of a framing system as the container acts as a rigid frame in and of itself. The framing elements which it would omit from the construction process are the following:

- Underfloor structural framing (containers are fitted with extensive underfloor support systems to allow them to be craned and lifted)
- Structural lateral wall framing (container wall sheets provide adequate strength and support and provide a weatherproof seal)
- Roof structure (no roofing structure required as the containers roof provides all roofing elements).

Additionally, essential building elements not accounted for include:

- Subfloor (container has pre-fitted plywood flooring)
- External Cladding (the external container walls provide a rigid weatherproof seal from the elements)
- Roof Cladding (the roof of the container is weatherproof and negates the application of roofing elements and cladding).

Therefore, the elements which must be assessed against each other are the shipping container, the framing, cladding, and subfloor. For the basis of these calculations, a 40-foot shipping container has been used as a basis (see Fig. 2). The usable space inside the container was recorded, and a traditionally built building was constructed

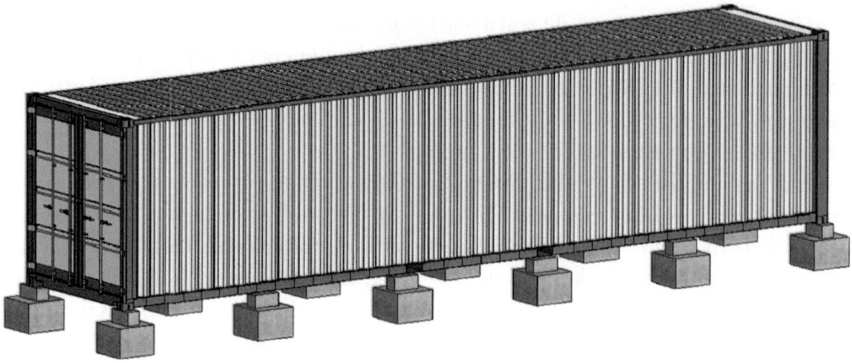

Fig. 2 Shipping container building, external 3D model. *Source* Developed by the authors

to the same dimensions. The traditional building was outfitted in terms of common building practices stipulated by the Australian Government [6]. This includes:

- A structural LVL constructed floor
- A particleboard subfloor
- Structural timber framing
- Exterior hardwood weatherboard cladding
- A timber truss roof
- Corrugated steel roof cladding.

4.2 CAD Models Used for Data Analysis

This section presents the CAD models that the bill of quantities was created from and will thereafter include all relevant details and information. Figures 2, 3 and 4 represent the shipping container information with the 3D model, floor plan and elevation details, respectively.

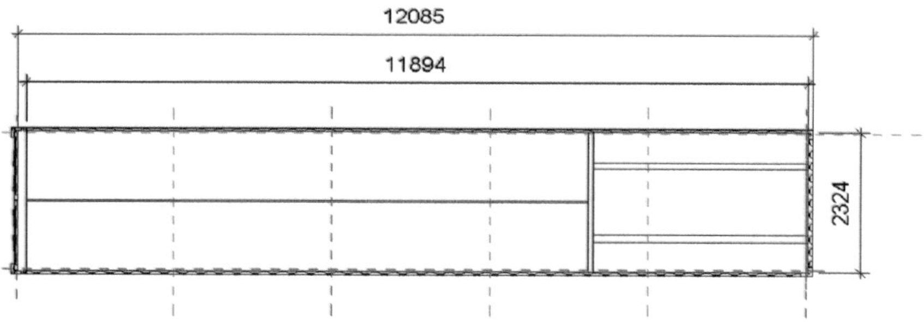

Fig. 3 Shipping container building, floor plan. *Source* Developed by the authors

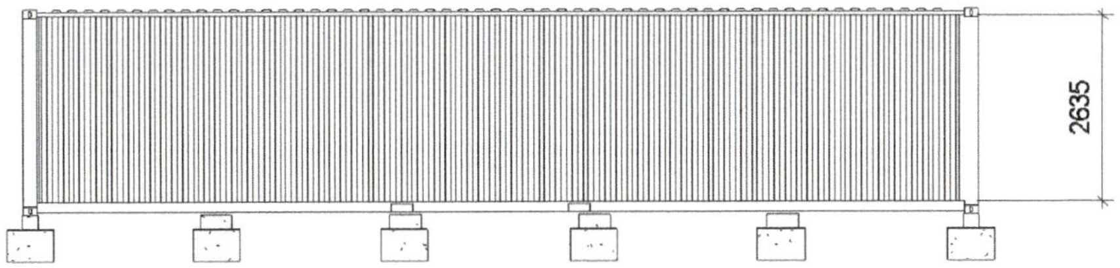

Fig. 4 Shipping container building, elevation. *Source* Developed by the authors

The shipping container is designed to have approximately 28m^2 gross floor area with a floor height of 2.64 m.

In addition to the shipping container, the CAD model of a traditional building was also designed. Their details are presented in Figs. 5, 6 and 7 with the 3D model, floor plan, and the elevation details, respectively.

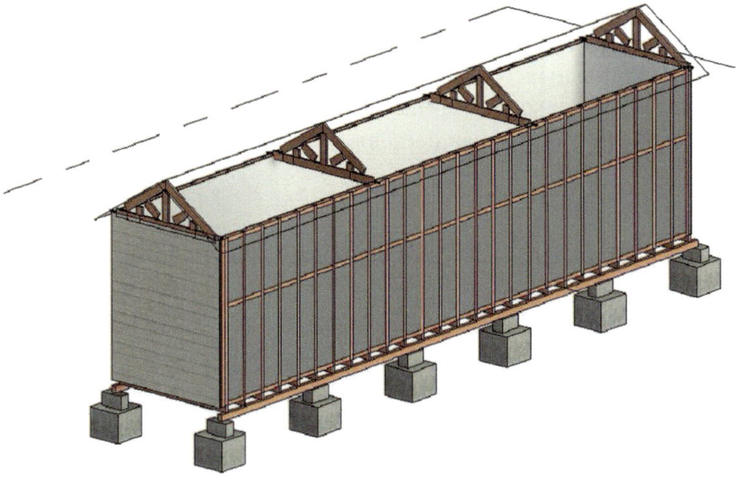

Fig. 5 Traditionally constructed building, 3D model. *Source* Developed by the authors

Fig. 6 Traditionally constructed building, floor plan. *Source* Developed by the authors

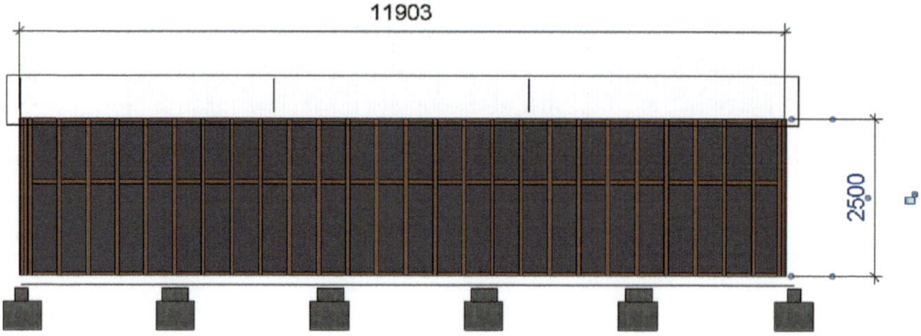

Fig. 7 Traditionally constructed building, elevation. *Source* Developed by the authors

Similar to the shipping container, the traditional building has a gross area of 28m^2 and a floor height of 2.50 m.

Firstly, to address the differences between the two subjects, the dimensions differ slightly based upon various factors. The slight difference in floor area dimensions is due to the container structure not yet being fitted with internal walls, which would be necessary for housing to incorporate any services and finishes. The height difference is to account for in-roof services as the space in the roof of the traditional building allows for this while the container does not.

4.3 Embodied Energy Calculations and Discussion of Findings

The timber structure of the traditional building has been constructed according to the Timber Framing Code, AS 1684 [31]. This includes the spacing of underfloor bearer and joint construction, timber stud frame wall construction, and roofing truss systems. This encompassed 450 mm upright stud spacing for external walls, a dual bearer floor construction, and 4 timber trusses supporting the roof structure [31]. These calculations were then used to generate a Bill of quantities for the traditional building elements (Table 2).

From these figures, a weight was assigned to each material to be able to determine the embedded energy value of the building. These were based on identical products found on the Bunnings Warehouse website which defines weights for all products [8]. Based on these figures and the values presented within Table 2, the embodied energy in the traditionally constructed building was estimated as presented in Table 3.

This brings the total embodied energy of the traditionally constructed building to 74,532.5 Megajoules of embodied energy. The next step was to determine the value of the container to assess which option is more sustainable. This can be completed in a much simpler manner than the traditional building due to the container having minimal differing materials. The same process was undertaken starting with the weight of the shipping container. [30] defined the tare weight of a 40ft shipping

Table 2 Bill of quantities for traditionally constructed building

Element	Description	Quantity	Unit	Weight (kg)
Floor structure	100 × 75 mm Underfloor bearers (LVL)	24.08	m	77.1
	90 × 45 mm Underfloor joists (LVL)	62.7	m	169.3
	19 mm particleboard subfloor	24.7	m^2	326.8
Wall structure	90 × 45 mm timber studs	247.9	m	527.2
	External weatherboard cladding	71.2	m^2	1265
Roof structure	90 × 45 mm timber truss	26.44	m	100.4
	70 × 35 mm roofing battens	47.2	m	31.2
Roof cladding	Corrugated steel sheet	39.4	m^2	155.1

Source Developed by the authors

Table 3 Embodied energy calculations for traditionally constructed building

Material	Mj/kg	Weight (kg)	Embodied energy (Mj)
100 × 75 LVL	34.3	77.1	2,644.5
90 × 45 LVL	34.3	169.3	5,807
19 mm particleboard	22	326.8	7,189.6
90 × 45 mm timber	19	627.6	11,924.4
Weatherboard	26.9	1265	34,028.5
Roofing battens	19	31.2	592.8
Corrugated steel	79.6	155.1	12,346
Total			74,532.5

Source Developed by the authors

container to be approximately 3750 kg, and it was also denoted that the plywood flooring that comprises the shipping container is approximately 40 mm thick. These numbers were then cross-referenced with product listings with associated weights at [8] and the CAD models to produce estimates:

- The total floor area of the container = 26.47m^2
- 40 mm Plywood flooring = 679.52 kg.

Minus the flooring weight from the total container weight and the answer is as follows:

- Corten/weathering steel shipping container = 3,070.48 kg.

The authors then determined the embodied energy in the shipping container as presented in Table 4.

When compared to a traditional building, a container building has immensely more embodied energy than a traditionally constructed building. The embodied energy value is, however, an estimate and may not be accurate in the slightest. The authors

Table 4 Embodied energy calculations for shipping container building

Material	Mj/kg	Weight (kg)	Embodied energy (Mj)
40 mm Plywood	42.9	679.52	29,151.40
Weathering steel	79.6	3,070.48	244,410.21
Total			273,561.61

Source Developed by the authors

Table 5 Further embodied energy calculations for shipping container building

Material	Mj/kg	Weight (kg)	Embodied energy (Mj)
40 mm Plywood	42.9	679.52	29,151.40
Weathering steel	23	3,070.48	70,621.04
Total			99,772.44

Source Developed by the authors

acknowledge that the calculations are bound to assumptions that were made for this calculation process. Recalculations were also undertaken with the 23Mj/kg of embodied energy in weathering steel presented by Makeitfrom [26] (see Table 5) to determine whether that would skew the result in the shipping container building's favour.

Even when the shipping container is assessed to have a much lower embodied energy for weathering steel (23 Mj/kg in Table 5 against 79.6 Mj/kg in Table 4), it still presents as less sustainable than a traditionally constructed building. It would then stand to reason that these container buildings are not as sustainable as they initially were considered. Against the previous conclusions of Abrasheva et al. [1], Dara et al. [10], Islam et al. [19] and Bongiorno et al. [7], this research findings present a myth buster for the typical assumption that shipping container buildings are inherently more sustainable compared to traditionally constructed buildings.

5 Implications for Research Findings

The research findings hold significant implications for various stakeholders that are involved in the construction industry. The findings provide insights to architects and designers who are involved in the design process of shipping container buildings. Careful consideration of embodied energy and their related environmental implications over the long run should be considered when selecting materials and building elements that are deemed to be "environmentally friendly" due to their social implications and adherence to circular economy principles. For example, Kucukvar et al. [23] conducted a life cycle impact assessment of the FIFA World Cup stadium in Qatar, which was made out of 974 recycled shipping containers, to conclude that circular design principles reduce human health impacts and significantly reduce the material footprint. While those research findings confirm the positive impacts of using shipping containers in mega scale, it should be noted that the carbon calculations and embodied energy-based evaluations require specific analysis that are particularly relevant for each scenario. As such, one of the current research implications is to be cautious when selecting well-known recycled or upcycled materials for building construction.

Currently, Australia is facing a national housing crisis which creates avenues for selecting quick and easy fixes such as shipping container buildings for affordable social housing that also fall under the umbrella category of sustainable offsite construction solutions. Therefore, policymakers can leverage these research findings to inform policy decisions by understanding the environmental impact and feasibility of shipping container buildings. The research findings highlight the importance of raising environmental awareness within the construction industry and among the general public. By highlighting the embodied energy of construction materials, the study contributes to the growing body of knowledge that mitigates the environmental impact of construction projects.

6 Conclusions

This research shows that trends are not all that they may appear at first glance. Throughout the research, many avenues were discussed on how to arrive at its results and the simplest route was the most effective. This research aim was to scale the sustainability of contemporary construction practices compared to traditional ones to gauge the industry's progress towards a more sustainable future, particularly with reference to shipping containers that are being used as modules or complete building structures under offsite construction. The research findings indicate how a comparison of a traditional building, and a shipping container building shows a higher embodied energy utilisation by the shipping container building. In comparison, the traditional building and the shipping container building had a comparative embodied energy requirement of 74,532.5 Megajoules and 99,772.44 Megajoules, respectively.

Shipping container buildings that are recognised as a type of offsite construction, while they may appear to promote a more sustainable industry, in fact, do the opposite. They appear to not aid the industry in its sustainability goals but actively push against them. It is possible that it is trends like these which have seen the industry struggle to meet emissions targets and fall behind other industries in terms of 'cutting back'. It is a trap that the industry must be aware of moving forward. Although projections for carbon output in the sector do not appear to be moving forward as fast as some might have hoped, it is still an active movement championed by contemporary leaders in construction. It is a goal that appears attainable; however more research must be conducted before wide-scale trends such as shipping container buildings take hold and cause a "one step forward two steps back routine". Although the goals of this line of research were met, simply identifying the problem is only the first step in a long routine of rectification work which only becomes a larger process when trends are left unresearched.

The research can be further expanded to include more complex building structures that include several shipping containers and various building elements including typical finishes. In addition, empirical data-based research can be conducted to evaluate both operational and embodied energy of these two different types of buildings over the building life cycle, from inception till demolition. In particular, circular economy principles can be incorporated to improve such studies. As the embodied energy calculations were done using the simplest methods and a selected data source, the research findings may limit the accuracy if a different database was used. Given the unavailability of a universal standard method for carbon estimations, this is acknowledged as a limitation of the current study.

References

1. Abrasheva G, Senk D, Häußling R (2012) Shipping containers for a sustainable habitat perspective. Revue de Métallurgie 109:381–389
2. Al-Atesh EA, Rahmawati Y, Zawawi NAWA, Utomo C (2021) A decision-making model for supporting selection of green building materials. Int J Constr Manag 23:922–933
3. Alshikh Z, Trepci E, Rodriguez-Ubinas E (2023) Sustainable off-site construction in desert environments: zero-energy houses as case studies. Sustainability 15
4. Australian Government (2020) Embodied energy. Australian Government. https://www.yourhome.gov.au/materials/embodied-energy. (Online)
5. Australian Government (2022) National inventory by economic sector: annual emissions. Department of Climate Change, Energy, the Environment and Water. https://www.dcceew.gov.au/climate-change/publications/national-greenhouse-accounts-2020/national-inventory-by-economic-sector-annual-emissions. (Online)
6. Australian Government (2024) Your home-materials. Australian Government. https://www.yourhome.gov.au/materials. (Online)
7. Bongiorno E, Borgia C, Detommaso M, Nocera F (2021) Sustainable and green building design: shipping container as Passivhaus. New Metrop Perspect 1423–1432
8. Bunnings (2023) Bunnings warehouse. https://www.bunnings.com.au/products. (Online)
9. Churkina G, Organschi A, Reyer CPO, Ruff A, Vinke K, Liu Z, Reck BK, Graedel TE, Schellnhuber HJ (2020) Buildings as a global carbon sink. Nat Sustain 3:269–276
10. Dara C, Hachem-Vermette C, Assefa G (2019) Life cycle assessment and life cycle costing of container-based single-family housing in Canada: a case study. Build Environ 163
11. Department of Climate Change Energy the Environment and Water (2022) National inventory by economic sector: data tables and methodology. Australian Government (Online)
12. Department of Climate Change Energy the Environment and Water (2024) Net zero. Department of Climate Change Energy the Environment and Water. https://www.dcceew.gov.au/climate-change/emissions-reduction/net-zero. (Online)
13. El Ghonaimy I, Habbab AH (2020) Sustainability and the recycle of the portable shipping containers in offering health care services in COVID 19 circumstances. In: 2020 Second international sustainability and resilience conference: technology and innovation in building designs (51154), pp 1–5
14. Emissions Database for Global Atmospheric Research (Edgar) (2023) Community GHG database, a collaboration between the European Commission, joint research centre (JRC), the international energy agency (IEA), and comprising IEA-EDGAR CO2, EDGAR CH4, EDGAR N2O, EDGAR F-GASES version 8.0. European Commission

15. Ginigaddara B, Perera S, Feng Y, Rahnamayiezekavat P (2021) Development of an offsite construction typology: a delphi study. Buildings 12
16. Green Building Council Australia (2024). Exploring green star. Green Building Council Australia. https://new.gbca.org.au/green-star/exploring-green-star/. (Online)
17. Huang L, Krigsvoll G, Johansen F, Liu Y, Zhang X (2018) Carbon emission of global construction sector. Renew Sustain Energy Rev 81:1906–1916
18. Hurlimann AC, Browne GR, Warren-Myers G, Francis V (2018) Barriers to climate change adaptation in the Australian construction industry–impetus for regulatory reform. Build Environ 137:235–245
19. Islam H, Zhang G, Setunge S, Bhuiyan MA (2016) Life cycle assessment of shipping container home: a sustainable construction. Energy Build 128:673–685
20. Jayawardana J, Kulatunga AK. Jayasinghe JASC, Sandanayake M, Zhang G (2023) Environmental sustainability of off-site construction in developed and developing regions: a systematic review. J Arch Eng 29
21. Karlsson I, Rootzén J, Johnsson F (2020) Reaching net-zero carbon emissions in construction supply chains–Analysis of a Swedish road construction project. Renew Sustain Energy Rev 120
22. Kiani Mavi R, Gengatharen D, Kiani Mavi N, Hughes R, Campbell A, Yates R (2021) Sustainability in construction projects: a systematic literature review. Sustainability 13
23. Kucukvar M, Kutty AA, Al-Hamrani A, Kim D, Nofal N, Onat NC, Ermolaeva P, Al-Ansari T, Al-Thani SK, Al-Jurf NM, Bulu M, Al-Nahhal W (2021) How circular design can contribute to social sustainability and legacy of the FIFA World Cup Qatar 2022™? The case of innovative shipping container stadium. Environ Impact Assess Rev 91
24. Living Future Institute Australia (2024) The living building challenge. Living Future Institute Australia. https://living-future.org.au/living-building-challenge/. (Online)
25. Madkour WM (2017) Shipping containers as an approach to increase the quality of economic sustainable buildings in Egypt. In: The 1ST international conference on towards a better quality of life. El Gouna, Red Sea Region Egypt
26. Makeitfrom (2020) EN 1.8961 (S235J2W) weathering steel. https://www.makeitfrom.com/material-properties/EN-1.8961-S235J2W-Weathering-Steel. (Online)
27. OECD (2021) Australia-progress in the net zero transition
28. Pacific Marine & Industrial (2024) Shipping container parts. Pacific Marine & Industrial. https://www.pacificmarine.net/marine-deck/shipping-container-parts.htm. (Online)
29. Sieffert Y, Huygen JM, Daudon D (2014) Sustainable construction with repurposed materials in the context of a civil engineering–architecture collaboration. J Clean Prod 67:125–138
30. Sizirici B, Fseha Y, Cho CS, Yildiz I, Byon YJ (2021) A review of carbon footprint reduction in construction industry, from design to operation. Mater (Basel) 14
31. Standards Australia (1992) National timber framing code. Standards Australia
32. Tafesse S, Girma YE, Dessalegn E (2022) Analysis of the socio-economic and environmental impacts of construction waste and management practices. Heliyon 8:e09169
33. Teng Y, Pan W (2020) Estimating and minimizing embodied carbon of prefabricated high-rise residential buildings considering parameter, scenario and model uncertainties. Build Environ 180
34. Tumbaga J, Vimonsatit S, Lopez JR, Holschemacher K, Quapp U, Singh A, Yazdani S (2022) Trends in sustainable practices in the construction industry using green star certified buildings data. Proc Int Struct Eng Constr 9
35. United Nations (UN) (2024) The Paris agreement. UN. https://unfccc.int/process-and-meetings/the-paris-agreement. (Online)
36. Yu M, Wiedmann T, Crawford R, Tait C (2017) The carbon footprint of Australia's construction sector. Procedia Eng 180:211–220

Chapter 12
The Impact of Indoor Air Quality in Higher Education Institutions

Mahnoor, Nirodha Gayani Fernando, and Hemantha Sandaruwan Bandara Rathnaweera Patabandige

Abstract Indoor Air Quality (IAQ) is a critical aspect of the educational environment, significantly influencing the well-being of students in higher education institutions (HEIs). This chapter is dedicated to presenting the impact of IAQ on students' well-being in a HEI, with a specific focus on carbon dioxide (CO_2) Particulate Matter less than 2.5 μm in diameter ($PM_{2.5}$) concentrations and the Air Quality Index (AQI). A university in Edinburgh has been selected as a case for this study. Various locations within the case study, such as the library, central computer lab, reception hallway, cafeteria, and two classrooms, were used to measure IAQ. In the case study, fifty questionnaire surveys were conducted among students to gain insights into how IAQ impacts students' well-being and their awareness. The data were analysed using frequency analysis as part of the statistical techniques. The case study revealed that the AQI and $PM_{2.5}$ values in 2 classrooms exceeded acceptable thresholds due to reduced ventilation caused by closed doors and windows during the summer. In contrast, other places maintained a favourable AQI, CO_2, and $PM_{2.5}$ range, likely attributed to natural ventilation through open windows and doors. The chapter presents several health impacts of the IAQ, including reduced cognitive performance, headaches, dizziness, dry and irritated throats, fatigue, and tiredness. Allergy symptoms such as sneezing, flu-like symptoms, and coughing were also prevalent among some respondents. The study's practical implications highlight the need for policies that manage CO_2 and $PM_{2.5}$ levels through effective ventilation systems. It also recommends the implementation of indoor plants and mechanical

Mahnoor · N. G. Fernando (✉)
School of Computing, Engineering & the Built Environment, Edinburgh Napier University, Edinburgh, UK
e-mail: N.Fernando@napier.ac.uk

Mahnoor
e-mail: 40522185@live.napier.ac.uk

H. S. B. Rathnaweera Patabandige
Department of Civil Engineering, Faculty of Engineering, Sri Lanka Institute of Information Technology, Malabe, Sri Lanka
e-mail: hemantha.b@sliit.lk

T. Ramachandra et al. (eds.), *Concepts, Policies and Technology for Sustainable Buildings*, Advances in Sustainability Science and Technology,
https://doi.org/10.1007/978-981-96-3086-8_12

ventilation systems, the setting of standardised occupancy levels, and the regular monitoring and management of IAQ.

Keywords Indoor air quality · CO_2 · Indoor particulate matter · Student well-being · Higher education institution

1 Introduction

According to the EPA's definition, IAQ is the air quality within and around buildings and structures, especially concerning the health and comfort of building occupants [58]. The adverse effects of IAQ on health manifest in heightened hospitalisation rates, affecting respiratory, nervous, and cardiovascular systems. Additionally, it is associated with carcinogenicity and endocrine disruption [18, 29, 37]. According to Korsavi et al. [28], three main factors influence IAQ: Context, Occupation, and building. Contextual factors include macro-level elements like climatic conditions [9] and micro-level factors such as regional temperature [49] and window draughts [21]. Occupant-related factors involve behaviours [6], system maintenance and operation [6], occupancy levels [34], activity levels, duration of occupancy, previous occupancy [56], age, diet [49], and individual thermal comfort [27, 53]. Building-related factors involve considerations such as airtightness [6], school location, classroom and window design [21], ventilation type and rate [6], various sources of outdoor air pollution (e.g., traffic, industrial activities, construction, and combustion activities), CO_2 exhalation rate, and room volume [30].

Gaining insight into the impact of occupant-related factors within educational institutions is crucial for understanding IAQ [2]. In environments such as classrooms and libraries, where population density is high, occupants' exhalation of CO_2 is more pronounced than in other work settings [34]. The accumulation of CO_2 in these spaces harms IAQ, leading to discomfort and impaired cognitive function [27]. Particulate Matter (PM) are carbonaceous particles associated with adsorbed organic chemicals and reactive metals [55]. Fine Particulate Matter ($PM_{2.5}$) is a significant pollutant regulated under the Clean Air Act by the United States Environmental Protection Agency (EPA). $PM_{2.5}$ has been associated with adverse health outcomes, and recent studies have revealed that even low levels of exposure can contribute to increased mortality rates [37]. Epidemiological research consistently shows that exposure to $PM_{2.5}$ negatively affects human health.

Proper ventilation is essential for improving IAQ as it helps circulate fresh air, reduce levels of CO_2 and $PM_{2.5}$, and eliminate excessive heat and moisture. Natural ventilation is crucial in creating a conducive academic environment in educational institutions. However, inadequate AQI within these HEIs, primarily caused by high levels of CO_2 and $PM_{2.5}$, along with heat accumulation due to a large number of students, increases their exposure to indoor air pollution and can lead to various health issues [37, 50].

Previous studies have focused on residential or uninhabited spaces, examining temperature, humidity, and outdoor CO_2 and $PM_{2.5}$ concentrations. Haddad et al. [17] conducted a study investigating IAQ and thermal conditions in secondary school classrooms in Australia's subtropical region. Turunen et al. [57] investigated IAQ and its impact on students' health in Finland. Additionally, lower ventilation rates and higher temperatures were associated with more reports of unsatisfactory indoor air quality by students [57]. A lack of research has been conducted specifically in higher education settings on how students' performance impacts IAQ. Yuhe et al. [64] emphasised the insufficient research in quantifying the effects of CO_2 and $PM_{2.5}$ levels within school premises on student activities.

In this context, this chapter aims to explore the connection between AQI, particularly the levels of CO_2 and $PM_{2.5}$, and their influence on students' health and academic performance in HEIs. By analysing the factors affecting IAQ and studying the potential health consequences, this chapter offers valuable insights to educational authorities and policymakers. Based on the case study findings, the recommendations are to enhance IAQ standards and establish healthier learning environments.

2 Literature Review

2.1 Indoor Air Quality

IAQ index encompasses four main aspects: thermal comfort, indoor air quality (IAQ), visual comfort, and acoustic comfort. It plays a crucial role in ensuring the comfort, health, and well-being of occupants in buildings [27]. Many school buildings face various challenges regarding the thermal environment and indoor air quality. A report from the Public Health Agency of Sweden indicated that 15% of Swedish schools were classified as having a 'bad' or 'poor' IAQ Index [8]. Inadequate IAQ can result in Sick Building Syndrome (SBS), which presents symptoms such as feeling cold, experiencing headaches, dizziness, confusion, nausea, fatigue, respiratory problems, and irritation of the eyes, throat, nose, and skin. Furthermore, AQI significantly impacts buildings' lifecycle costs and energy consumption [27]. Sufficient IAQ can be achieved through incorporated ventilation systems [14]. The primary objective of ventilation is to remove or diminish the concentration of pollutants in the air, which can originate from building materials, occupants, and indoor activities [33]. Ventilation rates specific to indoor environments are defined by international and national standards and legal regulations [8]. According to Montazami et al. [38], it's essential to ensure that classrooms provide maximum comfort in every aspect, as the learning experience involves multiple senses like hearing, speaking, and visualising. Hence, improving the Indoor Air Quality (IAQ) index is vital for boosting productivity and performance. Failing to address thermal comfort issues may compromise learning and academic activities due to discomfort or distractions [52].

De Giuli et al. [13] conducted a study on Indoor Environment Quality in primary schools in Italy. Their findings highlighted significant issues such as excessively high levels of CO_2, inadequate lighting, uncomfortable thermal conditions, and acoustic discomfort as the primary concerns observed in the studied schools. Further, Turanjanin et al. [56] investigated IAQ and its impact on students' health in Finland. They found a significant statistical correlation between temperature and self-reported poor IAQ. Additionally, lower ventilation rates and higher temperatures were associated with more reports of unsatisfactory indoor air quality by students [56]. The study by Ghita and Catalina [16] aimed to assess the indoor AQI in three different types of rural schools (old, new, and renovated) over an extended period from autumn 2013 to spring 2014. The researchers used long-term measurements and spot recordings to evaluate Indoor Environment Quality (IEQ). The findings revealed that the new and renovated buildings rank Class C on the IEQ index scale, whereas the old building is rated D Class (IEQ). Catalina and Iordache [10] carefully selected IEQ index thresholds to represent different IEQ classes optimally. Class A classification is assigned to IEQ index values exceeding 90, while Class B includes IEQ index values falling between 67.5 and 90. For Class C, IEQ index values range from 32.5 to 67.5, Class D contains IEQ index values between 10 and 32.5, and any IEQ index below 10 is categorised as Class E. Furthermore, indoor air pollutants emanate from various indoor sources, such as smoking, cooking, building materials, furnishings, space heaters, pesticides, personal care products, office equipment, cleaning agents, and paints [46]. Additionally, outdoor sources, including fuel combustion processes in power plants and industrial facilities, vehicular traffic, biomass burning, and the re-suspension of road dust, contribute to indoor air pollutants [62]. Table 1 shows the common indoor air pollutants, their potential sources, and their effects on human health.

2.2 *Impact of CO_2 and $PM_{2.5}$ on IAQ*

CO_2 is an invisible and scentless gas emitted by humans through metabolic processes and generated by the combustion of fossil fuels. In indoor environments, where CO_2 levels are usually low, it remains ordinary and is generally considered harmless by individuals. The primary contributor of CO_2 in classrooms is the exhaled breath of the occupants [12]. Myhrvold et al. [40] found that classrooms experiencing CO_2 concentrations exceeding 1500 ppm could result in symptoms such as headaches, dizziness, fatigue, difficulty concentrating, and unpleasant odours.

Several studies have examined the impact of CO_2 levels on the health and well-being of students in educational settings. Madureira et al. [35] found that highly occupied classrooms exceeded 1,000 ppm CO_2 concentration, reducing indoor air quality and potential health issues such as allergies and asthma. Similarly, Simoni et al. [51] highlighted respiratory health in students and reported that children exposed to CO_2 levels above 1,000 ppm had a higher risk of experiencing a dry cough. Furthermore, elevated values beyond recommended levels were associated with reduced

Table 1 Indoor air pollutants, their potential sources, and their effects on human health

Pollutants	Source	Health effects	References
Particular Matter (PM)	Transport, smoking, cooking, heating, candles and incense burning, cleaning, and personal care products	Allergies, discomfort, eye, nose and throat irritation, bronchitis, cardiovascular diseases, respiratory diseases, Asthma, lung cancer, and heart attacks	[1, 19, 31]
Carbon Monoxide (CO)	Cooking stoves, smoking, traffic emissions, vehicle exhaust	Headache, nausea, fatigue, chest pain, reduced brain function, impaired vision and coordination, dizziness, and reduced brain functioning	[23, 36]
Carbon Dioxide (CO_2)	Combustion emissions, metabolic activities, and smoking, cooking, and heating	Headache, sleepiness, fatigue, and concentration difficulty	[4, 65]
Nitrogen Dioxide (NO_2)	Gas-fuelled cooking and heating appliances	Increased asthmatic reactions, respiratory damage, and airway inflammation	[48, 61]
Ozone (O_3)	Outdoor sources, photocopiers, disinfecting devices, air purifiers, and printers	DNA damage, chest pain, cough, asthma, lung inflammation, airway narrowing, and decreased respiratory functions	[26, 47]
Volatile Organic Compounds (VOCs)	Paints, varnish, wood preservatives, aerosol sprays, cleaning products, consumer products, cosmetics, pesticides, adhesives, paints, dyes, polishes, dry-cleaned clothes, lubricants, smoking, perfumes, office equipment such as laser printers and photocopiers	Eye, nose and throat irritation, nausea, loss of coordination, asthma, bronchial hyper-reactivity, damage to the liver, kidney and central nervous system, respiratory illness, and cancer	[25, 54]

nasal patency in exposed children [51]. The measurements and calculations on the classroom's CO_2 concentration revealed inadequate ventilation system effectiveness, resulting in poor air quality [41]. The influence of CO_2 on cognitive function remains diverse, and inhaling CO_2 leads to its accumulation in the body, which subsequently triggers psychological responses like heightened arousal or increased sleepiness level; it may impact cerebral blood flow or alter the oxygen partial pressure in the brain, thereby affecting brain functionality [42, 65].

Hänninen et al. [20] found significantly low ventilation rates, high CO_2 concentrations, and a sensation of stuffiness in Albanian school classrooms during the winter, primarily due to low temperatures and high occupancy. In Kuwait, Al-Awadi [3] conducted research and found elevated CO_2 concentrations in classrooms, indicating poor indoor air quality and increased health risks for students, negatively affecting their educational performance. Insufficient and inadequate ventilation rates, emissions from cleaning products, and chemicals released by building materials and furnishings were identified as the main contributors to indoor air pollution, according to the study [7]. Due to the heightened susceptibility, children and young adults are particularly vulnerable to the negative consequences of poor indoor AQI [43]. Schools face a multitude of challenges as they often accommodate a high number of occupants. These challenges encompass the buildup of harmful gases such as CO_2 and particulate matter (PM) and the potential health risks associated with these elements [43]. Vouriot et al. [59] utilised CO_2 concentration data to assess the prevalence of airborne infectious diseases in classrooms across schools in the UK. Their research demonstrated that variations in CO_2 concentration and ventilation rates influenced the risk of infection, with the highest risk observed in January 2021. The significance of indoor air quality, specifically the concentration of CO_2 in schools, for children's health and academic performance is widely acknowledged.

In 2021, the World Health Organisation (WHO) released revised global air quality guidelines encompassing outdoor (ambient) and indoor air recommendations. These guidelines outline acceptable levels of PM, including PM_{10} and $PM_{2.5}$. The recommended daily average for $PM_{2.5}$ concentration is 35 μg per cubic meter of air (μg/m^3). This signifies the threshold level below which the concentration of $PM_{2.5}$ is deemed within acceptable limits for maintaining good air quality [62]. "Micrograms per cubic meter" (μg/m^3) is a unit used to measure the concentration of particulate matter with a diameter of 2.5 μm or smaller, commonly known as $PM_{2.5}$. This unit indicates the mass of $PM_{2.5}$ particles in each air volume [62]. The Environmental Protection Agency (EPA) standards, similar to those of the UK's Health and Safety Executives, were used to determine indoor air quality. The corresponding CO_2, $PM_{2.5}$ concentration, and AQI levels per the EPA standards are shown in Table 2.

Particulate Matter (PM) originates from various sources, such as chalk dust, soil dust, furniture, cleaning procedures, particles stirred up by children's movements, combustion sources like heaters, gas, and wood stoves, smoking in permitted areas, as well as outdoor traffic and industrial emissions [12]. "Particulate Matter" combines solid and liquid particles in the air. These particles are small enough to remain suspended in the atmosphere rather than settle on the Earth's surface due to gravity [62]. According to Reddy [45] fine particulate matter ($PM_{2.5}$) is identified as the

Table 2 Standards for AQI, $PM_{2.5}$, and CO_2

Standard	Air quality index	$PM_{2.5}$ (μg/m^3)	CO_2 (ppm)
Good	0–50	0–12	0–700
Moderate	51–100	12.1–35	700–1000
Unhealthy for sensitive groups	101–150	35.1–55	1000–1500
Unhealthy	151–200	55.1–75	
Very unhealthy	201–300	75.1–110	>1500

Source Adapted from EPA [15]

fifth-leading cause of death worldwide. The fatalities associated with $PM_{2.5}$ exposure primarily resulted from diseases such as lower respiratory tract infections, lung cancer, ischaemic heart disease, cerebrovascular disease, and chronic obstructive pulmonary disease [39]. Murray et al. [39] revealed that PM pollution had escalated to the fourth position in the global rankings of risk factors based on the total number of deaths. Air pollution is a significant worldwide issue contributing to various health complications, including heightened mortality rates, hospitalisations, and outpatient visits, particularly cardiovascular and respiratory diseases [63].

Several studies have highlighted that the air quality in classrooms can contain a variety of pollutants, including CO_2 and $PM_{2.5}$. Among these environmental concerns, $PM_{2.5}$ has received considerable public attention due to their detrimental impact on students' health [32]. Researchers examined the relationship between mortality rates and the size of particles and discovered a strong correlation between mortality and $PM_{2.5}$. They reported that for every 10 μg/m3 increase in the average concentration of $PM_{2.5}$, mortality rates would rise by approximately 1.5 [44]. In Japan, research indicated a positive link between sickness-related absences and $PM_{2.5}$ [60]. Individuals who do not engage in cooking or smoking activities, including vulnerable groups like children and older adults, can still be exposed to $PM_{2.5}$ inhalation as these particles disperse within enclosed spaces.

A recent study revealed that indoor sources account for more than 50% of the overall $PM_{2.5}$ exposure experienced by urban residents [24]. A study on school buildings revealed that $PM_{2.5}$ and CO_2 were identified as the primary indoor pollutants, surpassing the established Chinese standards [24]. The influence of indoor sources, such as electronic devices, on $PM_{2.5}$ levels was minimal. A strong association was observed between indoor and outdoor $PM_{2.5}$ levels, with outdoor pollution predominantly contributing to indoor contamination. Buildings in urban areas with high outdoor $PM_{2.5}$ concentrations displayed elevated levels of indoor $PM_{2.5}$.

3 Case Study

A university building in Edinburgh was selected as a case for the study. This case study university has three campuses in Edinburgh offering advanced facilities for learning, teaching, and research. The selected campus is the largest in the University for teaching, research, and knowledge transfer activity. There are 4,000 students and several hundred full-time academic and professional support staff on campus. The campus includes two schools: Arts & Creative Industries, Computing, and Engineering & the Built Environment. The study included various locations within the campus, such as a 500-seat computer centre, reception hallway, cafeteria, and lecture rooms.

Data collection for this study was done in two stages. First, portable air quality monitors are utilised to collect the occupancy levels, indoor temperature, humidity, concentrations of CO_2 and $PM_{2.5}$, and AQI values within the campus-selected premises to evaluate the IAQ, CO_2 and $PM_{2.5}$ levels. To ensure accurate readings, the monitoring apparatus was positioned 500 mm from doors, windows, and individuals, enabling precise measurements of IAQ parameters and facilitating a comprehensive assessment of IAQ conditions within different campus spaces. The data collection process involved taking readings at 15-min intervals over 3 h at various locations. Subsequently, a questionnaire survey was conducted among 50 students to explore their awareness of IAQ guidelines and the health impacts of IAQ, their perception of thermal comfort, any changes they experienced in their well-being, actions taken to address temperature discomfort, factors contributing to poor indoor air quality, and their strategies for improvement. The data collected from the survey were analysed using statistical measures such as mode and mean.

4 Case Study Findings

4.1 Air Quality Measurements in the Selected Case Study

Indoor carbon dioxide levels are frequently used to indicate sufficient ventilation per occupant. Previous standards, such as ASHRAE 62-1989, established a maximum acceptable indoor CO_2 concentration of 1000 ppm and recommended a minimum ventilation rate of 8 lisp. More recent standards, like ASHRAE 62-1999, 62-2001, and 62-2004, propose that the disparity in concentration between indoor and outdoor environments should not surpass 700 ppm [49]. "ppm" stands for parts per million, a unit used to express the concentration of a specific gas, such as CO_2 in each environment.

The collected data were then analysed to calculate the mean for each location and presented in Table 3. The table includes occupancy levels, the number of doors and windows, outside temperature (minimum and maximum), inside temperature, humidity, CO_2 and $PM_{2.5}$ concentrations, and AQI values.

Case study findings indicated that the concentrations of CO_2 and $PM_{2.5}$ in all locations were within permissible limits, indicating acceptable indoor air quality. However, the AQI and $PM_{2.5}$ values in Classroom I were above 50 and 12, respectively. This observation is moderate, especially during the summer when the doors and windows of the classroom, operated manually, were closed for an extended period, potentially leading to stagnant air and reduced ventilation. In contrast, the AQI and $PM_{2.5}$ values in Classroom 2 lie within a suitable range because one window was opened, allowing for natural ventilation. Deviations in AQI and $PM_{2.5}$ from the good to moderate range indicate the presence of other pollutants beyond CO_2 and $PM_{2.5}$, such as volatile organic compounds, NO, PM_{10}, CO, and other gases. These pollutants may have contributed to the deterioration of indoor air quality, posing health risks for students within the HEI premises.

Table 3 Average air quality indicators at different locations in HEI

Time	Date	Location	No. of people in the room	No. of doors in the room	No. of windows in the room	Outside temperature (Min/max) °C	Inside temperature °C	Humidity (%)	CO_2 (ppm)	$PM_{2.5}$ (μg/m^3)	AQI
11.00–13.45	4 July	Library	18	3	20	9/17	25	73	400	3.2	13
14.00–16.45	4 July	Computer lab	9	1	0	9/16	24	64	400	0.7	2
12.15–15.00	7 July	Reception hallway	22	1	0	13/23	23	60	400	2.7	11
11.15–14.00	13 July	Cafe	6	1	3	11/18	23	45	400	2.6	10
11.00–13.45	20 July	Classroom 1	3	1	3	9/18	21	40	400	13.5	53
14.00–16.45	20 July	Classroom 2	1	1	4	9/17	23	38	400	7.1	30

4.2 Students' Exposure and Duration in the Case Study Buildings

Case study findings revealed that a majority of students (56%) reported spending 3–6 h inside the premises, indicating that a significant part of their day is spent in the indoor environment. The data further revealed that 32% of students spent 1–3 h within the case study building, which still represents a substantial amount of time spent indoors, even less than the first group. Additionally, 12% of students reported spending more than 6 h inside the building, indicating a notable proportion of students with extended exposure to the indoor environment.

The implications of these findings emphasised the importance of considering and improving the indoor air quality within the HEI building. With students spending significant amounts of time indoors, there is a higher likelihood of encountering indoor air pollutants or discomfort within the premises. Addressing these indoor air quality issues becomes essential for ensuring a healthier and more conducive learning environment for the students.

The findings identified a significant lack of awareness among students concerning guidelines for indoor air quality in educational environments. Specifically, 60% of the students were unfamiliar with the WHO guidelines, which are widely regarded as authoritative in this field and recognised internationally. This indicates that most students lacked knowledge about the specific recommendations set by the WHO to ensure a healthy indoor environment in educational settings. The findings highlight the importance of addressing this critical gap in understanding and awareness among the student population. Lack of awareness may hinder students' ability to recognise and address potential indoor air quality issues that could impact their health, comfort, and academic performance. Educating students on these principles is essential to foster a healthier and more positive learning atmosphere within educational settings.

Furthermore, the case study findings indicated that 60% of the students demonstrated awareness of the potential impacts of indoor air quality on human health. This implies that most participants understand the connection between indoor air quality and its effects on well-being. They likely know about the health consequences of poor indoor air quality exposure. To address this awareness gap, there is a need for educational interventions to enhance students' knowledge and understanding of the health consequences related to poor indoor air quality. Providing information about specific health risks, such as respiratory problems, allergic reactions, cognitive impairment, and other potential impacts, can help students better appreciate the importance of maintaining good indoor air quality for their well-being and academic performance. Empowering students with this knowledge can encourage them to take active measures to improve indoor air quality and create a healthier environment within educational institutions.

The case study results revealed that 24% of the students felt hot inside the premises, indicating that a significant proportion of participants perceived the indoor environment to be warmer than their desired comfort level. This finding suggests a potential issue with the indoor temperature control or ventilation system, leading to discomfort among these students. Furthermore, 50% of the students considered the indoor environment slightly hotter than usual. Although they did not report feeling excessively hot, they still perceived the indoor temperature as somewhat elevated compared to what they would consider ideal or standard. The presented data indicates room for improvement in maintaining an optimal temperature within the premises to meet students' expectations and comfort preferences. Enhancing thermal comfort can be achieved through various measures, including optimising heating and cooling systems, ensuring adequate insulation, and implementing proper ventilation strategies. By providing a comfortable indoor environment, the institution can create an atmosphere favourable to learning and promote students' overall well-being. Ensuring students' comfort contributes to their focus, concentration, and overall satisfaction, positively impacting their academic performance and educational experience.

Moreover, 26% of the students reported feeling fatigued while inside the premises. The findings underscore the presence of fatigue and tiredness among a considerable proportion of students within the indoor environment of the HEI premises. Approximately one-fourth of the students reported feeling fatigued during their time indoors. On the other hand, the fact that 28% of the students did not report any fatigue or tiredness indicates that they did not experience significant exhaustion or weariness. However, it is worth noting that a majority of the students, 46%, expressed a slight sense of fatigue and tiredness, suggesting a moderate level of tiredness experienced by this group.

The reported levels of fatigue and tiredness among students within the indoor environment of the HEI premises have crucial implications for their overall well-being and academic performance. Fatigue can negatively impact students' ability to concentrate, retain information, and actively participate in learning activities. Addressing concerns about fatigue and tiredness within the indoor environment is essential for creating a conducive learning environment that promotes students' well-being and academic success. Optimising ventilation, maintaining appropriate temperature and humidity levels, and encouraging regular breaks or physical activity can reduce fatigue and enhance students' overall experience within the HEI premises. By providing a comfortable and healthy indoor environment, the institution can positively influence students' focus, concentration, and engagement, ultimately fostering a more productive and fulfilling learning experience.

Of the students surveyed, 28 reported feeling fatigued after spending time indoors. This finding indicates that many students experienced exhaustion and weariness during their stay, potentially affecting their ability to concentrate and participate in academic activities effectively. Moreover, 11 students showed reduced cognitive performance, suggesting difficulties thinking, focusing, or processing information inside the building. Such cognitive impairments can hinder their learning and academic achievements. Additionally, six students reported experiencing headaches and dizziness, which may have caused discomfort and feelings of light-headedness or vertigo. Headaches and dizziness can significantly impact students' comfort, concentration, and overall well-being, potentially affecting their ability to engage in learning activities effectively.

Furthermore, four students mentioned a dry and irritated throat, indicating possible discomfort in their respiratory system. This observation raises concerns about indoor air quality or humidity levels, which could negatively affect students' respiratory health and overall comfort. Additionally, three students reported feeling shortness of breath, which is a concerning symptom indicating potential respiratory difficulties or discomfort. This highlights the need to assess indoor air quality and ventilation to ensure a healthy and breathable environment.

Notably, 14 students did not report any changes in their well-being after staying inside the building. This suggests that some participants did not experience noticeable symptoms or adverse effects on their physical or cognitive well-being during their indoor stay.

The findings underscore the importance of considering the impact of the indoor environment on students' well-being. Addressing the reported symptoms, such as fatigue, cognitive performance, headaches, and respiratory discomfort, and ensuring a comfortable and healthy indoor environment can promote students' overall well-being and academic success. The institution can enhance students' learning experience and overall quality of life by providing a conducive and supportive indoor environment.

Case study findings reported that most students, comprising 62% of the participants, reported not experiencing any respiratory issues. This indicates that many students did not encounter any difficulties or discomfort related to their respiratory system while indoors. This positive finding suggests that the indoor environment may generally be conducive to respiratory health for most students. However, some students did report respiratory symptoms. Specifically, 9% of the students mentioned experiencing coughing, which indicates irritation or inflammation of the airways leading to coughing episodes. This could be associated with poor indoor air quality or environmental irritants. Additionally, 20% of the students reported allergy symptoms, including sneezing and flu-like symptoms. These symptoms suggest some students may be sensitive to indoor allergens, resulting in allergic reactions. Sneezing and flu-like symptoms could indicate allergic rhinitis or other allergic responses from indoor allergens.

Moreover, 4% of the students experienced more severe respiratory symptoms, such as wheezing and chest tightness. These symptoms are characteristic of respiratory conditions like asthma or bronchial constriction, which can be triggered by irritants or allergens present in the indoor environment. The distribution of these responses, as depicted in Fig. 1, highlights the varying prevalence of respiratory symptoms reported by students. While a majority did not experience respiratory issues, some students' symptoms highlight the importance of addressing indoor air quality factors that may contribute to such discomfort.

Findings revealed that 45% of the participants opted to open windows in response to uncomfortable temperatures. This indicates that many students relied on natural ventilation to regulate indoor temperature. Opening windows allows fresh air to enter the space, potentially alleviating feelings of warmth or stuffiness. Additionally, 36% of students add just their clothing layers to deal with uncomfortable temperatures. This shows that students recognised the importance of modifying their clothing

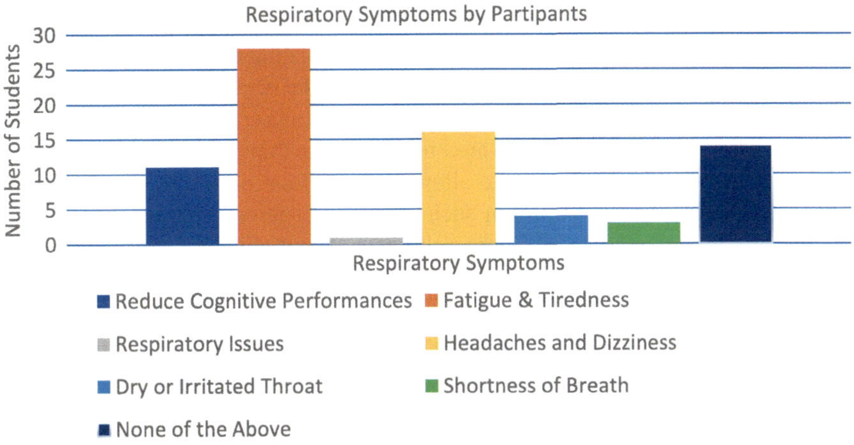

Fig. 1 Different respiratory symptoms by participants

to adapt to the prevailing indoor temperature. Adjusting clothing layers, such as adding or removing outerwear, provides a practical and immediate way to regulate personal comfort. Furthermore, 17% of the students reported taking breaks outside the building when faced with uncomfortable temperatures. This suggests that some students sought relief from the indoor temperature by temporarily stepping outside. Taking breaks outdoors allows students to experience a different climate and may provide a temporary respite from the uncomfortable indoor environment. These adaptive strategies employed by students demonstrate their resourcefulness and ability to adapt to their surroundings. Opening windows, adjusting clothing layers, and taking breaks outside the building are individual actions that allow students to personally address temperature-related discomfort.

The findings emphasise the importance of providing a comfortable indoor temperature within the campus premises and creating an environment that maintains an optimal temperature range that is responsive to the needs of students and can enhance their comfort, concentration, and overall well-being. It also highlights the significance of considering individual adaptive strategies when designing and managing indoor spaces within educational institutions. By understanding and accommodating students' responses to temperature discomfort, educational institutions can foster a more conducive and pleasant learning environment.

Case study findings highlighted that 25 students pointed out inadequate ventilation systems as a significant factor contributing to poor indoor air quality among the participants. This indicates that many respondents believed that the ventilation systems within the building were not effectively removing pollutants and providing a fresh air supply. Inadequate ventilation can lead to the buildup of contaminants, reduced air circulation, and increased indoor pollutant levels, resulting in compromised air quality. Moreover, 27 students emphasised inefficient heating and ventilation systems as a contributing factor. This suggests that the heating and ventilation systems in place within the building were not effectively maintaining an appropriate temperature and air distribution. Inefficient systems may cause uneven heating or cooling, inconsistent airflows, and difficulties in regulating the indoor environment, ultimately impacting the comfort and well-being of the occupants. Additionally, 21 students identified limited access to natural ventilation as a critical factor contributing to poor indoor air quality and elevated temperatures. Natural ventilation involves openings like windows and doors to allow fresh outdoor air to circulate within the building. The limited availability of such ventilation sources can hinder the inflow of fresh air and the removal of indoor pollutants, potentially leading to stagnant air and increased temperatures.

These findings underscore the importance of well-functioning ventilation systems and access to natural ventilation in ensuring a healthy indoor environment. Addressing the issues of inadequate ventilation, inefficient heating and ventilation systems, and limited access to natural ventilation is crucial for creating a comfortable and healthy learning environment for students. Improving indoor air quality and temperature conditions can enhance students' well-being, concentration, and academic performance.

4.3 The Ways to Improve IAQ

Case study findings incorporated valuable suggestions from students on improving indoor air quality within the HEI premises. These strategies offer insights into potential strategies and interventions.

One student proposal was to introduce indoor plants throughout the campus. Indoor plants are known for naturally purifying and filtering the air through photosynthesis. Having indoor plants can enhance air quality, as they help reduce indoor pollutants and increase oxygen levels, creating a more refreshing and pleasant indoor environment. This aligns with the concept of biophilic design, which emphasises integrating natural elements into built environments to improve well-being. Another significant highlight was the implementation of mechanical ventilation systems. Mechanical ventilation involves using fans or air conditioning systems to facilitate fresh air circulation and the removal of indoor pollutants. By employing efficient mechanical ventilation systems, the HEI can ensure a continuous supply of fresh air and maintain optimal indoor air quality, particularly in spaces where natural ventilation may be limited.

Additionally, students recommended standardising the number of occupants in each classroom at a given time. This suggestion aims to address overcrowding issues that could affect indoor air quality. By limiting the number of individuals in a space, the HEI can ensure that ventilation systems are adequately designed to accommodate the ventilation needs of the occupants. This can prevent the buildup of indoor pollutants and create a healthier indoor environment for students and faculty. Implementing these strategies can significantly improve indoor air quality within the HEI premises. Educational institutions must consider these proposals and explore feasible ways to implement them. Effective indoor air quality management promotes students' health and well-being and fosters a conducive learning and academic success environment.

5 Discussion

As the above case study findings reveal, some students expressed satisfaction with the indoor environment, while others reported adverse effects on their health. This diversity in perceptions could be attributed to awareness of indoor air quality, influencing how they interpret and react to the indoor environment's conditions. Similar findings were reported in a recent study that the environmental awareness of air pollution is still not very distinct in the general population, and young people especially often have only a diffuse and abstract idea of air pollutants [22]. Additionally, individual sensitivities to indoor conditions may differ, and other unmeasured air components could contribute to respiratory issues among specific individuals. These findings were supported by Chen et al. [11]. An important finding from the survey was that most participants lacked knowledge of responding to poor indoor air quality and were unaware of international protocols for addressing such issues. These findings align with previous studies [5, 22]. This indicates the need for educational programmes to raise awareness among students about the impact of indoor air quality on their health and well-being and how they can take proactive measures to improve IAQ. The statement suggests that overall, the indoor air quality in case study premises meets the standards set by the EPA, except for Classroom I. This highlights a potential issue with classroom ventilation systems, leading to suboptimal air quality. Adequate ventilation removes indoor air pollutants and ensures students and faculty a healthy and comfortable learning environment. These findings were supported by Haddad et al. [17].

The previous studies which examined the impact of ventilation on indoor air quality in educational settings Haddad et al. [17] and Turunen et al. [57] provided proper ventilation help to dilute and remove pollutants, including CO, volatile organic compounds (VOCs), and particulate matter, which can accumulate in enclosed spaces like classrooms. Similarly, the case study findings provided that inadequate ventilation can lead to a buildup of these pollutants, potentially causing health issues and reducing the overall well-being and productivity of students and staff. The findings stressed the importance of addressing indoor air quality issues in the HEI. Implementing valuable strategies such as introducing indoor plants, installing mechanical ventilation systems, standardising occupancy levels, and regularly monitoring and managing all air components can significantly create a healthier and more conducive learning environment. By raising student awareness through educational programmes about the significance of IAQ and the necessary actions to combat poor indoor air quality, the HEI can foster a positive and supportive atmosphere for all occupants. Therefore, most of the findings are similar to what has been established in the literature and could be regarded as generic issues among HEI indoor air quality on students' health and well-being.

6 Conclusions

This chapter investigated the potential students' health and well-being in a HEI associated with elevated CO_2, $PM_{2.5}$ concentrations, and the AQI. This was aided by a case study conducted at a university building in Edinburgh, UK. Findings revealed that students spend most of their time inside the HEI premises, highlighting the indoor environment's substantial influence on their well-being. However, a concerning aspect emerged as most students were unfamiliar with the WHO or EPA guidelines on indoor air quality, potentially hindering their ability to respond effectively to poor air quality situations despite being aware of the potential health impacts. Thermal comfort was identified as a crucial aspect, with most students perceiving indoor temperatures as slightly warmer or hotter than ideal. Additionally, a minority of students felt tired after spending extended periods indoors, suggesting that some aspects of the indoor environment can negatively affect their well-being. Findings revealed other health impacts of the IAQ, including reduced cognitive performance, headaches, dizziness, and dry, irritated throats. Some participants even experienced shortness of breath, which could be influenced by existing health conditions exacerbated by the indoor environment. Allergy symptoms, such as sneezing, flu-like symptoms, and coughing, were also prevalent among some respondents. However, a few students perceived the indoor environment as ideal, experiencing no adverse changes in their well-being. Additionally, inadequate heating and ventilation systems contributed to poor air quality within the premises. Maintaining acceptable indoor air quality is essential not only for the comfort and health of the occupants but also for optimising learning and working environments within the institution. The following recommendations were derived from the study.

- Implementation of Indoor Plants

Indoor plants are well-known for their natural ability to purify and filter the air through photosynthesis. Introducing indoor plants can enhance air quality by reducing indoor pollutants and increasing oxygen levels. This approach aligns with the concept of biophilic design, which emphasises incorporating natural elements into built environments to promote overall well-being.

- Mechanical Ventilation Systems

This involves fans or air conditioning systems facilitating fresh air circulation and removing indoor pollutants. By adopting efficient mechanical ventilation systems, the HEI can ensure a continuous supply of fresh air, especially in areas where natural ventilation may be limited or not feasible.

- Standardising Occupancy Levels

Overcrowding issues that could adversely impact indoor air quality. By limiting the number of individuals in a space, the HEI can ensure that ventilation systems are adequately designed to meet the occupants' ventilation needs, effectively preventing the accumulation of indoor pollutants and creating a healthier indoor environment for students and faculty.

- Regular Monitoring and Management

A regular indoor air quality monitoring system is essential to promptly identifying fluctuations and potential issues. A proactive management approach allows timely interventions to maintain a healthy indoor environment.

- Student Awareness Programmes

Awareness programmes should be conducted to educate students about indoor air quality guidelines and appropriate actions to take when the air quality is poor. This can empower students to actively contribute to maintaining a healthier indoor environment.

- Collaboration with Experts

With the implementation of hybrid ventilation systems, the HEI can take significant strides towards creating a healthier and more conducive indoor environment. These efforts will positively impact the well-being of students and faculty and foster a thriving educational environment that promotes learning and academic success. Moreover, future research should consider conducting the study in different seasons, mainly when occupancy levels are high, outdoor temperatures are low, and the open windows and doors are limited. This will provide a more comprehensive understanding of indoor air quality dynamics throughout the year and aid in developing targeted strategies to maintain optimal air quality under varying conditions.

References

1. Abdel-Salam, M. M. M. (2022). Indoor exposure of elderly to air pollutants in residential buildings in Alexandria, Egypt. Build Environ 219. https://doi.org/10.1016/j.buildenv.2022.109221
2. Abhijith KV, Kukadia V, Kumar P (2022) Investigation of air pollution mitigation measures, ventilation, and indoor air quality at three schools in London. Atmos Environ 289. https://doi.org/10.1016/j.atmosenv.2022.119303
3. Al-Awadi L (2018) Assessment of indoor levels of volatile organic compounds and carbon dioxide in schools in Kuwait. J Air Waste Manag Assoc 68(1):54–72
4. Azuma K, Kagi N, Yanagi U, Osawa H (2018) Effects of low-level inhalation exposure to carbon dioxide in indoor environments: a short review on human health and psychomotor performance. Environ Int 121:51–56
5. Baecher C, Sohui F, Ball L, Masson O (2020) Indoor air quality: a multifaceted public health problem. J Field Action, Special Issue 21. https://journals.openedition.org/factsreports/5986
6. Batterman S, Su F-C, Wald A, Watkins F, Godwin C, Thun G (2017) Ventilation rates in recently constructed U.S. school classrooms. Indoor Air 27(5):880–890. https://doi.org/10.1111/ina.12384
7. Bogdanovica S, Zemitis J, Bogdanovics R (2020) The effect of CO_2 concentration on children's well-being during the process of learning. Energies 13(22). https://doi.org/10.3390/en13226099

8. Cabovská B, Bekö G, Teli D, Ekberg L, Dalenbäck JO, Wargocki P, Psomas T, Langer S (2022) Ventilation strategies and indoor air quality in Swedish primary school classrooms. Build Environ 226. https://doi.org/10.1016/j.buildenv.2022.109744
9. Canha N, Almeida SM, Freitas MC, Täubel M, Hänninen O (2013) Winter ventilation rates at primary schools: comparison between Portugal and Finland. J Toxicol Environ Health-Part A Curr Issues 76(6):400–408. https://doi.org/10.1080/15287394.2013.765372
10. Catalina T, Iordache V (2012) IEQ assessment on schools in the design stage. Build Environ 49(1):129–140. https://doi.org/10.1016/j.buildenv.2011.09.014
11. Chen X, Li F, Liu C, Yang J, Zhang J, Peng C (2017) Monitoring, human health risk assessment and optimized management for typical pollutants in indoor air from random families of university staff, Wuhan City, China. Sustainability 9(7):1115. https://doi.org/10.3390/su9071115
12. Chithra VS, Nagendra SMS (2018) A review of scientific evidence on indoor air of school building: pollutants, sources, health effects and management. Asian J Atmos Environ 12(2):87–108
13. De Giuli V, Da Pos O, De Carli M (2012) Indoor environmental quality and pupil perception in Italian primary schools. Build Environ
14. Elsaid AM, Ahmed MS (2021) Indoor air quality strategies for air-conditioning and ventilation systems with the spread of the global coronavirus (COVID-19) epidemic: improvements and recommendations. Environ Res 199. https://doi.org/10.1016/j.envres.2021.111314. (Academic Press Inc.)
15. EPA (2023) EPA United States Environment Protection Agency. https://www.epa.gov/environmental-topics/air-topics
16. Ghita SA, Catalina T (2015) Energy efficiency versus indoor environmental quality in different Romanian countryside schools. Energy Build 92:140–154
17. Haddad S, Synnefa A, Ángel Padilla Marcos M, Paolini R, Delrue S, Prasad D, Santamouris M (2021) On the potential of demand-controlled ventilation system to enhance indoor air quality and thermal condition in Australian school classrooms. Energy Build 238:110838. https://doi.org/10.1016/j.enbuild.2021.110838
18. Halios CH, Landeg-Cox C, Lowther SD, Middleton A, Marczyl, T, Dimitroulopoulou S (2022) Chemicals in European residences–part I: a review of emissions, concentrations and health effects of volatile organic compounds (VOCs). Sci Total Environ 839. https://doi.org/10.1016/j.scitotenv.2022.156201. (Elsevier B.V.)
19. Hamanaka RB, Mutlu GM (2018) Particulate matter air pollution: effects on the cardiovascular system. Front Endocrinol 9:680
20. Hänninen O, Canha N, Kulinkina AV, Dume I, Deliu A, Mataj E, Lusati A, Krzyzanowski M, Egorov AI (2017) Analysis of CO_2 monitoring data demonstrates poor ventilation rates in Albanian schools during the cold season. Air Qual Atmos Health 10:773–782
21. Heebøll A, Wargocki P, Toftum J (2018) Window and door opening behaviour, carbon dioxide concentration, temperature, and energy use during the heating season in classrooms with different ventilation retrofits—ASHRAE RP1624. Sci Technol Built Environ 24(6):626–637. https://doi.org/10.1080/23744731.2018.1432938
22. Höfner S, Schütze A (2021) Air quality measurements and education: improving environmental awareness of high school students. Front Sens 2. https://doi.org/10.3389/fsens.2021.657920
23. Hoskins JA (2011) Health effects due to indoor air pollution. In Gökçekus H, Türker U, LaMoreaux JW (eds) Survival and sustainability: environmental concerns in the 21st century, pp 665–676. Springer, Heidelberg. https://doi.org/10.1007/978-3-540-95991-5_61
24. Hu Y, Zhao B (2022) Indoor sources strongly contribute to exposure of Chinese urban residents to $PM_{2.5}$ and NO_2. J Hazard Mater 426:127829
25. Huang Y, Ho SSH, Ho KF, Lee SC, Yu JZ, Louie PKK (2011) Characteristics and health impacts of VOCs and carbonyls associated with residential cooking activities in Hong Kong. J Hazard Mater 186(1):344–351
26. Huang Y, Yang Z, Gao Z (2019) Contributions of indoor and outdoor sources to ozone in residential buildings in Nanjing. Int J Environ Res Public Health 16(14):2587

27. Korsavi SS, Montazami A, Mumovic D (2020) Indoor air quality (IAQ) in naturally-ventilated primary schools in the UK: occupant-related factors. Build Environ 180. https://doi.org/10.1016/j.buildenv.2020.106992
28. Korsavi S, Montazami A, Brusey J (2018) Developing a design framework to facilitate adaptive behaviours. Energy Build 179:360–373. https://doi.org/10.1016/j.enbuild.2018.09.011. (Elsevier Ltd.)
29. Kumar P, Singh AB, Arora T, Singh S, Singh R (2023) Critical review on emerging health effects associated with the indoor air quality and its sustainable management. Sci Total Environ 872. https://doi.org/10.1016/j.scitotenv.2023.162163. (Elsevier B.V.)
30. Lazović IM et al (2016) Impact of CO_2 concentration on indoor air quality and correlation with relative humidity and indoor air temperature in school buildings in Serbia. Therm Sci
31. Li Z, Wen Q, Zhang R (2017) Sources, health effects and control strategies of indoor fine particulate matter (PM2.5): a review. Sci Total Environ 586:610–622. https://doi.org/10.1016/j.scitotenv.2017.02.029. (Elsevier B.V.)
32. Liang C-S, Duan F-K, He K-B, Ma Y-L (2016) Review on recent progress in observations, source identifications and countermeasures of $PM_{2.5}$. Environ Int 86:150–170
33. Liu G, Xiao M, Zhang X, Gal C, Chen X, Liu L, Pan S, Wu J, Tang L, Clements-Croome D (2017) A review of air filtration technologies for sustainable and healthy building ventilation. Sustain Cities Soc 32:375–396. https://doi.org/10.1016/j.scs.2017.04.011. (Elsevier Ltd.)
34. Luther MB, Horan P, Tokede O (2018) Investigating CO_2 concentration and occupancy in school classrooms at different stages in their life cycle. Archit Sci Rev 61(1–2):83–95. https://doi.org/10.1080/00038628.2017.1416576
35. Madureira J, Paciência I, Ramos E, Barros H, Pereira C, Teixeira JP, de Fernandes EO (2015) Children's health and indoor air quality in primary schools and homes in Portugal—study design. J Toxicol Environ Health Part A 78(13–14):915–930
36. Maynard RL (2019) Health effects of indoor air pollution. Indoor Air Pollut 48:196
37. Mendoza DL, Pirozzi CS, Crosman ET, Liou TG, Zhang Y, Cleeves JJ, Bannister SC, Anderegg W RL, Robert PIII (2020) Impact of low-level fine particulate matter and ozone exposure on absences in K-12 students and economic consequences. Environ Res Lett 15(11). https://doi.org/10.1088/1748-9326/abbf7a
38. Montazami A, Gaterell M, Nicol F (2015) A comprehensive review of environmental design in UK schools: history, conflicts and solutions. Renew Sustain Energy Rev 46:249–264. https://doi.org/10.1016/j.rser.2015.02.012. (Elsevier Ltd.)
39. Murray CL, Aravkin AY, Zheng P, Khatab K, Cristiana A, Ashkan A, Stephen SL (2020) Global burden of 87 risk factors in 204 countries and territories, 1990–2019: a systematic analysis for the global burden of disease study 2019. Lancet 396(10258):17–23
40. Myhrvold AN, Olsen E, Lauridsen O (1996) Indoor environment in schools–pupils health and performance in regard to CO_2 concentrations. Indoor Air 96(4):369–371
41. Nowak-Dzieszko K, Kisilewicz T (2020) Internal particulate matter pollution in educational building. E3S Web Conf 172:06008
42. Oman CM, Schneiderman J, Stankovic A, Alexander D (2016) A review of cognitive and behavioral effects of increased carbon dioxide exposure in humans
43. Ortiz Perez A, Bierer B, Scholz L, Wöllenstein J, Palzer S (2018) A wireless gas sensor network to monitor indoor environmental quality in schools. Sensors 18(12):4345
44. Pun VC, Kazemiparkouhi F, Manjourides J, Suh HH (2017) Long-term $PM_{2.5}$ exposure and respiratory, cancer, and cardiovascular mortality in older US adults. Am J Epidemiol 186(8):961–969
45. Reddy KS (2016) Global burden of disease study 2015 provides GPS for global health 2030. Lancet 388(10053):1448–1449. https://doi.org/10.1016/S0140-6736(16)31743-3. (Lancet Publishing Group)
46. Rivas I, Fussell JC, Kelly FJ, Querol X (2019) Indoor sources of air pollutants. Indoor Air Pollut 48:1
47. Salonen H, Salthammer T, Morawska L (2018) Human exposure to ozone in school and office indoor environments. Environ Int 119:503–514

48. Salonen H, Salthammer T, Morawska L (2019) Human exposure to NO_2 in school and office indoor environments. Environ Int 130:104887
49. Santamouris M, Synnefa A, Asssimakopoulos M, Livada I, Pavlou K, Papaglastra M, Gaitani N, Kolokotsa D, Assimakopoulos V (2008) Experimental investigation of the air flow and indoor carbon dioxide concentration in classrooms with intermittent natural ventilation. Energy Build 40(10):1833–1843. https://doi.org/10.1016/j.enbuild.2008.04.002
50. Shrestha M, Rijal H, Kayo G, Shukuya M (2022) An investigation on CO_2 concentration based on field survey and simulation in naturally ventilated Nepalese school buildings during summer. Build Environ
51. Simoni M, Annesi-Maesano I, Sigsgaard T, Norback D, Wieslander G, Nystad W, Canciani M, Sestini P, Viegi G (2010) School air quality related to dry cough, rhinitis and nasal patency in children. Eur Respir J 35(4):742–749
52. Singh MK, Ooka R, Rijal HB, Kumar S, Kumar A, Mahapatra S (2019) Progress in thermal comfort studies in classrooms over last 50 years and way forward. Energy Build 188–189:149–174. https://doi.org/10.1016/j.enbuild.2019.01.051. (Elsevier Ltd.)
53. Stabile L, Dell'Isola M, Russi A, Massimo A, Buonanno G (2017) The effect of natural ventilation strategy on indoor air quality in schools. Sci Total Environ 595:894–902. https://doi.org/10.1016/j.scitotenv.2017.03.048
54. Tang X, Misztal PK, Nazaroff WW, Goldstein AH (2015) Siloxanes are the most abundant volatile organic compound emitted from engineering students in a classroom. Environ Sci Technol Lett 2(11):303–307
55. Tran VV, Park D, Lee YC (2020) Indoor air pollution, related human diseases, and recent trends in the control and improvement of indoor air quality. Int J Environ Res Public Health 17(8):2927
56. Turanjanin V, Vučićević B, Jovanović M, Mirkov N, Lazović I (2014) Indoor CO_2 measurements in Serbian schools and ventilation rate calculation. Energy 77:290–296. https://doi.org/10.1016/j.energy.2014.10.028
57. Turunen M, Toyinbo O, Putus T, Nevalainen A, Shaughnessy R, Haverinen-Shaughnessy U (2013) Indoor environmental quality in school buildings, and the health and wellbeing of students. Int J Hyg Environ Health 217(7):733–739. https://doi.org/10.1016/j.ijheh.2014.03.002
58. United States Environment Protection Agency (2024) Introduction to indoor air quality. https://www.epa.gov/indoor-air-quality-iaq/introduction-indoor-air-quality. Last Accessed 16 June 2024
59. Vouriot CVM, Burridge HC, Noakes CJ, Linden PF (2021) Seasonal variation in airborne infection risk in schools due to changes in ventilation inferred from monitored carbon dioxide. Indoor Air 31(4):1154–1163
60. Watanabe M, Noma H, Kurai J, Kato K, Sano H (2021) Association with ambient air pollutants and school absence due to sickness in schoolchildren: a case-crossover study in a provincial town of Japan. Int J Environ Res Public Health 18(12):6631
61. Weichenthal S, Dufresne A, Infante-Rivard C (2007) Indoor nitrogen dioxide and VOC exposures: summary of evidence for an association with childhood asthma and a case for the inclusion of indoor ultrafine particle measures in future studies. Indoor Built Environ 16(5):387–399
62. World Health Organization (2021) WHO global air quality guidelines: particulate matter ($PM_{2.5}$ and PM_{10}), ozone, nitrogen dioxide, sulfur dioxide and carbon monoxide. World Health Organization
63. Xu D, Chen Y, Wu L, He S, Xu P, Zhang Y, Luo J, Ye X, Chen Z, Wang X (2020) Acute effects of ambient $PM_{2.5}$ on lung function among schoolchildren. Sci Rep 10(1):4061
64. Yuhe Z, Guangfei Y, Xianneng L (2021) Indoor $PM_{2.5}$ concentrations and students' behavior in primary school classrooms. J Clean Prod
65. Zhang X, Wargocki P, Lian Z, Thyregod C (2017) Effects of exposure to carbon dioxide and bioeffluents on perceived air quality, self-assessed acute health symptoms, and cognitive performance. Indoor Air 27(1):47–64

Chapter 13
Minimising Waste in Construction Projects: A Focus on Ceramic Tile Waste

Udesh Perera, Nguyet Tong, Wajiha Shahzad, and Niluka Domingo

Abstract Construction waste poses a significant challenge in New Zealand, with the construction industry accounting for over one-third of the waste deposited in the country's landfills. Addressing this issue requires effective management of waste generated during construction projects. Among the various materials contributing to this waste stream, ceramic tiles stand out due to their fragile nature, susceptible to damage during handling and installation. However, research on minimising ceramic tile waste in construction projects, particularly in New Zealand, remains limited. This chapter aims to fill this gap by investigating current waste management practices related to ceramic tiles. Through a combination of literature review and semi-structured interviews with experienced tilers, site managers, and project coordinators, the study identifies the causes of ceramic tile waste and proposes measures for its minimisation. The research objectives are achieved by categorising these measures into five stages of construction projects: design and documents, material procurement, transportation and handling, and management and onsite operation. By analysing challenges associated with minimising ceramic tile waste, the study provides valuable insights into the New Zealand context. This research contributes to enhancing understanding and implementation strategies for waste minimisation in construction projects, ultimately supporting the industry's efforts to reduce its environmental impact and achieve sustainability goals.

Keywords Ceramic tiles · Construction waste · Waste minimisation · New Zealand

U. Perera · N. Tong · W. Shahzad · N. Domingo (✉)
School of Built Environment, Massey University, Albany, New Zealand
e-mail: n.d.domingo@massey.ac.nz

T. Ramachandra et al. (eds.), *Concepts, Policies and Technology for Sustainable Buildings*, Advances in Sustainability Science and Technology,
https://doi.org/10.1007/978-981-96-3086-8_13

1 Introduction

The construction industry has contributed to a major proportion of waste in landfills, about 35% of construction and demolition (C&D) waste [16]. This results in adverse impacts on the environment, natural resources, and the depletion of landfill sites. In addition, construction waste generation leads to significant increases in project costs due to tons of material losses, and additional costs for waste handling, transportation, and landfills [1]. In New Zealand, the construction industry contributes 30–50% of total waste sent to landfills, and proper waste management can reduce the waste on building sites by at least a half [6].

Construction waste can be classified into eight categories, including (1) concrete, bricks, tiles, and ceramics; (2) wood, glass, and plastic; (3) coal and asphalt; (4) metals; (5) soil; (6) insulation materials and materials containing asbestos; (7) gypsum-based materials; (8) other construction waste. Tiles are one of the most essential materials in building projects but also one of the main sources of waste generated in construction sites [5, 30]. Ceramic tile waste accumulation is a big environmental hazard if they are not recycled, however, the common method of disposal for ceramic tiles is landfilling due to a lack of standards, as well as a lack of experience and awareness in dealing with ceramic waste [7].

Minimising ceramic tile waste can make great contributions to addressing the negative impacts of C&D waste and supporting the zero-waste target. However, there are limited studies on waste minimisation in the New Zealand construction industry, particularly in ceramic tile waste. Previous research in other countries has identified the causes of construction waste, but only a few studies were conducted on ceramic tile waste generated in construction projects. Sources of waste generation in one country or region may not coincide with those of other locations [17]. Therefore, this study aims to investigate the current practice in waste management to minimise ceramic tile waste in construction projects from extant literature and from practitioners' perspectives. Accordingly, the research objectives are developed as follows:

- To identify existing causative sources of waste generation in construction projects including ceramic tile waste from the extant literature.
- To determine measures that can be taken to reduce ceramic tile waste generated in construction projects in New Zealand.
- To understand the challenges that the construction industry can encounter when implementing the minimisation of ceramic tile waste in New Zealand.

2 Literature Review

Sources of Construction Waste Generation

Understanding the waste sources is fundamental to effective waste management. Many previous studies were conducted to identify the sources for construction waste generation. The impacts of those sources depend on project size, location, and related activities [20]. In Thailand, Luangcharoenrat et al. [20] found that design issues, improper storage, and incompetent labour have the most significant impacts on waste generation. Meanwhile, improper management skills, insufficient environmental awareness, and lack of legislative enforcement are the major causes of waste in Chinese construction [30]. In Ireland, improper handling, material damage, rushed programmes, odd-sized materials, and design issues are the main reasons for waste generation on construction sites [21]. Understanding waste sources is fundamental to implementing effective measures towards waste minimisation. Based on a review of the previous research, this study categorised sources of construction waste into four groups, including design and documents, material procurement, transportation and handling, and management and onsite operation, as shown in Table 1.

Design problems rank at the top as construction waste sources [20]. In the extant literature, design changes and design errors are mentioned with the highest frequency as common causes of waste generation in construction projects (Table 1). Design changes or mistakes during the construction stage will lead to rework and material waste due to the demolition of completed work failing to meet design requirements. In addition, complex design and lack of knowledge of standard sizes on the market can lead to a number of off-cuts which mostly end up as waste. To solve the design problem, previous studies agreed that good coordination and effective communication will help to improve design quality and, therefore, can avoid waste during the construction stage [11, 29]. Furthermore, recent researchers have recognised that non-consideration of reusable and recyclable materials from the design stage is a causative source of construction waste [4, 22].

Accurate material ordering is important to prevent unwanted construction waste; however, many sources have negative impacts on the material ordering process. Purchasing materials not complying with specifications is the most acknowledged cause in the literature, followed by ordering errors, mistakes in material quantity take-off, and lack of possibility to order small quantities (Table 1). When ordering materials more than the actual amount used, the materials that cannot be used in the project may be kept on-site and later turned into waste [20]. This is due to the impossibility of small order quantity or inappropriate methods used to estimate the amount of needed material. Hence, quantity surveying is a significant contributor to preventing material leftovers [15].

Transportation, unloading, and storage processes are further cited as sources of waste generation in construction sites. As mentioned at the highest frequency in the literature, storage problems involve improper storage methods, insufficient space and facilities, and long distances from storage areas to application points [11]. Hence, Marinelli et al. [21] and Naji et al. [22] suggested that Just-In-Time (JIT) deliveries can reduce material damage due to improper storage as materials will be delivered in the right quantity, in the right specification, at the right time to be used in the construction process. Material damage also occurs during transportation and unloading due to insufficient protection, difficulties for vehicles to access construction sites, narrow sites, and inefficient methods of unloading [8, 17].

Table 1 Causative sources of construction waste (CW) generation in extant literature

Causative sources of CW generation	Previous research											Frequency
	(1)	(2)	(3)	(4)	(5)	(6)	(7)	(8)	(9)	(10)	(11)	
Design and documents												
Design changes	✔	✔	✔	✔		✔	✔	✔	✔	✔	✔	10
Design errors		✔	✔	✔				✔	✔	✔		6
Errors in construction drawings	✔			✔				✔		✔		4
Complex design	✔	✔		✔		✔				✔		5
No design standardisation	✔	✔										2
Unclear/unsuitable specification		✔			✔	✔	✔			✔		5
Document problems				✔	✔	✔		✔		✔		5
Poor coordination	✔	✔			✔	✔					✔	5
Non-consideration of reusable materials											✔	1
Uneconomic design or shapes						✔			✔			2
Material procurement												
Ordering errors		✔	✔	✔			✔		✔			5
Mistakes in quantity take-offs		✔					✔		✔	✔		4
Not comply with specification		✔		✔	✔	✔	✔			✔		6
Lack of possibility of ordering small quantities					✔	✔	✔		✔	✔		5
Suppliers' errors		✔	✔							✔		3
Poor schedule of material procurement							✔					1
Causative sources of CW generation	Previous research											Frequency
	(1)	(2)	(3)	(4)	(5)	(6)	(7)	(8)	(9)	(10)	(11)	
Transportation and handling												
Improper material storage	✔	✔	✔	✔		✔	✔	✔	✔	✔	✔	10

(continued)

Table 1 (continued)

Causative sources of CW generation	Previous research											Frequency
	(1)	(2)	(3)	(4)	(5)	(6)	(7)	(8)	(9)	(10)	(11)	
Improper material handling	✔			✔			✔	✔	✔	✔	✔	7
Improper unloading methods	✔	✔			✔	✔	✔					5
Transporting problems	✔	✔	✔	✔			✔	✔	✔	✔		8
Packaging problems			✔	✔			✔	✔				4
Management and onsite operation												
Misused material	✔	✔		✔						✔	✔	5
Incompetent workers	✔	✔	✔	✔	✔	✔		✔	✔	✔	✔	10
Improper planning for required material		✔			✔	✔						3
Accident due to negligence		✔	✔		✔	✔					✔	5
Equipment misuse/ malfunction		✔	✔	✔	✔				✔	✔		6
Time pressures	✔	✔					✔			✔		4
Lack of onsite waste management plans		✔	✔	✔	✔			✔		✔	✔	7
Lack of onsite supervision		✔	✔	✔			✔	✔	✔	✔		7
Lack of onsite quality control		✔	✔	✔		✔		✔	✔	✔	✔	8
Improper construction methods				✔		✔		✔	✔	✔		5
Cutting uneconomical shapes		✔				✔	✔			✔		4
Poor coordination	✔			✔	✔	✔		✔	✔	✔	✔	8
Rework			✔	✔	✔		✔		✔			5
Poor planning and schedule	✔					✔	✔	✔			✔	5
Lack of training					✔	✔					✔	3

(continued)

Table 1 (continued)

Causative sources of CW generation	Previous research											Frequency
	(1)	(2)	(3)	(4)	(5)	(6)	(7)	(8)	(9)	(10)	(11)	
Lack of waste knowledge							✔		✔	✔	✔	4
Poor site layout						✔			✔	✔		3
Lack of waste segregation							✔					1

(1)—Domingo [11], (2)—Umar et al. [29], (3)—Arshad et al. [3], (4)—Luangcharoenrat et al. [20], (5)—Kolaventi et al. [17], (6)—Sweis et al. [27], (7)—Chibuike et al. [8], (8)—Almusawi et al. [2], (9)—Naji et al. [22], (10)—Saad et al. [24], (11)—Bajjou and Chafi [4]

According to Umar et al. [29] and Saad et al. [24], onsite operation and management are cited as the most critical sources contributing to waste generation. First, using incompetent workers causes the misuse of equipment and materials, errors, and reworks on-site, this results in a significant amount of material damage and waste. Furthermore, the lack of quality control and supervision has negative impacts on the quality of completed construction as well as the effectiveness of waste minimisation measures onsite [17]. It is also noted that onsite waste management plans play a crucial role in implementing waste minimisation during the construction stage [22, 24]. Arshad et al. [3] suggested that a full-time waste manager in larger construction projects is also an effective solution to control and minimise waste. Equipment problems are further causes of material waste, including equipment misuse, malfunction, and scarcity [20].

During the construction stage, waste can be generated due to poor coordination and communication; meanwhile, effective coordination can address most of the waste generation sources [20]. For example, good coordination between clients and designers can prevent design changes during the construction stage, and adequate communication between supervisors and workers can avoid errors in construction activities. In addition, design drawings are normally changed during the construction stage and construction sites contain several subcontractor layers. Appropriate communication can reduce conflicts among parties, which will help to prevent rework and unwanted waste on sites [30].

Minimising Ceramic Tile Waste in Construction Projects

Apart from general sources of construction waste as mentioned in Table 1, the major causes of each type of material waste can be different. For example, over-ordering and formwork leakage are the major causes of concrete waste, while the main sources of tile waste are design changes and damage during installation and transportation [30].

Ceramic tiles are essential construction materials in both commercial and residential buildings [31]. Global ceramic tile production is witnessing an increased rate worldwide today, particularly in 2021, the annual global production of ceramic tiles reached approximately 18.3 billion square meters [12]. Accordingly, ceramic tile waste has also increased worldwide due to the fragile nature of ceramic which easily breaks during transportation, handling, and installation [13]. Such a significant amount of waste will inevitably create growing loads to landfills as well as additional costs due to material losses and waste disposal costs.

Furthermore, the manufacture of ceramic tiles also causes adverse impacts on the environment. Normally starting from extracting raw material, blending, mixing, pressing, glazing, and firing, the production of such materials consumes natural resources and involves energy-intensive processes that result in the depletion of natural resources and significant pollution. Ceramics is also one of the construction materials that account for the most impact on the environment [28]. Ceramic tile waste results in not only the depletion of landfills but also additional environmental impacts raising from the production process, including pollution and exploration of natural resources. Hence, minimising such waste in construction projects is essential to protect the environment and natural resources.

Ceramic tiles are fragile materials which easily break during transportation and can generate a lot of waste when cutting to fit the shape and size of floors. Preferable prevention methods for tile waste can be different from others. However, there are limited studies on minimising ceramic tile waste in construction projects.

The C&D waste management hierarchy is considered an effective strategy for minimising construction waste overall. The hierarchy is divided into five levels of priority (Fig. 1). Waste prevention is the highest priority for waste minimisation, which is often related to appropriate design and planning to reduce construction waste, followed by reuse and recycling materials. Recovery and disposal are the least preferred but will be used when waste cannot be reused or recycled [19].

Minimising Ceramic Tile Waste in Construction Projects

Apart from general sources of construction waste as mentioned in Table 1, the major causes of each type of material waste can be different. For example, over-ordering

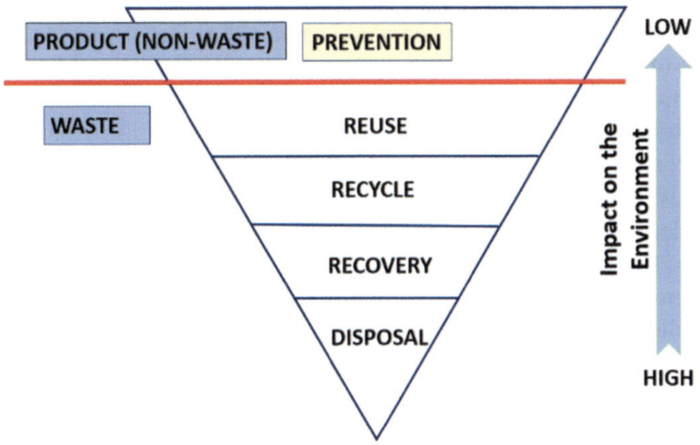

Fig. 1 Waste management hierarchy [10] (*Source* Directive 2008/98/EC)

and formwork leakage are the major causes of concrete waste, while the main sources of tile waste are design changes and damage during installation and transportation [30].

Ceramic tiles are essential construction materials in both commercial and residential buildings [31]. Global ceramic tile production is witnessing an increased rate worldwide today, particularly in 2021, the annual global production of ceramic tiles reached approximately 18.3 billion square metres [12]. Accordingly, ceramic tile waste has also increased worldwide due to the fragile nature of ceramic which easily breaks during transportation, handling, and installation [13]. Such a significant amount of waste will inevitably create growing loads to landfills as well as additional costs due to material losses and waste disposal costs.

Furthermore, the manufacture of ceramic tiles also causes adverse impacts on the environment. Normally starting from extracting raw material, blending, mixing, pressing, glazing, and firing, the production of such materials consumes natural resources and involves energy-intensive processes that result in depletion of natural resources and significant pollution. Ceramics is also one of the construction materials that account for the most impact on the environment [28]. Ceramic tile waste results in not only the depletion of landfills but also additional environmental impacts raising from the production process, including pollution and exploration of natural resources. Hence, minimising such waste in construction projects is essential to protect the environment and natural resources.

Ceramic tiles are fragile materials which easily break during transportation and can generate a lot of waste when cutting to fit the shape and size of floors. Preferable prevention methods for tile waste can be different from others. However, there are limited studies on minimising ceramic tile waste in construction projects.

The C&D waste management hierarchy is considered an effective strategy for minimising construction waste overall. The hierarchy is divided into five levels of priority (Fig. 1). Waste prevention is the highest priority for waste minimisation, which is often related to appropriate design and planning to reduce construction waste, followed by reuse and recycling materials. Recovery and disposal are the least preferred but will be used when waste cannot be reused or recycled [19].

In terms of waste prevention, appropriate design decisions can reduce up to 30% of construction waste [23]. However, designers tend to not pay attention to reducing tile waste, such as non-consideration of minimising tile off-cuts in the design stage, or lack of details in reusing tile off-cuts in shop drawings. Hence, construction practitioners must make decisions in terms of reusing and cutting based on their experience. This leads to rework and significant wastage of materials, labour, and project investment [18]. To aid the design process, Wu et al. [32] proposed a BIM-based framework to optimise floor tile design to reuse off-cuts and prevent tile waste, as a result, reducing 12% of waste, 5% of cut tile numbers, and saving up to 5% of the investment. Furthermore, ordering materials in the right quantity and size will contribute to minimising ceramic tile waste [5], however, purchasing tiles is often based on industry benchmarks and exceeds than actual requirements by 10–20% [32]. Białko et al. [5] suggested further prevention methods for ceramic tile waste, including appropriate transportation and handling, timely delivery, and sufficient storage.

To date, recycling methods are not commonly employed to divert ceramic tile waste from landfills; hence, little attention is paid to waste segregation on-site [5]. Due to a lack of standards and knowledge of using ceramic waste, ceramic tile waste is normally disposed to landfills instead of recycling, even though researchers have reported the potential solutions to recycle ceramic tile waste, such as using it as coarse aggregate in concrete, alternative materials for road pavement subgrade Cabalar et al. [7] and Deboucha et al. [9], or an eco-friendly material in sustainable mortars [25]. Hence, more attempts should be made in ceramic recycling as a promising approach to lower the adverse impacts of ceramic tile waste on the environment and landfills.

3 Methodology

The New Zealand Context and Challenges

In New Zealand, the major proportion of construction waste was sent to clean fills without any waste disposal levy; as a result, this did not push the building industry to improve waste management and minimise construction waste. Construction waste was often co-mingled and sent to landfills or clean fills [14]. Towards zero-waste target, from July 2021, the government imposes a new levy on construction waste disposal to divert construction waste from landfills and clean fills. In addition, BRANZ [6] issued the guideline on designing out waste, developing a site waste management plan, and managing waste on-site in order to enhance the implementation of waste minimisation in the construction industry.

Nevertheless, according to Low et al. [19], the New Zealand construction industry is facing challenges when implementing waste minimisation. Firstly, due to high labour costs in New Zealand, sorting waste on-site will create more burdens on the project budget, although onsite separation brings more advantages than off-site separation, such as requiring fewer efforts, contributing to cost-savings on material purchases, transportation, and landfill cost, as well as facilitating material reuse and recycle process. To solve the problem, Low et al. [19] suggested using temporary workers when more waste is generated, such as on the days of deliveries and unloading materials. Secondly, the New Zealand construction sector has been facing a significant shortage of skilled labour. Relying on foreign and migrant workers also causes several disadvantages, such as language barriers and high labour turnover. Furthermore, due to the imbalance between the low disposal costs or levies and high labour costs for sorting on-site, construction organisations tend to send construction waste to landfills or clean fills rather than reusing and recycling it.

To thoroughly understand the causes of ceramic tile waste and current efforts to reduce it in New Zealand, we employed semi-structured interviews to gather detailed insights into the sources contributing to tile waste and the strategies being employed to minimise it in construction projects. Participants are selected based

Table 2 Interviewees' profiles

Respondent number	Position	Number of years in the role
P1	Project coordinator	2 years
P2	Site manager	20 years
P3	Tiler	More than 10 years
P4	Project coordinator	12 years
P5	Tiler	6 years
P6	Tiler	10 years

on their experience and expertise in tiling work and ceramic tile waste, including tilers, site managers, and project coordinators. As shown in Table 2, two-thirds of the participants possess over a decade of experience in tiling work, enabling them to offer trustworthy insights into the waste management practices concerning ceramic tile waste in construction projects.

The interview question included four main parts: (1) Participants' background, (2) Causes of tile waste generation during the construction stage, (3) Current waste treatment methods onsite for tile waste, and (4) Measures for minimising tile waste in construction projects. Information collection was stopped when reaching the saturation point of the data, which was achieved after six interviews. Interviewees' profiles were coded to protect their anonymity as in Table 2. Interviews were transcribed and imported into an Excel file to identify the themes.

4 Results and Discussion

According to interview findings, ceramic tiles in New Zealand are imported from other countries. China is the major tile supplier from Asia, whereas, in Europe, Italy is the leading tile supplier to New Zealand. Ceramic tiles are commonly used in both commercial and residential buildings due to various benefits, such as good appearance, high durability, ease of cleaning and drying, different grading of the tiles (size, colour, R-value, finishes), less expensive than stone tiles, fire-resistant, and easy to install. Various causes of ceramic tile waste are identified in different project stages from design, procurement to transportation, handling, and installation. Accordingly, measures for minimising ceramic tile waste and challenges in implementing waste minimisation are discussed in the following sections.

4.1 Causes of Ceramic Tile Waste in Construction Projects

4.1.1 Design and Documents

According to Sweis et al. [27], design changes during the construction stage are the most significant source of construction waste due to the rework of completed work. This is consistent with the interview findings that changes in tile layout during the construction stage will result in rework and significant material waste because it is difficult to remove and reuse tiles once they are installed (P2, P3, and P6). Participants added that when it comes to changes in tile colour, prior ordered glue and grout will turn into waste (P1), or when specifications are changed, ordered materials including tiles, glue, cement, and silicon can be wasted (P2 and P6). Furthermore, the selection of tile size directly affects the amount of generated waste, as one of the interviewees mentioned that the use of large tile sizes generates more waste compared to small sizes (P4). Poor tile layouts are another source of design waste as this leads to a significant number of off-cuts (P3).

4.1.2 Material Procurement

Errors in the procurement process are commonly mentioned in the literature as a causative source of construction waste generation, such as errors in quantity take-offs or not complied with specifications [8, 24, 29]. Meanwhile, from the interviewee's point of view, the common cause of waste generation is due to clients' requirements. Clients often preserve 5% of the overall project quantity of floor and wall tiles for future maintenance. If there are any tile leftovers after a few years, they can be reused in other projects, especially in residential projects. However, clients often prefer the latest tile trends rather than reusing tiles in new projects, this means tile leftovers will be wasted due to old fashion. In addition, leftovers of related materials, such as silicone, grout, glue, and waterproofing chemicals, can become waste because they must be used once taken out of the packaging or unable to reuse in other projects due to different colour requirements (P3).

4.1.3 Transportation and Handling

Ceramic tiles are easy to crack during the transportation and handling process. Improper material storage is acknowledged by most researchers as the source of construction waste generation [3, 4, 11]. Meanwhile, it is noted in the interview that humans are a critical source in construction waste generation, including using unskilled drivers, crane operators, and handling labour (P1, P2, P3, and P6). For example, truck drivers fail to use the proper belts to secure the packages, or tiles drop due to poor handling labour. In terms of transportation, the interviewees and literature agreed that poor protection and inappropriate methods will lead to material damage [29], such as improper tile arrangement (P1), and low-quality packaging, and pallets (P6). Apart from the sources mentioned in the literature, one of the experts added that bad weather also has negative effects on the transportation and handling process of ceramic tiles (P6).

4.1.4 Management and Onsite Operation

The respondents believed that the installation phase contributes to the greatest proportion of tile waste on-site. During the interviews, unskilled tilers are emphasised as the main cause of tile waste. Workers who are not trained and lack skills in assigned tasks will negatively impact the quality of work then leading to rework, repairs, and material wastage [20]. For instance, unskilled tilers contribute to the generation of construction waste due to improper cutting (P2) or installing the wrong tile types (P1).

Another causative source is due to poor communication and coordination among project parties, such as wrong instructions from main contractors to subcontractors, and from supervisors to workers (P6). Misunderstanding instructions, drawings, and specifications may result in mistakes in the selection of tile types, as well as errors in cutting and installation. In addition, ineffective coordination with other subcontractors may result in clashes and rework. In terms of tiling work, conflicts commonly occur with plumbers and electricians (P1, P2, and P3). This finding is aligned with [4] who suggested that onsite coordination and communication become a serious challenge when it comes to a project with multiple levels of subcontractors.

4.2 Measures for Minimising Ceramic Tile Waste

To reduce the ceramic tile waste generated from construction projects, measures are investigated and categorised into four main groups, as illustrated in Fig. 2. Detailed explanations are given in the following sub-sections.

4.2.1 Design and Documents

In line with the literature, the interview findings also confirm that changes and errors in design are critical waste sources [20]. Hence, drawings and specifications should be finalised before ordering materials and installing tiles on-site. Furthermore, architecture layouts can make great contributions to the reduction of waste generation by setting out floor tiles correctly and planning for minimum tile cutting (P2). One expert added a design measure of using small-size tiles to reduce off-cuts and wastage (P3). Effective coordination with clients is important to optimal tile layout to minimise off-cuts and still ensure the expected appearance (P4). According to Wu et al. [32], innovative technology can be employed to enhance the quality of design, such as BIM

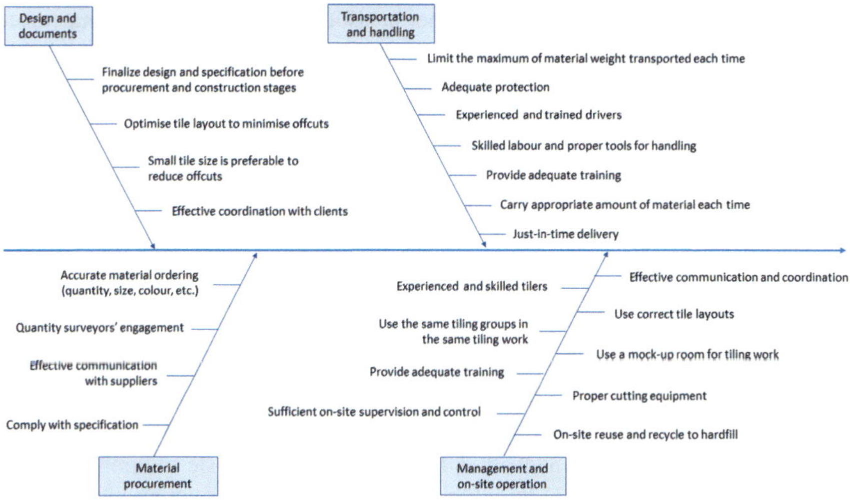

Fig. 2 Measures for minimising ceramic tile waste

being used to avoid design errors and support the layout design process to minimise tile cutting and reuse off-cuts.

4.2.2 Material Procurement

In the procurement process, accurate material ordering plays an important role in preventing unwanted waste, including correct quantity, colour, size, and other related specifications (P2). Quantity surveying plays a crucial role to estimate the needed materials for order, including not only tiles but also related materials, such as glue and grout [15]. Umar et al. [29] suggested that good communication with suppliers can avoid misunderstandings and errors in ordering.

4.2.3 Transportation and Handling

Due to the fragile characteristic of ceramic tiles, proper methods need to be applied during the transportation and handling process. The transportation process needs experienced drivers and adequate protection, including using good quality packaging and pallets. If there are any opened or damaged packages, they must be adequately covered or repackaged (P2). It is advised in the interviews that the maximum weight of tile transported at one time is limited to five tons for safer transportation to avoid unwanted tile damage (P1).

During the handling process, interviewees agreed that the measures for avoiding material damage include employing experienced labours for loading and unloading materials, and using proper handling tools, such as pallet jacks and custom-built trolley for moving tiles. Training is another measure for improving the competence of handling labours (P3). When loading and unloading materials, workers should carry an appropriate amount to avoid tile breakage (P1). Both the interview results and literature indicated that materials should be delivered at the time needed for installation to avoid material damage during storage periods on-site [22]. One respondent added that if tiles are carried to the application points too early, other labours can break them when sitting on them or moving them to other places (P3).

4.2.4 Management and Onsite Operation

In order to reduce tile waste during the installation stage, the interview results show that employing experienced and skilled tilers as well as using the same tiling groups in the same tiling work are prioritised. Adequate training should be provided to improve tilers' competence as well as their awareness of waste minimisation and environmental protection (P1). In addition, onsite supervisors can provide instructions for workers carrying out tiling work properly. Naji et al. [22] suggested that clear roles and responsibilities of the project team member will improve control and supervision to monitor and guide workers.

Furthermore, participants believed that effective communication would help to eliminate misunderstandings and errors during the installation process (P6 and P2). Tilers can clearly understand drawings, specifications, and instructions from supervisors to implement their work properly. In addition, construction sites often have multiple subcontractor layers, hence, keeping proper coordination among the main contractor and its subcontractors will help to avoid conflicts, and reduce rework and unwanted material waste. Project managers play an important role in ensuring transparent communication and effective coordination among parties [4], especially when it comes to design changes, information should be updated to related parties before they commence their work.

Interview findings also confirm that tile layouts and other requirements should be checked carefully and confirmed by the main contractors and clients before carrying out tiling work. An interviewee added that a mock-up room should be tiled first, if the tile waste is significant, setting out floor tiles should be discussed with the main contractors and clients to reduce tile waste (P4). Another method mentioned in the interviews is using proper cutting equipment, including high-quality electric tile cutters, sharp cutting blades, and renewing cutting wheels frequently (P3, P4, and P5). Moreover, tile off-cuts are encouraged to be reused on-site; otherwise, they can be crushed and mixed with soil and then used for hard filling (P3).

4.3 Challenges in Minimisation of Ceramic Tile Waste

4.3.1 Government Policies and Standards

In New Zealand, most of the construction waste, including ceramic tile waste, used to be sent to clean fills without any charges. Construction organisations did not attempt to reuse and recycle construction waste due to higher costs compared to sending to clean fills. Cabalar et al. [7] explained this is due to the lack of guidelines, standards, and experience in reusing and recycling tile waste, it is co-mingled into onsite bins with other construction waste and sent to clean fills instead of recycling. However, to achieve the zero-waste target, reusing and recycling waste is crucial when waste generation cannot be prevented. To maximise diversion from landfills as a key part of the construction industry's move to achieve a net-zero carbon economy by 2050, a new levy was introduced in July 2024 [6]. This will force the industry to make more efforts to reduce waste generated in construction projects. The government and industry still need to develop guidelines and standards on reuse and recycling tile waste to increase the waste diversion rate.

4.3.2 Client Requirements

Once construction starts, any design changes as client requirements may lead to reordering materials or rework on-site, such as changes in tile layouts or types when they are already ordered or installed will result in significant tile waste. Furthermore, clients expect to have tile layouts with a good look, even if they may lead to complicated tile layouts, more off-cuts, and tile waste. Therefore, it is important to raise clients' awareness of waste minimisation and environmental protection and coordinate with clients early from the design stage to clearly understand their requirements.

4.3.3 Labour Market

Skilled and experienced tilers are highlighted in the interviews as a crucial factor in construction waste minimisation. Nevertheless, the construction sector in New Zealand is struggling with labour shortages, especially the shortage of skilled workers. Immigration is considered a solution to supplement skilled tilers; however, language barriers are the main cause of misunderstanding and errors on-site. High labour cost is another barrier to construction waste management. Although onsite separation can support the reuse and recycling process, a high labour cost will contribute to higher handling costs compared to the clean filling option.

4.3.4 Limited Market Demand for Reused and Recycled Materials

Market demand is essential to sustain material reuse and recycling; however, this is still restricted in New Zealand [19]. There are no tile recycling facilities in New Zealand, tile waste is sent to clean fills. When it comes to tile leftovers, it is encouraged to reuse them in other projects by selling or giving them for free. If leftovers cannot be reused in other projects, they will also be sent to clean fills. To solve the problem, sharing centres or online sharing platforms can be developed as a place where tile leftovers and off-cuts can be bought and sold widely. Moreover, recycling ceramic waste as alternative raw materials in construction and other industries is a promising solution to maximise diversion from landfills.

5 Conclusions

In conclusion, the chapter has provided comprehensive insights into the sources contributing to ceramic tile waste generation in construction projects and proposed various measures for its minimisation. Design changes, errors in procurement, transportation mishaps, and onsite mismanagement were identified as key drivers of waste generation. These findings underscore the importance of meticulous planning, effective coordination, and skillful execution throughout the construction process.

One significant aspect highlighted is the need for accurate design and documentation prior to material procurement and installation. By finalising designs early and employing innovative technologies like BIM, potential errors and layout inefficiencies can be minimised. Moreover, establishing clear communication channels with clients can help align their expectations with waste reduction goals.

Efficient material procurement practices, including precise quantity estimation and good supplier communication, were emphasised as crucial for waste prevention. Transportation and handling procedures also require careful attention to avoid tile damage, with measures such as employing experienced drivers and utilising proper packaging and handling tools.

Furthermore, onsite operations management plays a pivotal role in waste minimisation. Investing in skilled labour, providing adequate training, and fostering effective communication among project stakeholders are essential steps. Additionally, implementing quality control measures and conducting mock-ups can help identify and address potential issues before full-scale installation.

The chapter also highlights broader industry challenges, such as limited market demand for reused and recycled materials and labour shortages. Addressing these challenges will require collaborative efforts from industry stakeholders, policymakers, and the government to develop guidelines, standards, and infrastructure to support waste diversion and recycling initiatives.

Overall, the findings underscore the need for a holistic approach to ceramic tile waste minimisation, encompassing design optimisation, efficient procurement and logistics, effective onsite management, and industry-wide collaboration. By implementing these measures, the construction sector can move closer towards achieving sustainability goals and minimising its environmental footprint.

Acknowledgements This research is a part of the "Building Capability to help the Construction Industry Transition to Zero Carbon" research programme, which is funded by the Building Research Association New Zealand (BRANZ) under Building Research Levy, grant number LR12965.

References

1. Ajayi SO, Oyedele LO (2018) Critical design factors for minimising waste in construction projects: a structural equation modelling approach. Resour Conserv Recycl 137:302–313
2. Almusawi MBH, Karim ATBA, Ethaib S (2022) Evaluation of construction and demolition waste management in Kuwait (Article). Recycling 7(6):Article 88. https://doi.org/10.3390/recycling7060088
3. Arshad H, Qasim M, Thaheem MJ, Gabriel HF (2017) Quantification of material wastage in construction industry of Pakistan: an analytical relationship between building types and waste generation. J Constr Dev Ctries 22(2):19–34. https://doi.org/10.21315/jcdc2017.22.2.2
4. Bajjou MS, Chafi A (2022) Exploring the critical waste factors affecting construction projects. Eng Constr Archit Manag 29(6):2268–2299. https://doi.org/10.1108/ECAM-12-2020-1097
5. Białko M, Hoła B, Topolski M, Woźniak Z (2022) Waste reduction methods used in construction companies with regards to selected building products. Appl Sci 12(11):5387. https://doi.org/10.3390/app12115387
6. Bulletin 671 reducing construction and demolition waste (2022)
7. Cabalar AF, Hassan DI, Abdulnafaa MD (2017) Use of waste ceramic tiles for road pavement subgrade. Road Mater Pavement Des 18(4):882–896. https://doi.org/10.1080/14680629.2016.1194884
8. Chibuike C, Booth C, Lamond J (2021) A review of the causes of construction waste generation in Nigeria and recommendations. Proc Inst Civ Eng-Waste Resour Manag 1–20. https://doi.org/10.1680/jwarm.20.00020
9. Deboucha S, Aissa Mamoune SM, Sail Y, Ziani H (2020) Effects of ceramic waste, marble dust, and cement in pavement sub-base layer. Geotech Geol Eng 38(3):3331–3340. https://doi.org/10.1007/s10706-020-01211-x
10. Directive 2008/98/EC of the European Parliament and of the Council of 19 November 2008 on waste and repealing certain Directives (2008)
11. Domingo N (2015) Assessment of the impact of complex healthcare features on construction waste generation. Buildings 5(3):860–879. https://doi.org/10.3390/buildings5030860
12. Garside M (2023) Global ceramic tile production volume 2018–2021. https://www.statista.com/statistics/1275511/global-ceramic-tile-production-volume/
13. Goyal RK, Agarwal V, Gupta R, Rathore K, Somani P (2021) Optimum utilization of ceramic tile waste for enhancing concrete properties. Mater Today Proc
14. Hernandez G, Low J, Nand A, Bu A, Wallis SL, Kestle L, Berry T-A (2022) Quantifying and managing plastic waste generated from building construction in Auckland, New Zealand. Waste Manag & Res 0734242X221105425. https://doi.org/10.1177/0734242X221105425
15. Jahan I, Zhang G, Bhuiyan M, Navaratnam S, Shi L (2022) Experts' perceptions of the management and minimisation of waste in the Australian construction industry. Sustainability 14(18):11319. https://doi.org/10.3390/su141811319
16. Kabirifar K, Mojtahedi M, Wang C, Tam VW (2020) Construction and demolition waste management contributing factors coupled with reduce, reuse, and recycle strategies for effective waste management: a review. J Clean Prod 263:121265

17. Kolaventi SS, Momand H, Tadepalli T, Siva Kumar M (2020) Construction waste in India: a structural equation model for identification of causes. Proc Inst Civ Eng-Eng Sustain
18. Liu H, Sydora C, Altaf MS, Han S, Al-Hussein M (2019) Towards sustainable construction: BIM-enabled design and planning of roof sheathing installation for prefabricated buildings. J Clean Prod 235:1189–1201. https://doi.org/10.1016/j.jclepro.2019.07.055
19. Low JK, Wallis SL, Hernandez G, Cerqueira I, Steinhorn G, Berry T-A (2020) Encouraging circular waste economies for the New Zealand construction industry: opportunities and barriers. Front Sustain Cities 2:35. https://doi.org/10.3389/frsc.2020.00035
20. Luangcharoenrat C, Intrachooto S, Peansupap V, Sutthinarakorn W (2019) Factors influencing construction waste generation in building construction: Thailand's perspective. Sustainability 11(13):3638. https://doi.org/10.3390/su11133638
21. Marinelli M, Dolan M, Spillane JP, Konanahalli A (2014) Material waste in the Northern Ireland construction industry: on-site management causes and methods of prevention
22. Naji KK, Gunduz M, Hamaidi MF (2022) Major factors affecting construction waste management in infrastructure projects using structural equation model. J Constr Eng Manag 148(10):04022101. https://doi.org/10.1061/(ASCE)CO.1943-7862.0002358
23. Osmani M, Glass J, Price A (2006) Architect and contractor attitudes to waste minimisation. Proc Inst Civ Eng-Waste Resour Manag
24. Saad A, Bal M, Khatib J (2022) The need for a proper waste management plan for the construction industry: a case study in Lebanon (Article). Sustain (Switz) 14(19):Article 12783. https://doi.org/10.3390/su141912783
25. Samadi M, Huseien GF, Mohammadhosseini H, Lee HS, Abdul Shukor Lim NH, Tahir M.M, Alyousef R (2020) Waste ceramic as low cost and eco-friendly materials in the production of sustainable mortars. J Clean Prod 266. https://doi.org/10.1016/j.jclepro.2020.121825
26. Saunders M, Lewis P, Thornhill A (2016) Research methods for business students, 7th edn. Pearson Education
27. Sweis GJ, Hiari A, Thneibat M, Hiyassat M, Abu-Khader WS, Sweis RJ (2021) Understanding the causes of material wastage in the construction industry (Article). Jordan J Civ Eng 15(2):180–192. https://www.scopus.com/inward/record.uri?eid=2-s2.0-85108652020&partnerID=40&md5=ede18c728699b5b4a3caf6939b56b5f6
28. Tikul N, Srichandr P (2010) Assessing the environmental impact of ceramic tile production in Thailand (Article). J Ceram Soc Jpn 118(1382):887–894. https://doi.org/10.2109/jcersj2.118.887
29. Umar UA, Shafiq N, Malakahmad A, Nuruddin MF, Salihi IU (2016) Ranking of principal causes of construction waste for Malaysian residential project. In: Engineering challenges for sustainable future-proceedings of the 3rd international conference on civil, offshore and environmental engineering, ICCOEE 2016
30. Wang JY, Kang XP, Wing-Yan Tam V (2008) An investigation of construction wastes: an empirical study in Shenzhen. J Eng, Des Technol 6(3):227–236. https://doi.org/10.1108/17260530810918252
31. Wattanasiriwech D, Saiton A, Wattanasiriwech S (2009) Paving blocks from ceramic tile production waste (Article). J Clean Prod 17(18):1663–1668. https://doi.org/10.1016/j.jclepro.2009.08.008
32. Wu S, Zhang N, Luo X, Lu W-Z (2022) Multi-objective optimization in floor tile planning: coupling BIM and parametric design. Autom Constr 140:104384. https://doi.org/10.1016/j.autcon.2022.104384

Chapter 14
Integrating Building Information Modelling with Modern Methods of Construction for Sustainable Construction in the UK

Devindi Geekiyanage and Mo Maleki Sadabad

Abstract The integration of Building Information Modelling (BIM) with Modern Methods of Construction (MMC) not only streamlines construction processes but also catalyses the achievement of sustainability benchmarks across a building's life cycle. While existing literature predominantly focuses on the overall impact of mainstreaming BIM in MMC, a notable gap exists in understanding how BIM integration specifically contributes to the success of diverse MMC applications. This chapter bridges the identified gap by systematically analysing the integral role of BIM within various MMCs, highlighting key principles, methodologies, benefits, and challenges, as well as emerging technologies. It reveals BIM's role in seven MMCs, underscoring its role in bolstering efficiency, collaboration, and sustainability, as evidenced in UK construction case studies. Furthermore, strategic solutions for BIM-MMC integration are discussed for key challenges, including limited standardisation, fragmentation and interoperability challenges, skill and knowledge gaps, financial implications, data security and privacy concerns, and cultural and organisational barriers. The study further outlines the contribution of emerging technologies, including innovative materials, Internet of Things, big data, automation, artificial intelligence, and blockchain technology, in advancing BIM-MMC integration, driving sustainable construction practices and shaping a more resilient built environment. By emphasising sustainable construction and drawing insights from relevant case studies, particularly within the UK construction industry, the chapter offers a nuanced account of key technologies integral to MMC.

Keywords Building Information Modelling (BIM) · Modern Methods of Construction (MMC) · UK · Construction industry · Sustainability

D. Geekiyanage (✉) · M. M. Sadabad
School of Science, Engineering and Environment, University of Salford, The Crescent, Manchester, UK
e-mail: d.geekiyanage@salford.ac.uk

M. M. Sadabad
e-mail: m.n.malekisadabad1@salford.ac.uk

T. Ramachandra et al. (eds.), *Concepts, Policies and Technology for Sustainable Buildings*, Advances in Sustainability Science and Technology,
https://doi.org/10.1007/978-981-96-3086-8_14

1 Introduction

In the rapidly evolving landscape of the twenty-first century, the construction industry stands at a critical juncture, poised to redefine its role in shaping the world we inhabit. Modern Methods of Construction (MMCs) encompass a diverse array of innovative building processes, transforming traditional construction methodologies. These methods span off-site, near-site, and on-site pre-manufacturing, as well as process improvements and technology applications. While the term MMC is often interchangeably used with off-site or modular construction, it is crucial to recognise the broader scope of MMC, covering advancements in both off-site and on-site processes [26]. One facet of MMC involves off-site construction methods, where components or modules are fabricated in controlled factory environments. This includes Precast Flat Panel Systems, 3D Volumetric Construction, and Insulating Concrete Formwork (ICF) [14, 42, 58]. MMC extends beyond off-site methods such as Tunnel Form construction, to encompass on-site process improvements [53]. Hybrid Concrete Construction (HCC) is another noteworthy category within MMC, seamlessly integrating traditional cast-in-place concrete with off-site produced precast elements [21].

MMCs have improved the sustainability landscape in the construction industry, offering multifaceted benefits across various dimensions. Firstly, MMC often results in a reduction in construction time, contributing to a decrease in the environmental impact associated with extended construction periods [14]. This aligns with the industry's increasing emphasis on sustainable practices and resource efficiency. Additionally, studies indicate substantial reductions in embodied energy and greenhouse gas emissions associated with modular construction [14, 25, 47]. The controlled environments of off-site fabrication contribute to precise material use, reducing waste, and promoting sustainability [58].

As of the latest available data from Build off-site (2021), the adoption of MMC in the UK construction industry has shown steady progress but varies across different segments. MMC accounts for approximately 3.5% of all construction activities in the UK, with a higher prevalence in new constructions (5% of all projects) compared to refurbishments (2% of renovation projects). This increase is particularly notable in residential developments, where MMC is employed in approximately 8% of new housing projects. The government's push for housing innovation and sustainability has contributed to this upward trajectory, encouraging the integration of MMC in new builds. While the application of MMC in refurbishments is gradually gaining traction, the complexities involved in retrofitting existing structures pose challenges compared to integrating MMC in new developments. Notably, the type of buildings also plays a role in the adoption of MMC. Residential construction leads in MMC usage, with modular and off-site construction methods being popular choices for housing projects. In contrast, MMC adoption in commercial and institutional buildings hovers around 3%, reflecting a slower but emerging trend in these sectors.

Moreover, the integration of Building Information Modelling (BIM) with MMC emerges as a key driver for sustainability. BIM is a sophisticated digital technology transforming the architecture, engineering, and construction (AEC) industry. Unlike traditional 2D drawings, BIM involves creating and managing digital representations of physical and functional characteristics of places while offering a collaborative platform to design, construct, and manage buildings and infrastructure. BIM facilitates visualisation, simulation, and analysis of project aspects like spatial relationships, quantities, and building component properties, fostering improved communication, coordination, and decision-making. This optimises designs for energy efficiency, aids in accurate material estimation, and supports Life Cycle Assessment (LCA) for informed decisions across a building's entire life cycle [1, 76].

The combination of BIM and MMC has been shown to significantly enhance collaboration, error reduction, and efficiency in construction projects [67]. Recent studies suggest that this integration not only streamlines the construction process but also contributes to achieving sustainability goals throughout a building's life cycle. For instance, Alwan et al. [4] found that BIM adoption in MMC projects led to a 30% reduction in construction time and a 20% decrease in material waste. Moreover, the Construction Industry Institute [15] reported that BIM implementation alongside MMC resulted in a 25% increase in productivity and a 15% improvement in project cost predictability. This evidence underscores the transformative impact of combining BIM with MMC on construction practices and sustainability outcomes.

While existing literature predominantly focuses on the overall impact of mainstreaming BIM in MMC, there is a notable gap in understanding how BIM integration has specifically contributed to the success of various types of MMC. Most discussions tend to centre around off-site constructions, overlooking the nuanced differentiations among MMC types that warrant a more contextualised approach. This chapter aims to delve into the multifaceted role of BIM across diverse dimensions of MMC, examining its underlying principles, methodologies, and the profound influence it exerts on the built environment. By navigating through key technologies integral to MMC and emphasising sustainable construction, the chapter will provide a nuanced exploration, drawing insights from relevant case studies, particularly within the context of the UK construction industry.

2 Background of the Inquiry

2.1 Application of MMC in the UK

The spectrum of MMC ranges from the construction of entire homes using factory-built volumetric modules to the implementation of innovative methods for laying concrete blockwork directly on-site [26]. Table 1 provides an overview of commonly employed MMC types within the UK construction industry.

Table 1 Key features of diversified types of MMCs

MMC type	Key features	UK application	Sources
Precast flat panel system and foundations	• Off-site production in a controlled factory environment • Assembled on-site for robust structures • Ideal for repetitive cellular projects • Load-bearing elements with insulation and cladding • Tailored for diverse functions • Accommodates services, windows, doors, finishes, and foundational support • Improves productivity, especially in adverse weather conditions • Reduces excavation, advantageous for contaminated ground	• Structures up to 22 storeys have been completed in the UK using crosswall construction	[33, 40, 42, 44, 58]

(continued)

Table 1 (continued)

MMC type	Key features	UK application	Sources
Tunnel form	• Specialised formwork system • Focus on repetitive structural elements like walls and floors • Blends speed, quality, and accuracy of factory-produced ready-mixed concrete • Maintains flexibility and cost-effectiveness of cast in situ construction • Suitable for projects with repetitive structures, such as hotels, apartment blocks, and student accommodations • Offers economy, speed, quality, and accuracy in construction • Utilises inherent benefits of concrete, including fire and sound resistance • Particularly popular in countries with a demand for mass housing and high-rise structures	• Contributes to meeting the demand for housing in a timely manner • Aligns well with the need for rapid construction in urban areas	[2, 13, 38, 48, 53, 75]

(continued)

Table 1 (continued)

MMC type	Key features	UK application	Sources
3D Volumetric Construction (often referred to as modular or off-site construction)	• Prefabricating 3D modules in a controlled factory environment • Modules include finishes, fixtures, and fittings • Versatility highlighted by the production of basic structures to fully finished units (e.g., concrete stair, lift cores, and bathroom pods) • Precision and rapid on-site assembly emphasised • Inherently embodies concrete benefits, including thermal mass, sound resistance, and fire resistance • Ensures factory quality and accuracy	• Emerging as a pragmatic solution for housing shortages • Major projects include residential developments, student accommodations, and healthcare facilities • Embraced to realise efficiency gains and cost-effectiveness in construction	[2, 3, 14, 27, 28, 38, 42, 55, 63, 71, 75]

(continued)

Table 1 (continued)

MMC type	Key features	UK application	Sources
Flat slabs	• A widely adopted structural engineering method • Directly supports floor slabs on columns without the use of beams or girders • Notable for incorporation of column heads and drop panels • Prioritises reinforcing slab-column connections, ensuring overall structural integrity • Accelerated project timelines • Increased flexibility in architectural layouts • Especially beneficial in seismic regions • Allows for unrestricted positioning of services and partitions, providing occupants with the flexibility to modify internal layouts according to their needs • The use of post-tensioning enhances structural efficiency • Enables the creation of longer and thinner slabs with reduced reinforcement	• Aligns with industry demands for creative and efficient interior spaces, showcasing its adaptability across various types of structures, including residential, commercial, and institutional buildings	[6, 32, 36, 73]

(continued)

Table 1 (continued)

MMC type	Key features	UK application	Sources
Hybrid Concrete Construction (HCC)	• Integrate cast-in-place and precast concrete • Columns, beams, and slabs manufactured in controlled environments for high-quality and precision • Accelerates project schedules • Minimises on-site waste through efficient material use • Allows on-site customisation and adaptation for various architectural designs and project requirements	• Addressing housing shortages, reducing environmental impact, and enhancing overall construction efficiency	[21, 45, 64]

(continued)

Table 1 (continued)

MMC type	Key features	UK application	Sources
Thin joint masonry	• Involves using thin mortar joints, typically 1–3 mm thick • Utilises precision-engineered blocks made from materials like autoclaved aerated concrete (AAC) or lightweight aggregate concrete • The mortar used is often a pre-mixed adhesive that sets rapidly, facilitating quick construction • Adoption of large-format concrete blocks, equivalent to two traditional blocks, for efficient walling • Rapid mortar curing achieves full bond strength within one to two hours, addressing the challenge of 'floating' • Allows for laying more courses per day	• Aligns seamlessly with the industry's emphasis on sustainability and energy efficiency	[29, 56, 66]

(continued)

Table 1 (continued)

MMC type	Key features	UK application	Sources
Insulating concrete formwork	• Utilises twin-walled panels or blocks made of expanded polystyrene for swift assembly to establish formwork for building walls • The formwork is filled with pre-manufactured, quality-guaranteed, ready-mixed concrete, creating a resilient structure • The enduring expanded polystyrene blocks deliver exceptional thermal insulation, enhancing energy efficiency and reducing heating and cooling costs • Combination of concrete and foam forms ensures structural integrity, robustness, and durability • Offers resistance against severe weather conditions • Excels in sound insulation, contributing to a more comfortable indoor environment	• While not as ubiquitous as in some regions, ICF has garnered recognition in the UK for its insulation properties and durability • Studies conducted by BRE in the UK have contributed to the acceptance of ICF construction, particularly in projects prioritising energy efficiency and sustainable design	[11, 57, 60]

2.2 Contributions of MMC to Sustainable Construction

As the global construction industry increasingly embraces MMC, the statistics speak volumes about the transformative impact on sustainability. The contributions of MMC towards construction sustainability are multifaceted. One pivotal aspect is the reduction in construction time [14, 37, 51, 54]. In particular, Bertram et al. [14]'s research indicates that modular construction slashes construction timelines by up to 50%, thereby not only accelerating project delivery but also minimising the environmental impact associated with extended construction periods. Similarly, it has been proved that tunnel form can reduce the cost of the frame of a typical hotel or similar building by about 15% when compared with traditional construction, chiefly due to the increased speed of construction [60]. In addition, Thin Joint Masonry Construction method significantly reduces construction time, with statistics indicating a potential speed increase of up to 13.5% when using large-format concrete blocks [58]. Overall, savings in construction time of up to 25% over traditional frame construction methods can be achieved; for low-rise construction, which would traditionally be in brick or block, savings of 45% on the construction time have been achieved [58]. This reduction in time is attributed to the efficiency gains achieved through off-site construction processes.

Focusing on the environmental impact, studies shed light on the substantial reductions in embodied energy and greenhouse gas emissions associated with modular construction. Glass and Baiche [25] report a remarkable 56% and 26% reduction in embodied energy for small and large structures, respectively. Quale et al. [47] expand the discussion, highlighting that while modular construction may increase water usage, it significantly decreases eutrophication and various pollutants, showcasing a holistic approach to sustainability. Furthermore, Aye et al. [8] demonstrate that a steel prefabrication system can reduce material consumption by up to 78% compared to conventional reinforced concrete construction, emphasising the resource efficiency of MMC. Lawson et al. [30]'s findings on modular construction align with this efficiency narrative, showing reductions in safety incidents, noise, delivery vehicle frequency, and embodied energy. In addition, Thin Joint Masonry aligns with energy efficiency goals through the utilisation of lightweight and thermally insulating materials. On the contrary, Dabirian et al. [16] note that prefabrication facilities may produce more noise than on-site construction, underscoring a potential trade-off in sustainability aspects. However, Lawson et al. [30] counterbalance this concern by demonstrating that modular construction reduces overall noise by 30–50%, indicating a nuanced consideration of noise-related sustainability factors.

Regarding financial implications, Baldwin et al. [12] suggest little difference in overall costs between modular and conventional construction, whereas Ferdous et al. [19] argue that despite higher initial investment costs, modular construction leads to reduced labour costs, ultimately resulting in a 20% overall cost reduction.

2.3 BIM and MMC Integration

The statistical evidence presented above has positioned MMC as a cornerstone in the construction sector's journey towards a more sustainable and resilient future. It has been evident that BIM stands at the forefront of advancing sustainability within MMC. Through comprehensive 3D modelling and simulation, BIM optimises designs for energy efficiency, minimising environmental impact [1, 74]. BIM's role extends to efficient resource management, aiding accurate material estimation and reducing wastage, with studies indicating potential waste reductions of up to 30% [24]. Moreover, BIM facilitates Life Cycle Assessment (LCA), enabling informed decisions for sustainable practices across a building's entire life cycle [76]. In the context of MMC, BIM fosters collaboration and coordination, reducing errors and improving efficiency [67]. In off-site construction, the integration of BIM ensures precision in prefabricated components, contributing to sustainable construction by minimising waste and enhancing speed, with reports showcasing up to 50% reduction in construction timelines [14]. Furthermore, BIM supports energy performance analysis, allowing for the assessment of energy efficiency, aligning with sustainability goals [61]. As buildings move into the operational phase, the digital models created by BIM continue to contribute to sustainability by aiding efficient facility management and maintenance [34, 50].

3 Analysis and Results

The following synthesis comprises a desk study encompassing case studies, industry reports, and discussion briefs from construction practitioners regarding the integration of BIM with diverse approaches to MMC, with a specific focus on insights extracted from UK construction project scenarios. The analysis and synthesis provide a detailed discussion on specific benefits and challenges associated with integrating BIM across seven types of MMCs. Furthermore, the synthesis includes an in-depth examination of three prominent BIM- and MMC-involved major construction projects, shedding light on their methodologies and outcomes. Additionally, the synthesis explores industry proposals aimed at mainstreaming the integration of BIM within MMC practices.

3.1 Benefits and Challenges in Integrating BIM with MMC in the UK Construction Industry

In the dynamic landscape of the UK's construction industry, the strategic integration of BIM within various MMCs has emerged as a game-changer. This comprehensive exploration delves into the profound benefits and consequential impacts witnessed through the seamless amalgamation of BIM in different MMC scenarios, underscoring its role in bolstering efficiency, collaboration, and sustainability, as evidenced by practitioners in the UK construction industry.

3.1.1 Precast Flat Panel System

The utilisation of BIM in precast flat panel systems has transformed the design process by enabling efficient design coordination. Through precise alignment among precast elements, BIM significantly reduces errors, enhancing overall design efficiency. Automated clash detection in BIM minimises conflicts during precast assembly, ensuring a seamless fit and reducing errors [31]. Moreover, BIM's parametric modelling capabilities allow for quick adjustments to precast designs, optimising flexibility in the construction process. In the context of the UK, where efficiency in construction projects is paramount, BIM has transformed the precast flat panel system. The intricacies of projects like the One Nine Elms development in London, leveraging precast elements, underscore the critical role of BIM in precise design coordination receiving the Best Commercial Project in the 2020 Tekla Global BIM Awards [62]. The reduction in errors and enhanced design efficiency achieved through BIM align with the industry's demand for streamlined processes.

In the realm of the precast flat panel system, ensuring seamless coordination among elements demands heightened precision in the BIM model to minimise errors during assembly, a fact exemplified in projects like the Battersea Power Station redevelopment, where despite extensive use of precast elements, precise coordination through BIM faced challenges [39]. The parametric modelling capabilities of BIM, while offering quick adjustments, introduce complexities in mastering intricate precast designs. Simultaneously, on the foundation front, exemplified by the Wembley Park Regeneration project, challenges arise in accurately modelling structural elements within BIM, given varying soil conditions and complex geometries [49]. The need for an accurate representation of foundation details for precise structural analysis adds a layer of complexity. Furthermore, effective communication and coordination challenges persist across both systems, as highlighted by statistics from the Crossrail project [43], necessitating a strategic and concerted effort to overcome barriers and fully capitalise on the benefits of BIM integration in these critical components of UK construction projects.

3.1.2 Tunnel Form

BIM's 3D visualisation and simulation capabilities aid in better decision-making during planning, offering stakeholders a realistic preview of tunnel form structures [53]. Furthermore, the automated clash detection in BIM remains critical in identifying and resolving conflicts, reducing errors and rework during construction [31]. Amidst the bustling urban projects in the UK, exemplified by the Crossrail initiative, the integration of BIM in Tunnel Form construction has streamlined collaboration among multidisciplinary teams [43]. The visualisation and simulation capabilities of BIM have proven pivotal in the decision-making processes of projects like Crossrail, ensuring a unified approach among architects, engineers, and contractors.

One prominent challenge lies in the complexity of coordinating the repetitive nature of Tunnel Form structures with the detailed modelling requirements of BIM, demanding a nuanced approach to capture the intricacies of each component while maintaining efficiency. A case in point is the Crossrail project, where the integration faced hurdles in aligning the continuous casting process of Tunnel Form with the need for precise information modelling, leading to clashes and inefficiencies [43]. Additionally, the resistance to change within traditional construction practices, as evidenced in projects like the One Nine Elms development [62], underscores the need for comprehensive training and cultural shifts to fully exploit BIM's potential. Furthermore, the absence of standardised protocols and clear regulatory frameworks, exemplified by challenges encountered in the Thames Tideway Tunnel project [43], hinders seamless collaboration and data exchange.

3.1.3 3D Volumetric Construction

In the realm of 3D volumetric construction, BIM ensures precision in off-site construction by accurately modelling volumetric modules. The data-rich environment of BIM promotes seamless collaboration among stakeholders, facilitating a unified understanding of designs [38]. Moreover, BIM's support for the entire life cycle of 3D volumetric construction projects, from design to operation and maintenance, enhances project management and sustainability [28]. The off-site construction trend, exemplified by projects like the Wembley Park regeneration, witnesses BIM's impact on precision [49]. BIM's role in modelling volumetric modules contributes to the accuracy demanded by off-site construction methodologies, aligning with the UK's push for innovation and sustainability in construction practices.

One significant hurdle involves the complexity of coordinating volumetric modules within a BIM framework, as these 3D units demand meticulous detailing and precise alignment for successful integration [59]. A case in point is the Modular Cancer Centre at Guy's Hospital, where the incorporation of BIM into volumetric construction faced challenges related to ensuring accurate parametric modelling and seamless coordination of the modular components. Moreover, the industry-wide struggle with standardisation and interoperability persists, hindering the seamless exchange of data between various stakeholders and platforms in volumetric construction projects [59]. This is evident in the Crossrail project, where different teams using distinct BIM software encountered difficulties in consolidating data, leading to inefficiencies and communication gaps [43].

3.1.4 Flat Slabs

The integration of BIM in flat slab construction enhances structural analysis, optimising coordination and ensuring integrity. Furthermore, 3D visualisation capabilities in BIM contribute to a deeper understanding of flat slab designs, facilitating informed decision-making [23]. In iconic projects such as The Shard, BIM's role in enhancing structural analysis and facilitating effective communication in flat slab construction has been instrumental [68]. The visualisation capabilities of BIM have allowed stakeholders in projects like The Shard to make informed decisions, ensuring the integrity and efficiency of flat slab elements.

Integrating BIM into flat slab construction methods in the UK encounters several challenges, notably stemming from the complex nature of structural design and coordination. The intricate detailing required for flat slabs demands precise BIM modelling, and a notable challenge lies in the potential discrepancies between the virtual model and the physical construction. The intricacies of structural analysis and coordination in flat slab construction can be exemplified by projects like The Leadenhall Building in London, where deviations between the BIM model and on-site construction were identified, leading to adjustments during the construction phase [5]. Additionally, a study by the Construction Industry Council (CIC) in the UK revealed that while BIM adoption is increasing, there is a need for further standardisation and training within the industry to ensure seamless integration with construction methods like flat slabs [17].

3.1.5 Hybrid Concrete Construction

BIM's role in HCC lies in promoting integration between traditional and modern methods. The clash detection features in BIM identify and resolve conflicts between different construction elements, minimising rework [21]. Additionally, BIM optimises material use, contributing to economic and environmental sustainability in hybrid construction projects. In the evolving urban landscape, projects like Battersea Power Station highlight the integration of traditional and modern construction methods. BIM's clash detection features are particularly relevant in such scenarios, reducing rework and optimising material use New Steel Construction [39], a vital consideration for the economic and environmental sustainability goals of UK construction.

Integrating BIM into HCC methods in the UK presents several challenges rooted in the dynamic interplay of traditional and modern construction approaches. A noteworthy challenge is the intricate coordination required between diverse construction elements within hybrid projects, demanding a seamless integration of BIM to ensure effective collaboration and minimise clashes. The Battersea Power Station redevelopment, a prominent UK project, encountered challenges in reconciling the intricate balance between traditional and modern construction methods, reflecting the complexities of BIM integration in hybrid contexts [39]. Additionally, a notable hurdle lies in optimising material use, critical for economic and environmental sustainability in hybrid construction. The application of BIM's clash detection features and efficient material utilisation was exemplified in the Crossrail project [43], illustrating the potential solutions BIM offers to mitigate challenges within the unique realm of HCC in the UK.

3.1.6 Thin Joint Masonry

The utilisation of BIM in thin joint masonry construction is instrumental in optimising material usage through meticulous design representation, thereby minimising waste. BIM's collaborative platform ensures streamlined design coordination, mitigating errors in the construction process. Additionally, the 3D visualisation capabilities of BIM contribute to enhanced comprehension and communication regarding thin joint masonry designs. This strategic application of BIM is exemplified in residential developments like the Greenwich Peninsula, where the emphasis on material optimisation and efficient design coordination aligns seamlessly with BIM's capabilities (Design F9). The unique attributes of Thin Joint Masonry, requiring precision in design representation and economical material utilisation, necessitate customised BIM solutions tailored to the specific requirements of this construction method. A case study conducted during the execution of the Greenwich Peninsula project highlighted the complexities of coordinating diverse elements within Thin Joint Masonry using BIM, underscoring the importance of advanced BIM features to facilitate seamless collaboration among stakeholders and reduce errors throughout the construction phase.

3.1.7 Insulating Concrete Formwork

For Insulating Concrete Formwork (ICF), BIM contributes to energy-efficient design through analysis and coordination [57]. BIM supports resilient planning for ICF construction, ensuring structures that withstand various challenges [57]. Collaborative features in BIM enhance communication and coordination among stakeholders involved in ICF projects, fostering a unified approach [60]. As the UK strives for energy-efficient structures, exemplified by projects like the Eco Park in Gloucestershire, BIM plays a crucial role. The energy analysis and design coordination facilitated by BIM align with the national agenda for sustainable construction, contributing to resilient planning and enhanced collaboration among stakeholders.

One of the challenges lies in the complexity of aligning diverse stakeholders within the ICF construction process to a unified BIM platform, affecting the seamless flow of information. Case studies, such as the Eco Park project in Gloucestershire, have highlighted difficulties in achieving optimal collaboration and coordination due to varying levels of BIM adoption among project participants. Additionally, the absence of standardised protocols and guidelines specific to ICF within the current BIM framework contributes to interoperability issues. As exemplified by experiences in projects like the Greenwich Peninsula, the industry faces hurdles in ensuring a standardised approach to data exchange and collaboration in ICF projects (Design F9).

3.2 *Case Studies in BIM Integrated Sustainable MMC Projects*

This section examines three exemplary UK-based case studies—the Bloomberg London Office, the Dalston Works, and One Angel Square—each a beacon of sustainable construction practices. These projects were chosen not only for their successful implementation of BIM and MMC but also for their diverse contexts and scales, offering a broad perspective on sustainable practices across different types of construction projects. Each case study provides unique insights into how sustainable goals can be achieved in various settings from a high-profile corporate office and an urban residential complex to a large-scale commercial development, thus presenting a well-rounded exploration of sustainable construction in the contemporary urban landscape. These case studies collectively demonstrate the transformative impact of BIM and MMC in sustainable construction, offering valuable insights into practical applications, challenges, and future potentials in this field.

3.2.1 Case Study 1: The Bloomberg London Office

The Bloomberg London Office, located in the heart of the City of London, is a standout example of modern sustainable architecture. Designed by Foster + Partners, it serves as Bloomberg's European headquarters. The building is a marvel of contemporary design, blending seamlessly into the historic urban fabric while embodying the ethos of technological innovation and environmental stewardship. The project's aim was to establish a new benchmark in sustainable building design, culminating in receiving the highest BREEAM rating for an office building [11]. This recognition was a result of its state-of-the-art sustainable features, such as integrated ceiling panels that use 40% less energy and water conservation systems saving millions of litres annually and water systems, setting a precedent for future sustainable urban architecture [11] (Fig. 1).

Implementation of BIM and MMC

The Bloomberg London Office's construction showcases the cutting-edge application of BIM and MMC. This approach ensured that every aspect of the building, from its energy-efficient systems to its responsive architectural elements, was optimised for sustainability. The implementation of BIM and MMC exemplifies a forward-thinking approach to construction, marrying technological advancements with a deep commitment to environmental stewardship and operational efficiency [10].

Fig. 1 Bloomberg London Office: a model of sustainable architecture with innovative energy and water systems [72] (*Source* Wainwright [65])

BIM in Design and Planning: From the outset, BIM facilitated intricate design and efficient planning, allowing for a harmonious blend of innovative architectural features and environmental sustainability. The building's unique features, like it's 'breathing' facade, which responds to weather conditions and reduces energy consumption, were carefully modelled, and optimised using BIM [10]. This advanced modelling facilitated the integration of intricate design elements and sustainability features, ensuring a cohesive and efficient construction process.

MMC Techniques in Construction: The construction phase saw the application of various MMC techniques. A significant challenge was applying the bronze cladding to the building's spiral ramp, which was six floors high with minimal design tolerance. This was accomplished using sophisticated 3D laser scanners, allowing for pinpoint accuracy and efficient on-site assembly [52]. Further, the project's complexity necessitated continuous improvement processes, with weekly sessions where specialists tackled specific design elements, enhancing overall construction standards [52].

Contribution to Sustainable Construction: The design of the Bloomberg London Office was driven by a commitment to sustainable development and corporate responsibility. This ethos permeated every aspect of the project, from initial design concepts to the selection of materials and construction methods. The building's innovative design not only aimed to reduce its environmental footprint but also to enhance the well-being of its occupants and the surrounding community. Features like smart airflow systems and 'breathing' building technology demonstrate a forward-thinking approach to workplace environments, prioritising both ecological sustainability and human-centric design [11]. Furthermore, the integration of BIM and MMC was key to achieving the building's sustainability targets. The efficient use of resources, such as the adaptive reuse of the existing basement, showcased the flexibility and effectiveness of MMC. This approach ensured minimal downtime and optimised use of equipment and materials [52]. Additionally, the focus on safety measures, like comprehensive health and safety training for all site personnel, underscored the commitment to sustainable and responsible construction practices [52].

Outcomes and Lessons Learned

The Bloomberg London Office, hailed as the "world's most sustainable office building," represents a significant achievement in sustainable construction. Its design and construction resulted in a remarkable 73% saving in water consumption and a 35% reduction in energy consumption and associated CO_2 emissions compared to typical office buildings. This was accomplished through innovative features like a 'breathing' facade, integrated ceiling panels, and smart CO_2 sensors for airflow management. These features not only saved resources but also created a healthier and more inspiring work environment [22].

Michael Bloomberg, the founder of Bloomberg L.P., emphasised that environmentally friendly practices incorporated in the building are as beneficial for business as they are for the planet. This project demonstrates that sustainable design can go hand-in-hand with creating an exciting and inspiring workplace. The Bloomberg building sets a new standard for office environments, balancing ecological responsibility with employee well-being [22].

A key lesson from this project is the potential for buildings to positively impact both the environment and the occupants. The Bloomberg London Office showcases how sustainability and functionality can be integrated seamlessly. The project has set a benchmark in sustainable construction, proving that ambitious environmental goals are attainable in complex urban settings. It serves as a model for future developments, highlighting the importance of innovative design, careful planning, and the use of advanced technologies in sustainable construction [22].

The Bloomberg London Office's approach to sustainability extends beyond its physical features to influence corporate culture and operational practices. The building's environmentally conscious design inspires a deeper commitment to sustainability among employees and visitors alike. By proving that a sustainable office environment can enhance productivity and creativity, the project encourages other companies to adopt similar practices. This ripple effect of the building's success demonstrates the far-reaching impact sustainable construction can have on the broader business community, promoting an ethos of environmental responsibility across industries.

Furthermore, the project's success provides critical insights for urban planners and architects. It shows that integrating sustainability into the core design of buildings is not only feasible but also beneficial in densely populated urban areas. The Bloomberg London Office stands as a testament to the fact that environmental challenges in urban settings can be addressed through innovative construction methods and technologies. This case study serves as a valuable learning tool for future sustainable projects, offering a blueprint for achieving high sustainability standards in challenging environments.

3.2.2 Case Study 2: The Dalston Works, London

The Dalston Works in London, designed by Waugh Thistleton Architects, is renowned as the world's largest timber building. The building comprises 121 flats and over 3,460 square meters of commercial space, blending residential and commercial needs in a mixed-use development. This architectural marvel was designed with a focus on sustainability and reducing the urban carbon footprint (Fig. 2).

Implementation of BIM and MMC

The construction of the Dalston Works in London, a 10-storey carbon–neutral apartment complex, showcases a significant advancement in the application of BIM and MMC. The project, heralded as the world's largest cross-laminated timber (CLT) building, stands as a testament to the potential of MMC in transforming urban construction [46].

BIM in Design and Planning: While specific details on the use of BIM for the Dalston Works are limited, it is likely that BIM played a crucial role in the design and planning stages, especially in managing the complex logistics of using CLT. BIM would have been instrumental in optimising the design for sustainability, efficiency, and structural integrity, ensuring that all architectural and engineering components were seamlessly integrated.

MMC Techniques in Construction: The Dalston Works primarily utilised CLT, a method that embodies the essence of MMC. CLT is made by layering timber at right angles and glueing them together, creating a versatile and lightweight material that is less carbon-polluting to produce compared to traditional construction materials like steel or concrete.

Contribution to Sustainable Construction: The use of CLT significantly contributed to the Dalston Works being carbon-negative for the initial years of its use. This innovative approach to construction not only reduced the building's carbon footprint but also allowed for higher construction on the site due to the lighter core of the building, an essential factor considering its proximity to the Crossrail route [46]. This aligns with the broader goals of sustainable living and energy efficiency [20]. Furthermore, the use of timber not only enhances the building's environmental performance but also infuses aesthetic warmth and a connection to natural materials in the urban landscape [70].

Fig. 2 The Dalston Works: timber building showcasing sustainable mixed-use development (*Source* Waugh Thistleton Architects [69])

Outcomes and Lessons Learned

The Dalston Works project, upon its completion, not only became the world's largest CLT building but also marked a significant milestone in urban housing development. The 10-storey, 121-unit development is entirely made of CLT, from its external to core walls, floors, and stairs. This approach resulted in the building weighing a fifth of a comparable concrete structure and reduced construction deliveries by 80%, demonstrating the feasibility and efficiency of using timber in high-density urban housing [70]. The reduced weight of the building allowed for a taller structure on a previously neglected brownfield site, showcasing the potential of sustainable materials in transforming urban landscapes. The efficiency and reduced weight of the scheme allowed Regal Homes to increase the number of residential units by 25%, significantly contributing to tackling London's housing shortage. This project, therefore, exemplifies the viability of timber technology in high-density urban settings addressing London's need for high-quality, high-density housing while providing a natural and healthy living environment [70].

In addition, the rapid project delivery, completed in just 18 months, highlights the advantages of prefabricated CLT panels, setting a precedent for sustainable, efficient urban housing development [7]. Moreover, the innovative use of timber technology at the Dalston Works significantly lowered the carbon footprint of the building, both in material production and on-site energy consumption.

The Dalston Works has garnered international interest, positioning the London Borough of Hackney and the architectural and engineering teams involved as pioneers in timber construction. The project's success has fostered exciting partnerships and is inspiring a shift in the construction industry towards sustainable methods. It has become a case study in sustainable urban development, attracting attention worldwide and underscoring the importance of innovative approaches in addressing housing and environmental challenges in densely populated areas [70].

3.2.3 Case Study 3: One Angel Square, Manchester

Designed and built by architects 3DReid, One Angel Square in Manchester stands as one of Europe's most sustainable buildings, boasting a BREEAM 'Outstanding' rating and an A+ Energy Performance Certificate. This building exemplifies a commitment to sustainable design and operational efficiency, accommodating various offices in one energy-efficient structure and creating thousands of jobs during its construction [72] (Fig. 3).

Implementation of BIM and MMC

The construction of One Angel Square in Manchester notably utilised BIM and MMC, playing a crucial role in achieving its sustainability goals.

BIM in Design and Planning: BIM was instrumental throughout the project, from the initial design phase to construction and completion. In the design phase, BIM facilitated the detailed planning of One Angel Square's innovative features, such as its double-skinned façade and the bio-fuel cogeneration plant. It allowed for the precise

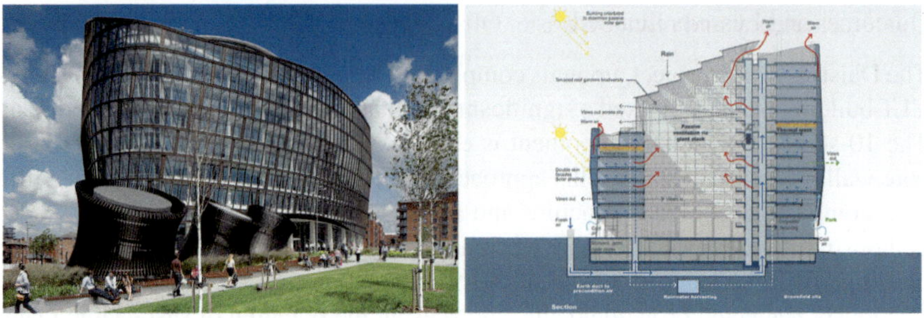

Fig. 3 One Angel Square: a leading sustainable building in Europe (*Source* Ongreening [41])

modelling of these elements, ensuring they would meet the strict sustainability targets set for the project. During the planning stage, BIM was crucial in identifying potential challenges and enabling the team to devise solutions ahead of time, minimising disruptions during construction. Overall, the use of BIM enabled a highly integrated and collaborative approach, ensuring that all aspects of the building's complex design and sustainability features were accurately modelled and executed.

MMC Techniques in Construction: The project's construction phase saw the application of various MMC techniques. These included prefabricated components and off-site construction methods, which significantly reduced on-site waste and improved construction efficiency. The use of MMC also contributed to a safer construction environment, with fewer on-site risks due to the reduced need for traditional construction processes.

Contribution to Sustainable Construction: The integration of BIM and MMC facilitated the successful implementation of One Angel Square's numerous eco-friendly innovations. The focus on sustainability was integral to every aspect of the building's development, from its initial location choice to the architectural design and construction process. The building features advanced engineering solutions, such as a double-skinned façade for minimised heating and cooling requirements and geothermal earth tubes for circulating fresh air. This comprehensive approach also influenced how the building is utilised, ensuring that its operations remain aligned with its sustainability ethos. One Angel Square is distinguished by its use of a pioneering bio-fuel cogeneration plant and an oil-fed combined heat and power (CHP) system, which utilises rapeseed oil grown on the Co-operative's farmland. This innovative energy solution, along with rainwater and greywater harvesting systems, ensures low water consumption. The project stands as a benchmark in sustainable construction, showing how the intelligent use of technology can lead to significant environmental benefits while also setting new standards in architectural design [35, 72].

Outcomes and Lessons Learned

One Angel Square in Manchester, achieving the BREEAM 'Outstanding' rating, stands as a testament to the practical viability of ambitious sustainability objectives in large-scale construction projects. The building's construction approach, involving the strategic use of long-span steelwork and exposed precast concrete, maximised resource efficiency and reduced waste. Off-site manufacturing, a key component of the project, led to a reduction in packaging waste and ensured 80% of site waste avoided landfill.

A key learning point from One Angel Square is the importance of integrating sustainable design principles right from the conceptual stage. The project illustrates how sustainability can be embedded in every aspect of construction, from material selection to waste management. The success of One Angel Square underscores the value of collaboration between architects, engineers, and construction teams, especially when adopting innovative techniques like BIM and MMC. This collaboration ensured precise planning, efficient resource usage, and effective risk management.

One Angel Square serves as a blueprint for future sustainable construction projects, demonstrating that high sustainability standards can be met without compromising on functional or aesthetic qualities. The project highlights the importance of considering long-term environmental impact in urban development, encouraging architects and developers to prioritise sustainable practices in their future projects [9].

4 Strategic Suggestions for BIM-MMC Integration

This section suggests strategic approaches to the BIM-MMC integration within the UK construction landscape, structured around three main areas. Initially, it suggests ways in which this integration can significantly enhance construction sustainability, assessing both immediate advantages and long-term implications for sustainability. Subsequently, it pinpoints core challenges that obstruct seamless integration of BIM with MMC, offering targeted suggestions for overcoming these barriers and advocating for industry-wide enhancements. Lastly, it highlights the potential of emerging technologies, suggesting their incorporation to further elevate construction sustainability through innovative BIM-MMC practices. By providing these strategic suggestions, this section aims to harness the full potential of BIM-MMC integration, propelling the construction industry towards a more innovative and sustainable future.

4.1 BIM-MMC Integration for Construction Sustainability

In discussing the long-term sustainability, unforeseen challenges, and industry adaptation in relation to the integration of BIM and MMC, it is essential to critically analyse how these aspects manifest across various construction methods.

Long-Term Sustainability: The integration of BIM with various MMC methods, such as precast flat panel systems, tunnel form, and 3D volumetric construction, shows immediate benefits in design efficiency and precision. However, the long-term sustainability of these benefits requires examination: will the efficiency gains and reduction in errors sustain over the life cycle of buildings, or will evolving construction technologies and environmental demands challenge them? Thus, the durability of these integrated methods in adapting to future sustainability benchmarks is a crucial factor to consider.

Unforeseen Challenges: While the initial benefits of integrating BIM with MMC methods are clear, potential unforeseen challenges may emerge over time. For instance, the complex coordination required in flat slab and hybrid concrete construction, or the detailed modelling in thin joint masonry and ICF, may lead to challenges in adapting to new environmental standards or changes in technology. The industry must be prepared to face and adapt to such challenges, ensuring the resilience of these integrated methods.

Industry Adaptation: The adaptation of the construction industry to the integration of BIM and MMC is another critical area. This includes assessing the readiness of the industry to embrace the paradigm shift towards more complex, technology-driven construction methods. Accordingly, there are two critical concerns to address: are there sufficient skills and training within the industry to effectively implement these methods? And how does the industry respond to the cultural and systemic changes required for this integration? The success of BIM and MMC integration depends heavily on the industry's ability to adapt to these new methods and technologies, ensuring that the benefits realised in initial projects can be sustained and expanded in future constructions.

In linking these points to the specific benefits and challenges discussed in Sect. 3.1, it becomes evident that while the integration of BIM and MMC offers significant immediate advantages, its long-term success, adaptability to unforeseen challenges, and the industry's capacity to embrace this integration will dictate its future efficacy and sustainability in the UK construction industry.

4.2 Challenges and Industry Proposals to Enhance MMC-BIM Integration in the UK

While the integration of BIM with MMC holds the promise of transforming the construction industry, the path to seamless implementation is fraught with barriers and challenges. This comprehensive discussion unravels the complexities and nuances surrounding the existing impediments, substantiating insights with real-world case study scenarios from the UK construction sector.

(1) Limited Standardisation, Fragmentation, and Interoperability Challenges

The absence of standardised protocols poses a significant challenge to the effective integration of BIM with MMC. Fragmentation and interoperability, that arose from limited standardisation, persist as one of the foremost obstacles to BIM integration across diverse MMC. In many cases, different software platforms, standards, and data formats create silos, hindering the seamless flow of information across the construction life cycle. The HS2 Rail Project encountered challenges due to the lack of standardised BIM protocols and similarly, the Thames Tideway

Tunnel project exemplifies the struggle with interoperability [43]. With multiple stakeholders utilising different BIM software and standards, the project encountered difficulties in consolidating data, leading to inefficiencies and communication gaps. In response, the industry has actively pursued initiatives to standardise protocols, promote collaboration, and enhance data exchange mechanisms. Some key protocols include:

- BIM Level 2 Standards: The UK government has mandated the adoption of BIM Level 2 standards for publicly funded projects. This protocol establishes a standardised framework for information exchange, collaboration, and project delivery, ensuring consistency across the industry.
- Industry Foundation Classes (IFC): IFC is an open standard for the exchange of BIM data among different software platforms. Adopting IFC allows interoperability between various BIM tools, mitigating issues related to incompatible software formats.
- UK BIM Alliance Guidance: The UK BIM Alliance provides guidance and best practices for BIM implementation. These resources offer industry-specific insights and recommendations to standardise BIM processes and enhance interoperability.
- COBie (Construction Operations Building Information Exchange): COBie is a standard for the exchange of facility information, ensuring that data is consistent and compatible across different stages of a building's life cycle. It enhances interoperability by defining a common format for sharing information.
- Common Data Environment (CDE): A Common Data Environment is a standardised platform where project information is stored and shared. It ensures that all stakeholders have access to the latest and most accurate data, promoting collaboration and reducing fragmentation.
- ISO 19650: ISO 19650, an international standard, provides a common framework for managing information throughout the entire life cycle of a built asset. This standard helps streamline data exchange and collaboration by offering guidelines for information management using BIM.
- BuildingSMART: BuildingSMART is an international organisation dedicated to developing and maintaining open BIM standards. Participation in BuildingSMART initiatives ensures alignment with global standards, promoting interoperability on an international scale.
- Classification Systems (e.g., Uniclass): Standardised classification systems like Uniclass help organise and categorise information within BIM models. Adopting such classification systems ensures consistency in data representation, aiding interoperability.

These measures aim to foster a more integrated and cohesive approach, mitigating inefficiencies and communication gaps prevalent in fragmented BIM integration scenarios.

(2) Skill and Knowledge Gaps

A shortage of skilled professionals proficient in both BIM and MMC methodologies poses a barrier to successful integration. The industry's rapid evolution demands a workforce equipped with interdisciplinary expertise; a need that often goes unmet. For instance, the Battersea Power Station project faced skill gaps, with professionals lacking the necessary expertise in both MMC and BIM [39]. This deficiency resulted in inefficiencies and hindered the project's ability to fully capitalise on the benefits of integration. In response to addressing skill and knowledge gaps in BIM and MMC integration, several targeted strategies include:

- Certified training and apprenticeship programmes in collaboration with higher-education institutes,
- Industry-specific workshops, seminars, and continuing professional developments (CPDs) (i.e., conducted by BRE Academy, Royal Institution of Chartered Surveyors (RICS), Chartered Institute of Building (CIOB), Institution of Civil Engineers (ICE), National Building Specification (NBS), and BIM Academy),
- Promotion of interdisciplinary collaboration,
- Government support and funding for training programmes and educational courses related to BIM and MMC (i.e., Construction Industry Training Board (CITB), Innovate UK, Government apprenticeship levy, and local enterprise partnerships),
- Integration of BIM and MMC in professional standards, and
- In-house training programmes by companies.

By implementing these targeted strategies, the UK construction industry aims to overcome skill and knowledge gaps, creating a proficient workforce capable of harnessing the full potential of BIM and MMC integration.

(3) Financial Implications

Financial considerations, including the costs associated with software, training, and transitioning to new workflows, act as a substantial barrier. Many organisations hesitate to invest in the necessary technology and resources, particularly in an industry traditionally resistant to change. The Crossrail 2 project faced financial challenges in transitioning to BIM-integrated MMC [43]. The upfront costs associated with software licenses, training, and restructuring workflows presented hurdles that slowed down the pace of integration. In this regard, the UK government offers Research and Development tax credits to incentivise companies engaged in innovative activities. Organisations implementing BIM and MMC technologies may be eligible for these tax credits, providing financial relief.

(4) Data Security and Privacy Concerns

The sensitive nature of construction project data raises concerns about security and privacy. Organisations may be hesitant to fully embrace BIM integration, fearing the potential vulnerabilities and risks associated with data breaches. The Shard project encountered data security concerns when considering BIM integration with MMC. The need to protect proprietary design and construction data prompted cautious implementation, delaying the realisation of full integration benefits. In the UK, several measures and regulations are in place to eliminate or control these concerns:

- Data Protection Legislation: The General Data Protection Regulation (GDPR) is a comprehensive data protection law in the UK and the European Union. GDPR governs the processing of personal data, ensuring that individuals' privacy rights are protected. Organisations integrating BIM and MMC must comply with GDPR requirements, including obtaining consent for data processing, implementing security measures, and ensuring transparent data practices.
- BIM Execution Plans (BEPs): BEPs outline the processes and procedures for information management throughout a construction project. They include considerations for data security and privacy, specifying how information, including personal data, will be managed, shared, and protected during different project stages.
- Information Security Standards: Adherence to information security standards, such as ISO/IEC 27001, is essential. ISO/IEC 27001 provides a framework for establishing, implementing, maintaining, and continually improving an information security management system.
- Contractual Agreements and Legal Frameworks: Clear contractual agreements between project stakeholders, including clients, contractors, and technology providers, should define responsibilities and liabilities concerning data security and privacy. Legal frameworks, such as the Common Data Environment (CDE) protocols, help establish the rules and responsibilities for information exchange in BIM projects.
- Cybersecurity Best Practices: Following cybersecurity best practices is crucial for protecting BIM and MMC systems from unauthorised access and cyber threats. This includes implementing firewalls, encryption, access controls, and regular security assessments to identify and address vulnerabilities.
- Data Encryption and Access Controls: Encrypting sensitive data and implementing access controls ensure that only authorised individuals have access to specific information. Encryption helps protect data during transmission and storage, while access controls limit who can view or modify data within BIM and MMC systems.
- Regular Audits and Assessments: Conducting regular audits and assessments of information security practices helps identify and address potential vulnerabilities. These assessments may include penetration testing, vulnerability assessments, and reviews of security policies and procedures.
- Data Residency and Sovereignty Considerations: Organisations should consider data residency and sovereignty when choosing where to store and process BIM data. Ensuring compliance with regional and national data protection laws helps mitigate risks associated with cross-border data transfers.

By incorporating these measures into their BIM and MMC practices, organisations in the UK can enhance data security and privacy, fostering responsible and compliant use of digital technologies in the construction industry.

(5) Cultural and Organisational Barriers

Resistance to change, especially in traditional construction sectors, often hinders the adoption of these innovative methods. There is also a challenge in aligning various stakeholders' interests and objectives, as BIM and MMC require collaborative approaches that may disrupt established practices.

Addressing these barriers necessitates a holistic approach that involves not only technological solutions but also organisational change management, stakeholder engagement, and policy reforms.

4.3 Emerging Technologies with MMC-BIM Integration for Elevated Construction Sustainability

In the UK, the detailed exploration of emerging technologies integrated with BIM and MMC is still a developing field. These technologies, including ***innovative materials*** like self-healing concrete and photovoltaic glass, promise to enhance the sustainability of various MMC methods, such as precast flat panel systems and 3D volumetric construction. ***Internet of Things (IoT) and big data*** are transforming tunnel form construction and thin joint masonry by improving precision through real-time data analysis, and similarly, ***3D printing, and automation*** are transforming flat slab and hybrid concrete construction, focusing on waste reduction and precision. The potential of ***Artificial Intelligence (AI) and Machine Learning (ML)*** is being tapped to optimise decision-making and resource allocation in complex scenarios like insulating concrete formwork. ***Blockchain technology*** is enhancing supply chain transparency, especially in MMC types like precast systems. ***Virtual and Augmented Reality (AR and VR)*** are improving design accuracy across different MMC methods, contributing significantly to construction sustainability.

By embracing these emerging technologies, the UK construction industry can not only meet but exceed current sustainability benchmarks, setting a new standard for future constructions. This strategic incorporation of innovative solutions into MMC-BIM integration offers a promising pathway towards achieving unparalleled efficiency, precision, and environmental stewardship in the construction sector. Therefore, these emerging areas present valuable opportunities for future research.

5 Conclusions

In conclusion, the varied types of MMC, ranging from off-site construction methods to on-site improvements and hybrid approaches, play a crucial role in advancing sustainability in the construction industry. The controlled environments, precision, and efficiency associated with MMC, coupled with the collaborative power of BIM, collectively contribute to a more sustainable and resilient built environment. Each

MMC type benefits uniquely from this integration, demonstrating its real-world applicability and setting new standards in construction management. While MMC-BIM integration accelerates construction timelines, enhances precision and quality, optimises resources, reduces waste, and promotes energy efficiency and eco-friendly construction, it is important to acknowledge the complexities and hurdles in their implementation.

Notably, regulatory and standardisation issues pose significant barriers, along with logistical challenges in adopting new technologies on a wide scale. Besides, the key challenges include high initial cost and investment needed, the requirement for specialised skills and training, data protection, and potential resistance to change from traditional construction methods. As the UK construction industry grapples with these barriers and challenges, it becomes evident that overcoming them requires a concerted effort from stakeholders, policymakers, and industry bodies. Collaborative initiatives to standardise protocols, invest in education and training, and develop clear regulatory frameworks will be crucial in fostering a more conducive environment for the integration of BIM with MMC. By learning from the experiences of past projects and adapting strategies to address these challenges, the industry can navigate the path towards a more innovative and integrated future unlocking the full potential of BIM in modern construction methods in the UK.

This chapter makes significant theoretical contributions to the field of sustainable construction by critically examining the integration of BIM with various types of MMC in the UK construction industry. It moves beyond simply outlining the benefits and challenges, offering an in-depth analysis of how this integration impacts construction practices, efficiency, and sustainability. The chapter elucidates the nuanced interactions between different MMC types and BIM, providing insights into long-term sustainability, unforeseen challenges, and the industry's adaptability to these emerging technologies. This analysis enriches existing literature by providing a detailed exploration of the specific implications and potential of BIM-MMC integration across different construction scenarios. Thus, the chapter significantly enhances our understanding of the complexities and dynamics involved in leveraging BIM and MMC for sustainable construction, setting a new benchmark for future research and practice in this domain.

The role of BIM and MMC in sustainable and smart city projects is also evolving, offering new avenues for integrating innovative materials and emerging technologies, including IOT, big data analytics, 3D printing and automation, AI, ML, Blockchain, VR, and AR, in urban development. Future research should explore the implementation challenges and benefits of these technologies, particularly in areas like real-time data analysis, supply chain optimisation, and sustainable resource management. This exploration will not only address current challenges but also pave the way for innovative solutions in the construction industry.

References

1. Abanda F, Tah J, Cheung F (2017) BIM in off-site manufacturing for buildings. J Build Eng 14:89–102
2. Agha A, Shibani A, Hassan D, Zalans B (2021) Modular construction in the United Kingdom housing sector: barriers and implications. J Arch Eng Technol 10:236
3. Ali AH, El-Mahdy GM, Ibrahim AH, Daoud AO (2022) Towards the adoption of modular construction in residential projects in Egypt: benefits, barriers, and enablers. In: Construction industry development board postgraduate research conference, pp 72–81. Springer
4. Alwan Z, Jones P, Holgate P (2017) Strategic sustainable development in the UK construction industry, through the framework for strategic sustainable development, using Building Information Modelling. J Clea Prod 140:349–358
5. Andy Y, Nigel A, Andy B, Brian S (2013) Case study: the Leadenhall building. CTBUH J 1
6. Apostolska R, Necevska-Cvetanovska G, Cvetanovska J, Mircic N (2008) Seismic performance of flat-slab building structural systems. In: The 14th world conference on earthquake engineering October, 2008
7. ArchDaily (2018, October 19) World's largest CLT building provides a model for high-density urban housing. Retrieved from https://www.archdaily.com/903839/worlds-largest-clt-building-provides-a-model-for-high-density-urban-housing
8. Aye L, Ngo T, Crawford RH, Gammampila R, Mendis P (2012) Life cycle greenhouse gas emissions and energy analysis of prefabricated reusable building modules. Energy Build 47:159–168
9. BAM Construction (n.d.) One angel square, Manchester - BAM case study. Retrieved 8 December 2023 from https://www.bam.co.uk/how-we-do-it/case-study/1-angel-square
10. BIM Plus (2017) Bloomberg's London HQ is 'most sustainable office building in the world'. BIM+. Retrieved 8 December 2023 from https://www.bimplus.co.uk/bloombergs-london-hq-most-sustainable-office-build/
11. BRE Group (n.d.) One of the world's highest BREEAM-rated major office buildings: Bloomberg, London. Retrieved 8 December 2023 from https://bregroup.com/case-studies/breeam-new-construction/one-of-the-worlds-highest-breeam-rated-major-office-buildings-bloomberg-london/
12. Baldwin A, Poon C-S, Shen L-Y, Austin S, Wong I (2009) Designing out waste in high-rise residential buildings: analysis of precasting methods and traditional construction. Renew Energy 34:2067–2073
13. Beheshti-Aval SB, Mohsenian V, Nikpour N (2018) A study on seismic performance of RC tunnel-form building structures with irregular plan. Modares Civ Eng J 17:57–68
14. Bertram N, Fuchs S, Mischke J, Palter R, Strube G, Woetzel J (2019) Modular construction: from projects to products. Capital Projects & Infrastructure, McKinsey & Company
15. Construction Industry Institute (2017) The future of building 2021 annual report. CII, The University of Texas at Austin, Texas. Retrieved 10 April 2024 from https://www.construction-institute.org/cii-2021-annual-report
16. Dabirian S, Han SH, Lee J (2020) Stochastic-based noise exposure assessment in modular and off-site construction. J Clean Prod 244:118758
17. Davies D, Tokbolat S (2023) Building information modelling implementation in the UK construction industry: a commercial perspective. Automation in construction toward resilience. CRC Press
18. Design F9 (n.d.) Greenwich Peninsula. https://www.designf9.com/greenwich-peninsula. Last Accessed 14 Oct 2023. (Online)
19. Ferdous W, Bai Y, Ngo TD, Manalo A, Mendis P (2019) New advancements, challenges and opportunities of multi-storey modular buildings–a state-of-the-art review. Eng Struct 183:883–893
20. Gerard (2018) Dalston works: the world's largest timber building. EBUILDING. Retrieved 8 December 2023 from https://ebuilding.blog/dalston-works-largest-timber-building

21. Glass J, Baiche B (2001) Perceptions of hybrid concrete construction in the UK construction industry. Eng Constr Archit Manag 8:67–77
22. Global Construction Review (2017) Bloomberg's London HQ is 'most sustainable office building in the world'. Global Construction Review. Retrieved 8 December 2023 from https://www.globalconstructionreview.com/bloombergs-london-hq-most-sustainable-office-build/
23. Goyal LK, Rai HS (2020) BIM approach for sustainable design of flat slab buildings: a review. IOP Conf Ser Mater Sci Eng 955:012012
24. Guerra BC, Bakchan A, Leite F, Faust KM (2019) BIM-based automated construction waste estimation algorithms: the case of concrete and drywall waste streams. Waste Manage 87:825–832
25. Hammad AW, Akbarnezhad A, Wu P, Wang X, Haddad A (2019) Building information modelling-based framework to contrast conventional and modular construction methods through selected sustainability factors. J Clean Prod 228:1264–1281
26. Hannah M, Hunter N (2018) Modern methods of construction: who's doing what. Milton Keynes, NHBC Foundation, UK
27. Hwang B-G, Shan M, Looi K-Y (2018) Key constraints and mitigation strategies for prefabricated prefinished volumetric construction. J Clean Prod 183:183–193
28. Jung J, Stachniss C, Ju S, Heo J (2018) Automated 3D volumetric reconstruction of multiple-room building interiors for as-built BIM. Adv Eng Inform 38:811–825
29. Kanyeto O, Fried A (2011) Flexural behaviour of thin joint concrete blockwork: experimental results. Constr Build Mater 25:3639–3647
30. Lawson RM, Ogden RG, Bergin R (2012) Application of modular construction in high-rise buildings. J Archit Eng 18:148–154
31. Liu P, Qi H, Liu J, Feng L, Li D, Guo J (2023) Automated clash resolution for reinforcement steel design in precast concrete wall panels via generative adversarial network and reinforcement learning. Adv Eng Inform 58:102131
32. Malviya V, Vyas J (2021) Comparison study of flat slab and conventional slab. Int Res J Eng Technol 8:2557–2560
33. Mapston M, Westbrook C (2010) Prefabricated building units and modern methods of construction (MMC). Materials for energy efficiency and thermal comfort in buildings. Elsevier
34. Matarneh ST, Danso-Amoako M, Al-Bizri S, Gaterell M, Matarneh RT (2020) BIM for FM: developing information requirements to support facilities management systems. Facilities 38:378–394
35. NBS (2014) One angel square. Retrieved 8 December 2023 from https://www.thenbs.com/knowledge/one-angel-square
36. Navyashree K, Sahana T (2014) Use of flat slabs in multi-storey commercial building situated in high seismic zone. Int J Res Eng Technol 3:439–451
37. Nawi MM, Hanifa FA, Kamar KM, Lee A, Azman MA (2014) Modern method of construction: an experience from UK construction industry. Aust J Basic Appl Sci 8:527–533
38. Nazir FA, Edwards DJ, Shelbourn M, Martek I, Thwala WDD, El-Gohary H (2021) Comparison of modular and traditional UK housing construction: a bibliometric analysis. J Eng Des Technol 19:164–186
39. New Steel Construction (2020) Steel power for Battersea. Kent, New Steel Construction, UK
40. Nightingale A, Ringrose J, Seaman R, Willett S, Ashdown S, Bowers K, Gandhi N, Waterson T, Hayward T (2014) Interrogating the technical, economic and cultural challenges of delivering the PassivHaus standard in the UK. In: Schoenefeldt H (ed) Canterbury. University of Kent, UK
41. Ongreening (2012) One Angel Square. https://ongreening.com/en/Projects/one-angel-square-1181. Last Accessed 22 July 2024
42. Pan W, Sidwell R (2011) Demystifying the cost barriers to offsite construction in the UK. Constr Manag Econ 29:1081–1099
43. Papadonikolaki E, Morgan B, Papachristos G (2023) Megaprojects as niches of sociotechnical transitions: the case of digitalization in UK construction. Environ Innov Soc Trans 48:100728

44. Paudel P, Dulal S, Bhandari M, Tomar A (2016) Study on pre-fabricated modular and steel structures. Int J Civ Eng (SSRG-IJCE) 3
45. Piek PJ (2014) An investigation into the time and cost factors for a decision between in-situ and hybrid concrete construction. Stellenbosch University, Stellenbosch
46. Planning, Building & Construction Today (2017) Dalston Works project showcases the future of cross-laminated timber. Retrieved 8 December 2023 from https://www.pbctoday.co.uk/news/planning-construction-news/dalston-works-project-showcases-the-future-of-cross-laminated-timber/36089/
47. Quale J, Eckelman MJ, Williams KW, Sloditskie G, Zimmerman JB (2012) Construction matters: comparing environmental impacts of building modular and conventional homes in the United States. J Ind Ecol 16:243–253
48. Ray P, Bera DK, Rath AK (2020) Comparison between the tunnel form system formwork and the MIVAN formwork system in a multi-unit building project. Recent Dev Sustain Infrastruct Sel Proc ICRDSI 2019:891–908
49. Roper L (2023) Knowledge foundations. University College of Estate Management, Reading, UK
50. Shalabi F, Turkan Y (2017) IFC BIM-based facility management approach to optimize data collection for corrective maintenance. J Perform Constr Facil 31:04016081
51. Shibani A, Agha A, Hassan D, Al-Hadeethi Y, Choudhury M (2021) Effectiveness of the modern methods of construction in terms of cost and time: a case study of the United Kingdom. J Civ Eng Res 11:19–28
52. Sir Robert McAlpine (n.d.) Construction of bloomberg's new London HQ. Retrieved 8 December 2023 from https://www.srm.com/projects/bloomberg-european-hq-london/
53. Sotoudeh Y, Salehi M, Moradzadeh S, Taghipoor H, Behboodi M (2013) Building technology for mass concrete tunnel form method. Adv Environ Biol 2190–2195
54. Švajlenka J, Kozlovská M, Spišáková M (2017) The benefits of modern method of construction based on wood in the context of sustainability. Int J Environ Sci Technol 14:1591–1602
55. Tay YWD, Panda B, Paul SC, Noor Mohamed NA, Tan MJ, Leong KF (2017) 3D printing trends in building and construction industry: a review. Virtual Phys Prototyp 12:261–276
56. Thamboo J (2020) Material characterisation of thin layer mortared clay masonry. Constr Build Mater 230:116932
57. The Concrete Centre (2012) Insulating concrete formwork: a guide to the performance benefits of ICF solutions. The Concrete Centre, London, UK
58. The Concrete Centre (2019) Offsite concrete construction: a guide to the design and construction of precast concrete in buildings. The Concrete Centre, London, UK
59. Thurairajah N, Rathnasinghe A, Ali M, Shashwat S (2023) Unexpected challenges in the modular construction implementation: are UK contractors ready? Sustainability 15:8105
60. Tovey A, Roberts J, Kilcommons M (2007) Design and construction using insulating concrete formwork. Published for the concrete centre by the concrete society. The Concrete Centre, Trowbridge, UK
61. Treeck CV, Wimmer R, Maile T (2018) BIM for energy analysis. Build Inf Model Technol Found Ind Pract 337–347
62. Trimble Incorporation (2023) One Nine Elms: a New London landmark built with Tekla. Espoo: Finland: Trimble Incorporation. https://www.tekla.com/resources/concrete/one-nine-elms-a-new-london-landmark-built-with-tekla. Last Accessed 22 Dec 2023
63. Tzourmakliotou D (2021) Modular disruption in construction industry—the environmental benefits. J Civ Eng Arch 15:1934–7359
64. Vagtholm R, Matteo A, Vand B, Tupenaite L (2023) Evolution and current state of building materials, construction methods, and building regulations in the UK: implications for sustainable building practices. Buildings 13:1480
65. Wainwright O (2017) Bloomberg HQ: A £1bn building that looks like a regional department store'. The Guardian, 25 October. https://www.theguardian.com/artanddesign/2017/oct/25/bloomberg-london-hq-norman-foster-architecture-review. Last Accessed 17 March 2025

66. Walliman N, Baiche B, Ogden R (2008) Thin-joint glued brickwork: building in the British context. Constr Build Mater 22:1081–1092
67. Wang J, Zhang S, Fenn P, Luo X, Liu Y, Zhao L (2023) Adopting BIM to facilitate dispute management in the construction industry: a conceptual framework development. J Constr Eng Manag 149:03122010
68. Waqar A, Othman I, Hayat S, Radu D, Khan MB, Galatanu TF, Almujibah HR, Hadzima-Nyarko M, Benjeddou O (2023) Building information modeling—empowering construction projects with end-to-end life cycle management. Buildings 13:2041
69. Waugh Thistleton Architects (2018) News: Waugh thistleton architects, dalston works featured in timber Rising exhibition. Waugh Thistleton Architects. https://waughthistleton.com/news/18/02/09/dalston-works-timber-rising/. Last Accessed 22 July 2024
70. Waugh Thistleton Architect. (n.d.) Dalston works. Retrieved 8 December 2023 from https://waughthistleton.com/dalston-works/
71. Weng Y, Li M, Ruan S, Wong TN, Tan MJ, Yeong KLO, Qian S (2020) Comparative economic, environmental and productivity assessment of a concrete bathroom unit fabricated through 3D printing and a precast approach. J Clean Prod 261:121245
72. World Construction Network (2013) One angel square. Retrieved 8 December 2023 from https://www.worldconstructionnetwork.com/projects/one-angel-square/
73. Yeomans D (2020) Concrete architecture: a lost opportunity? Archit Hist 63:257–282
74. Yin X, Liu H, Chen Y, Al-Hussein M (2019) Building information modelling for off-site construction: Review and future directions. Autom Constr 101:72–91
75. Young BE, Seidu RD, Thayaparan M, Appiah-Kubi J (2020) Modular construction innovation in the UK: the case of residential buildings. In: The 10th annual industrial engineering and operations management (IEOM) conference, 10–12 March 2020 Dubai. IEOM Society
76. Zimmermann RK, Bruhn S, Birgisdóttir H (2021) BIM-based life cycle assessment of buildings—an investigation of industry practice and needs. Sustainability 13:5455

GPSR Compliance

The European Union's (EU) General Product Safety Regulation (GPSR) is a set of rules that requires consumer products to be safe and our obligations to ensure this.

If you have any concerns about our products, you can contact us on ProductSafety@springernature.com

In case Publisher is established outside the EU, the EU authorized representative is:

Springer Nature Customer Service Center GmbH
Europaplatz 3
69115 Heidelberg, Germany

Batch number: 10704307

Printed by Printforce, the Netherlands